Current Themes in Theoretical Biology

Current Themes in Theoretical Biology

A Dutch Perspective

Edited by

Thomas A.C. Reydon
Leiden University,
The Netherlands

and

Lia Hemerik
Wageningen University,
The Netherlands

 Springer

A C.I.P. Catalogue record for this book is available from the Library of Congress.

ISBN 1-4020-2901-2 (PB)
ISBN 1-4020-2904-7 (e-book)

Published by Springer,
P.O. Box 17, 3300 AA Dordrecht, The Netherlands.

Sold and distributed in North, Central and South America
by Springer,
101 Philip Drive, Norwell, MA 02061, U.S.A.

In all other countries, sold and distributed
by Springer,
P.O. Box 322, 3300 AH Dordrecht, The Netherlands.

Printed on acid-free paper

springeronline.com

Printed in the Netherlands.

Table of Contents

Preface

The present volume originated in 2001 when we, together with our publishing editors at (then) Kluwer Academic Publishers, realized that the following year the 50th volume of our journal *Acta Biotheoretica* would see the light. We felt that this milestone should not pass unnoticed and that the appropriate way to mark it would be the publication of a special volume of papers on theoretical biology. While editing this book during 2003 and early 2004, we realized that another milestone was not far off: in 2005 it will be 70 years ago that the journal was founded. We hope that the book lying before you will serve well to mark both events.

The papers collected here have been written on invitation by representatives of the theoretical biology community in The Netherlands. They are intended to reflect the entire spectrum of topics on which *Acta Biotheoretica* publishes, ranging from philosophy of biology on one end to mathematical biology on the other. All chapters (except our own introductory one) have been peer reviewed according to the standards that are maintained with respect to regular submissions to *Acta Biotheoretica*.

We are indebted to many people for making this book possible: of course the authors and reviewers who kindly agreed to contribute to the book's contents; Charles Erkelens from Springer publishers and Peter de Liefde (formerly from Kluwer Academic Publishers) for suggesting the book and enabling its publication; and Elizabeth van Ast for her invaluable work in maintaining correspondence with authors and reviewers, copy editing of the chapters, preparing galley proofs and much more.

Thomas Reydon
Lia Hemerik
July 2004

1

1

The History of *Acta Biotheoretica* and the Nature of Theoretical Biology

Thomas A. C. Reydon, Piet Dullemeijer
and Lia Hemerik[1]

ABSTRACT

In this introductory chapter, the three most recent editors of *Acta Biotheoretica* briefly discuss the history, aims and nature of the journal in the context of the unique character of theoretical biology in The Netherlands. We stress that the broad conception of theoretical biology from which the journal has started almost 70 years ago has been maintained throughout all of its issues to the present day.

1.1 INTRODUCTION

Theoretical biology has existed as long as researchers have been interested in investigating the living world, although for most of the time it has not existed as an independent scientific discipline. In its earliest forms, theoretical biology can be understood as a general investigation into the nature of life and its development on Earth. In later stages, theoretical investigations became divided over several disciplines that study entities and phenomena on different levels of organization, from molecular biology to ecology.

Acta Biotheoretica, published for the first time in 1935, is one of the oldest international journals publishing in theoretical biology. Since its foundation, the journal has been devoted to the development and promotion of this field of investigation. As is illustrated in the following sections, the journal has continuously operated from a broad understanding on the nature of theoretical biology and has reflected the particular character of theoretical biology as it existed and still exists in The Netherlands. In this understanding, theoretical biology is seen as encompassing the entire spectrum of theoretical investigation of the living world, ranging from philosophy of biology to

---

[1]Piet Dullemeijer has been the Managing Editor of *Acta Biotheoretica* in the periods 1966 – 1969 and 1987 – 1998. Lia Hemerik has been the Managing Editor of the journal in the period 1999 – 2001. Since 2002 Thomas Reydon is the journal's present Managing Editor.

T.A.C. Reydon and L. Hemerik.,(eds.), Current Themes in Theoretical Biology,
1-8.
© 2005 *Springer. Printed in the Netherlands.*

mathematical biology. Consequently, the process of biological theory formation in the journal is allowed to range from purely verbal argumentation to the mathematical analysis of biological theory. This diverse nature of the journal's scope is also reflected in the articles published in the present book.

In 2002, the 50th volume of *Acta Biotheoretica* was published. Two of the editors, together with the publisher, felt that this jubilee should not pass unmarked and that an appropriate way to celebrate it would be to issue a book on the current state of theoretical biology in The Netherlands. The result of their co-operation together with the effort of the authors of the following ten chapters is now lying in front of you.

In this introductory chapter we briefly describe the history and scope of the journal *Acta Biotheoretica* in connection with the history of theoretical biology in The Netherlands.[2] We aim to provide an insight into the nature of theoretical biology as it is reflected in the articles published in *Acta Biotheoretica* and in the present book, and as it has been understood by the journal's editors from the beginning of the journal to the present day. In Section 2, we focus on the journal's early history; in Section 3, we discuss in more detail the perspective on theoretical biology that is taken in the journal.

1.2 THE JOURNAL'S ORIGIN AND EARLY YEARS

Acta Biotheoretica has its origins in the 1930s when the number of researchers working in theoretical biology was showing a strong increase. Due to this increase, in combination with the problematic situation regarding the mailing system and the travelling difficulties at the time, maintaining personal contacts among theoretical biologists became a complicated matter. This was the case not only within the field of theoretical biology itself but also with the rest of biological science and with areas outside biology. In 1934 three academic friends decided to issue a journal as they were aware of the increase in the number of researchers in the area of theoretical biology and the obstacles for good communication. These three professional friends were C. J. van der Klaauw (a zoologist at Leiden University), J. A. J. Barge (a medical anatomist at Leiden University) and A. Meyer (a biologist at the University of Hamburg). With this journal these scientists wanted to establish an open forum that would exist over an extended period of time and that, among other things, would enable students and scholars to always return to the study of questions and problems in theoretical biology.

To achieve their goal of issuing a journal the three friends approached many researchers spanning almost the entire academic field of biology. They received a very strong positive response, among other things because they

---

[2]However, we are no historians and we must consequently leave a more thorough discussion of this matter to specialists.

proposed to publish articles in the three modern European languages: French, English and German. Their idea was that every publication in the journal should be written in one of these three languages and be accompanied by summaries in the two other languages (Van der Klaauw *et al.*, 1935: 3).[3] Almost fifty researchers of a high standard responded in a positive way and agreed to support the three friends with respect to recommending new contributions and refereeing and editing submitted manuscripts. So, the three founders became the Board of the journal, and seven other scientists, each from a different country, were asked to form the Editorial Board. The others would constitute a main advisory team.

The new journal was called *Acta Biotheoretica* and started off with a support of the Jan van der Hoeven Foundation.[4] It was "(...) intended to be an international biological journal for the promotion of theoretical biology, being exclusively devoted to investigations on biological theories, particularly also the special mathematics and logic of biology." (*Acta Biotheoretica* 1 (1935): first printed page, no page number). In the years following the successful start of their journal, the editors supplemented *Acta Biotheoretica* with three additional series of publications. In 1936 *Folia Biotheoretica* was started, "(...) a series of introductory studies regarding theoretical biology. (...) Each number treats a certain subject and appears in connection with a symposium held at the University of Leyden." (*Acta Biotheoretica* 11 (1953-1956): inside front cover). In addition, to help students and scholars in gaining an overview of the available literature in their field, the editors collected references to typically theoretical papers and published these for subscribers in the *Bibliographia Biotheoretica*. This latter edition first appeared in 1938 and was discontinued in 1971, but lists of new titles in theoretical biology continued to appear in *Acta Biotheoretica* itself (Jeuken, 1971:1). This practice was discontinued when literature became available in Current Contents which itself later has been replaced by classified on-line literature search facilities covering almost all the literature in (theoretical) biology. Lastly, in 1941 the *Bibliotheca Biotheoretica* was established, "(...) a series of monographs on certain subjects of theoretical biology (...) [not aimed] at giving reviews in the

---

[3]From Volume 33 (1984) onward, the journal's official policy has become that papers should be written in English (Instructions for authors, *Acta Biotheoretica* 33 (1984): 63-66). Nevertheless, in later volumes papers in French have occasionally entered the journal's pages, for instance in special issues from the annual meeting of the *Société Francophone de Biologie Théorique*.

[4]Jan van der Hoeven was Professor of Zoology at Leiden University from 1826 to 1868. The Jan van der Hoeven Foundation for Theoretical Biology (the full Dutch name of which is '*Prof. Dr. Jan van der Hoeven Stichting voor Theoretische Biologie van Dier en Mensch, verbonden aan de Universiteit te Leiden*') was established in 1935 by C.J. van der Klaauw on the basis of a financial donation from the heirs of Jan van der Hoeven (Dullemeijer, 1976: 57) and exists to the present day.

manner of a pure compilation, but bearing the personal stamp of the author and having a scientific, objective and critical character." (*Acta Biotheoretica* 11 (1953-1956): inside front cover). At present, *Acta Biotheoretica* is the only one of these four series that still continues to be published on a regular basis. (See also Dullemeijer, 1976: 57-59.)

The researchers who began to publish in *Acta Biotheoretica* attempted to bring to the surface the specific character and value of biological investigation. In the view of the majority of these investigators biology could not be reduced to physics, chemistry or mathematics, even though many of them began to use these other scientific disciplines to support their own work. The journal aimed at achieving and maintaining a high quality by publishing on a wide variety of topics. From the first issues onward, the topics that were addressed in the articles published in the journal spanned the entire spectrum of theoretical biology as it was understood by the founding editors (see the next section), ranging from philosophy of biology (for example, there were two papers on the question whether biology could be considered an autonomous science; Sapper, 1935 and 1936) to the mathematical foundations of biological science (e.g. Volterra, 1937). Because it was felt that the specific character and value of biological investigation required the input from scientific disciplines outside biology (at the time, biologists frequently acquired training in fields outside biology), the task of the journal was adapted to include the publication of articles from other disciplines that could help biologists in their work.

Although the start of *Acta Biotheoretica* was very successful, unfortunately the main efforts to build bridges between various areas of scientific investigation began just before the outbreak of the Second World War. The journal barely survived the war because of reinforced language barriers and the particular situation of the founding editors: Van der Klaauw was imprisoned by the Germans; Meyer was physically unable to travel and, if he could, he was not allowed; Barge was in danger and was already advanced in years. Although young co-workers tried to consolidate the journal, the communication system was almost broken down fully. Because of these bad circumstances the journal was in need of a completely new editorial system. In the meantime, in different countries researchers in theoretical biology were becoming organised differently and were going their own way, among other things because they were working in different disciplines between which there was not much contact.

After the war a new start had to be made. Although the original intention had been to publish one volume of the journal annually (see *Acta Biotheoretica* 1 (1935): first printed page, no page number), most of the volumes that were published from 1939 onward covered two to four years per volume (Volumes 5 to 13). After 1960 the publication frequency increased, but it was not until the late 1960s that *Acta Biotheoretica* was again published annually. Moreover, not all of the previously established connections could be repaired immediately and certainly not to the original extent. Due to the severed connections, in the years following the war the journal's editors were unable to achieve the journal's aim "(...) to bring into the sphere of international thought theories and views which at present are too much restricted to certain schools and certain countries." (*Acta Biotheoretica* 1 (1935): first printed page, no page number), hoping to "(...) put an end to the terrible fragmentation of biotheoretical thought." (Van der Klaauw *et al.*, 1935: 2; translation TR). In stark contrast, the first issues of *Acta Biotheoretica* that were published after the war reflected the existing fragmentation of biological thought by containing mostly papers stemming from particular schools of theoretical research along with symposia reports of specialised disciplines or parts thereof. But as the situation improved a return to the original aims of the journal could be seen.

1.3 DUTCH THEORETICAL BIOLOGY AND THE SCOPE OF *ACTA BIOTHEORETICA*

In 1938, Van der Klaauw observed that theoretical biology " (...) has not yet developed into a universally acknowledged department of biology, a fact which gives rise to a good deal of confusion in the application of the term (...)" and expressed the view that theoretical biology should be understood "(...) on as broad a basis as possible, so that the subject may later be able to develop itself freely along its own lines." (Van der Klaauw, 1938: vii). Fifty years later, worldwide theoretical biology had become a mature field of investigation, with its Dutch branch occupying a special position in the international scholarly landscape. As the philosopher Michael Ruse acknowledged in an overview of the state of affairs at the time in philosophy of biology:

"Holland has a small subdiscipline which seems virtually unique to that country. So-called theoretical biology runs the gamut from hard-line mathematical modelling to serious study of the philosophical foundations of biology. To date, the subject's practitioners have perhaps been more successful at the mathematical end of the spectrum (...). Now, however, (...) a new generation of philosophically trained biologists is producing work on ethics, ecology, theory structure, and more (...)" (Ruse, 1988: 86).

Today, theoretical biology in The Netherlands still covers this entire spectrum.

While adopting the aim of the Jan van der Hoeven Foundation to develop "(...) a theoretical biology within, and strictly in the service of the science of practical biology" (Van der Klaauw, 1938: viii), *Acta Biotheoretica*'s founding editors did not see theoretical biology as subordinate to the gathering of empirical data but as a field of work that was of value in itself:

"Facts are regarded as parts of knowledge having eternal truth and
validity, while theory is regarded rather as a necessary evil, a means
by which facts may be deduced. (...) However, it is not true that
theory is a necessary evil. It is only to secure the joy of knowledge
that it [i.e. theory] gives us, that we search for the facts that can verify
it." (Van der Klaauw *et al.*, 1935: 1-2).[5]

Consequently, in their view theoretical biology should not only encompass the
foundations on which practical biology could rest, but also the development of
general theories that accounted for the phenomena studied in biological
practice. This stance is reflected in the conception of the character of
theoretical biology that the journal's founding editors adopted:

"By 'Theoretical Biology' is meant, in the first place theories and
views of a general character regarding the biological sciences, in the
second place the mathematical foundations of these sciences (...), in
the third place the philosophical and logical foundations of the
biological sciences (...)" (*Acta Biotheoretica* 11 (1953-1956): inside
front cover).[6]

While this broad view of the field was retained throughout the journal's
development to the present day, there were minor changes as new areas of
investigation were opened up. Some of these new areas were included into the
journal's scope, whereas others were not. When biochemistry became a
promising new area of investigation, for instance, researchers in this field
never used *Acta Biotheoetica* for publication of their theoretical results.
Similarly, physiological genetics and developmental biology did not find their
way to the journal's pages. Bioinformatics, however, was a field that did
explicitly enter into the journal's scope. As the editor, M. Jeuken, asserted in
an editorial in Volume 25 of the journal:

"Originally theoretical biology was only philosophy of biology.
About 1940 biomathematics came into the field. (...) Bioinformatics
is a new part of theoretical biology (...)." (Jeuken, 1976: i; cf.
Instructions for authors, *Acta Biotheoretica* 30 (1981)).

Similarly, the journal's scope was broadened to include studies into the
foundations of biomedical science (excluding, however, biomedical ethics).
One important new field of investigation that may in the near future explicitly
enter the scope of the journal is systems biology, a field in which experimental

---

[5]Van der Klaauw and co-authors wrote their 'Foreword by the editors' partly in
English, partly in German and partly in French. The first sentence of this quotation is
from the English part of the foreword, the sentences after '(...)' are from the German
part (translation by TR).
[6]This conception of theoretical biology was adopted from the constitution of the Jan
van der Hoeven Foundation. See also Van der Klaauw *et al.* (1935: 3) and Van der
Klaauw (1938: vii-viii).

biologists, biomathematicians and bioinformaticians have joined forces to understand how organisms work at various levels of organization (for recent overviews, see Kitano, 2001, 2002). In this field of investigation much co-operative work is currently being undertaken to enhance our understanding of living cells by means of modelling working cells *in silica* (Kitano, 2002; Nurse, 2003). (This, notably, notwithstanding the fact that until recently molecular biologists were not much interested in mathematical models of the cell.)

At present, the journal is conceived primarily as an international journal publishing on the mathematical and philosophical foundations of biological and biomedical science, as its subtitle since 2002 conveys (see Reydon, 2002: i). With respect to the philosophical end of the spectrum, the situation has changed considerably in the past decades. In 1976, *Acta Biotheoretica*'s editor observed that "As regards philosophy of biology, this journal is practically the only one existing in this special area of thinking (...)" (Jeuken, 1976: i). By now, there exist several high-quality journals that publish specifically on philosophy of biology, such as *Biology and Philosophy* and *Studies in the History and Philosophy of Biological and Biomedical Sciences*. Whereas these journals focus more on philosophy of science, *Acta Biotheoretica* aims to present philosophical investigation as an integral part of the whole spectrum of theoretical science.

1.4 CONCLUSION

The papers collected in the present book are intended to reflect the diverse nature of theoretical biology as it is understood from the perspective of *Acta Biotheoretica*. Although due to severe cutbacks in research funding, philosophy of biology in The Netherlands is at present in far from a good shape, we still believe that a mature theoretical biology should essentially encompass research into both the mathematical and philosophical foundations of biological science. Moreover, *Acta Biotheoretica* should in our opinion continue to publish in both fields of work, as well as on topics that lie in between these two.

We hope that the present book may serve well to mark the publication of the journal's 50th volume in 2002, as well as the journal's 70th birthday in 2005.

REFERENCES

Dullemeijer, P. (1976). Van Zoötomie tot Zoölogie: Een Historische Schets van de Leidse Algemene Dierkunde. Universitaire Pers Leiden, Leiden.
Jeuken, M. (1971). Note of the editor. Acta Biotheoretica 20: 1.

Jeuken, M. (1976). Note of the editor. Acta Biotheoretica 25: i.

Kitano, H. (Ed.) (2001). Foundations of Systems Biology. MIT Press, Cambridge, Mass.

Kitano, H. (2002). Computational systems biology. Nature 420: 206-210.

Nurse, P. (2003). Systems biology: understanding cells. Nature 424: 883.

Reydon, T. A. C. (2002). Editorial: Looking back, looking ahead. Acta Biotheoretica 50: i-ii.

Ruse, M. (1988). Philosophy of Biology Today. State University of New York Press, Albany, N.Y.

Sapper, K. (1935). Die Biologie als autonome Wissenschaft I. Acta Biotheoretica 1: 41-46.

Sapper, K. (1936). Die Biologie als autonome Wissenschaft II. Acta Biotheoretica 2: 12-18.

Van der Klaauw, C. J. (1938). Introduction. In: Van der Klaauw, C. J., J. A. J. Barge and A. Meyer (Eds). Bibliographia Biotheoretica 1925-1929. E. J. Brill, Leiden, pp. vii-xi.

Van der Klaauw, C. J., J. A. J. Barge and A. Meyer (1935). Foreword by the editors. Acta Biotheoretica 1: 1-4.

Volterra, V. (1937). Principes de biologie mathématique. Acta Biotheoretica 3: 1-36.

Thomas A. C. Reydon
Philosophy of the Life Sciences Group, Institute of Biology, Leiden University

Piet Dullemeijer
Mariahoevelaan 3, 2343 JA Oegstgeest, The Netherlands

Lia Hemerik
Biometris, Department of Mathematical and Statistical Methods,
Wageningen University

2

Images of the Genome: From Public Debates to Biology, and Back, and Forth[1]

Cor van der Weele

ABSTRACT

Public debates on genomics are still troubled by unclear images about the relationship between nature and nurture. The biological concept of norm of reaction can potentially play an important clarifying role here, but so far it has been a technical concept that is mainly used within biology. This essay starts from the assumption that metaphors play a large role in the understanding of scientific concepts. Since biology is not only a source of technical concepts, but also a rich source of images, the question arises which images it provides in association with reaction norms. For a long time the concept of reaction norms did not match dominant metaphors of DNA, but the metaphors are changing: a thorough innovation of images about DNA and the genome is currently going on.

The second part of the paper starts from the question of how we may expect metaphors to influence public debates. Recent findings suggest that we should critically assess common assumptions about the interpretation of genetic metaphors. Interpretation is not univocal but appears to depend on the context. Building on this, I suggest some further thoughts. First, in public contexts, the urgency of ethical questions requires innovative imagery about the genome. Second, because of the active cognitive and heuristic use of metaphors in scientific contexts, there is reason to look at biology as an especially productive source of such innovative imagery.

2.1 THE GENETICS THEATRE

Concerning the genome, large amounts of images[2] have come into use, and sometimes went out of use again, in recent decades, within biology as well as in society at large. DNA is a code, the secret of life, the blueprint of living

---

[1]This is an adapted version of an unpublished essay that was originally written for the program "Social Components of Genomics Research" of the Netherlands Organization for Scientific Research (NWO).

[2]I use the word "image" in a broad sense, which includes visual images as well as metaphors, or linguistic images, which are concepts that have been imported from other domains.

T.A.C. Reydon and L. Hemerik.,(eds.), Current Themes in Theoretical Biology,
9-31.

organisms, the holy grail of biology, the recipe for development, the list of parts of living beings, the language in which God created organisms, the lazy cigar smoking director of the cell, etc. Miltos Liakopoulos (2002) made an inventory of metaphors in public presentations of biotechnology. His subject is biotechnology and the images he is interested in therefore not only concern the understanding of DNA but also its manipulation. The latter generates hope and fear and keeps the imagination very busy: he had to create clusters and super-clusters in order to obtain some kind of overview of the large amount of metaphors he found. Since his inventory spans several decades, from the early seventies to the mid-nineties, he is also able to discern patterns in time. Metaphors expressing fear, such as Pandora's box, were relatively dominant in the seventies, when risks of genetic manipulation were emphasized. In the eighties, when biotechnology promised mountains of gold, metaphors of success and adventure flourished, while in the nineties "informative" metaphors dominated on what DNA is and how it works, such as code, program, blueprint or book. Those images are not new; they came into use soon after the structure of DNA had been elucidated in 1953.

This essay is mainly restricted to the informative metaphors, jumping back and forth between public and scientific contexts. Though it does not present a review of metaphor in biology, in the end the essay will turn more specifically to biology.

In her book *ImagEnation* (1998) José van Dijck presents science, at least its public aspects, as a theatre of representation. Within the theatre of genetics we see an ongoing contest in which nothing less is at stake than the definition of genetics. In this contest various groups defend their own interests with rivalling images of genetics, the "plays" in the theatre. The roles, plots and metaphors in these plays deserve our special attention, according to Van Dijck.

A theatre is a nice metaphor for the imagination in connection with science. It invites us to imagine that new plays are constantly being written. The definition of genetics may be contested *within* these plays but more importantly perhaps there is the contest *between* the plays, not in the form of direct confrontation but indirectly, as plays try to attract and entice the theatre audience as well as the press. The authors of these plays are genetic researchers but they are not the only playwrights. Ethicists, journalists, science fiction writers, business people and increasingly also visual artists try to persuade, provoke, seduce or shock the public with their "imagEnations".

Take Eduardo Kac, creator of the artworks *Genesis* and *GFP Bunny*. For *Genesis*, Kac took a passage of the bible book Genesis, translated it into Morse code, translated the Morse code to the four-letter code of DNA, had synthetic DNA created from this code, which was then mixed with DNA of Kac himself, put in a petri dish and under a microscope, and projected in purple colors on the walls of an exhibition room. *GFP Bunny* is about the genetically modified white rabbit Alba. Alba has a gene of a green fluorescent jellyfish inserted in its genome, which codes for Green Fluorescent Protein (GFP). Alba produces this green fluorescent protein in its body cells and lights up in the dark. When a transgenic organism is sitting in your lap and looks into your eyes, it is no longer a monster to be scared of according to Kac.

It does not require much effort to see these performances as plays in the theatre of genetics, meant, among other things, to play and provoke. Other people build on these performances with follow-up plays; *GFP Bunny* inspired the organization of ethics conferences with titles such as "Art, ethics and genetic engineering" (Madoff, 2002; and www.biota.org).

Nature versus nurture?

Genetics and genomics are developing so fast that ethics and public debates can hardly keep up. It is therefore not surprising that many plays in the theatre of genetics are outdated. Many old ways of imagining genetics and associated issues, such as relations between genes and their environment, or between nature and nurture, are not in line with new data, new images and new explanations within biology. This is a basic assumption of the present paper. It has been said before, for example by Evelyn Fox Keller (1995, 2000). José van Dijck, too, suggests that a discrepancy exists between new biological data about the fluidity and complexity of the genome on the one hand and old images that suggest fixity and rigid causality on the other. New concepts will be needed to bridge the gap between gene centric images of DNA and new developments in genetics and genomics.

A play that has been performed for innumerable years is the drama of nature versus nurture. The play has been declared outdated and worn out every now and then, but somehow it manages to survive in a constant series of revisions and updates. New authors, fascinated or irritated by the theme of the play, add new turns, twists and details, often hoping to fundamentally alter its plot and message, or better still, replace the play by something entirely new. Some of these authors are critical biologists who are worried about genetic determinism. Richard Lewontin (e.g. 2001) has been prominent among them for many decades, attacking the metaphors and models that define genetic determinism, such as blueprint or book of life, which suggest that DNA is a molecule that is in full control of the organism. Those metaphors are dangerous because they invite us to look for the definition as well as the solution of social problems in the genes. This is the wrong place to look because in fact DNA is not in control. It is a reactive molecule that does not do anything on its own; it can neither reproduce itself nor control the organism or even make proteins by itself. Biological processes are causally complex: in their causal models, critical biologists have long been urging that information does not only or primarily come from DNA but from many different sources. As Susan Oyama argues in *The Ontogeny of Information* (1985/2000) causal information consists of signals that make context dependent differences. The origin of such signals in biological processes is heterogeneous.[3]

These playwrights are advocating a biology that respects complexity, and their main worry concerns the reductionism found in the view that could be called "naturism", i.e. the view that the genetic nature of organisms fully determines its properties. This does not imply that they like the reductionism of the opposite, equally one-sided "nurturism" any better; they reject reductionism as such, including the polar dichotomy of nature and nurture. They hope to finally write a good play which is not simplistically black and white but which pictures a rich and complex reality that is biological as well as social through and through. Their third way is characterized by words such as interactionism, system approach or co-development. But their task is not an easy one, for the set-up of two opposing forces is an old and attractive scheme for a play, promising an exciting struggle. Though in fact many cherished dichotomies do not deal with real opposites at all, they are often associated with deep polarity; think of sun and moon, mind and body, man and woman. The tension between the elements of such pairs is further deepened by strong associations with hierarchy: the sun is stronger than the moon, the mind is higher than the body, etc.

The challenge is to design an exciting plot about complexity in the absence of hierarchically ordered pairs of opposites.

2.2 PHENOTYPIC PLASTICITY AND NORMS OF REACTION

In biology, the points of departure for alternative plays are promising enough. Let us start by looking in some detail at an old and still very good concept, *norm of reaction*[4], which has to do with phenotypic plasticity, the phenomenon that the characteristics of living beings vary to a certain degree under the influence of environmental variables. *Norm of reaction* is the main concept through which phenotypic plasticity is expressed by visualizing the interplay of genes and environment in a simple way.

Take a genotype or rather an organism with a certain genome. Take a lot of those organisms, let them develop in different environmental conditions and measure phenotypic outcomes. The diagram that can be drawn to visualize this

---

[3]There is much to be said about the concept of information in biology. Griffiths for example distinguishes between causal and intentional information (see Griffiths, 2001; Griffiths, in press). Causal information is about differences that make a difference, which makes information inherently context dependent. They contest the widespread idea that the role of DNA is best understood through an intentional view of information which implies a context independent relation between genotype and phenotype.

[4]The word "norm" is misleading since the concept is not normative.

procedure has on the x-axis some environmental variable, say temperature, and on the y-axis the phenotypic outcome, say the length of the organism. A flat norm of reaction (Figure 1, line a) means that this specific aspect of the environment, in this range of values, does not influence this specific phenotypic outcome for this genotype. In the figure the environment does influence the development of the organism with genotype b. The figure shows crossing norms of reaction, suggesting that there is not necessarily a simple relationship between the outcomes of different genotypes in different environments.

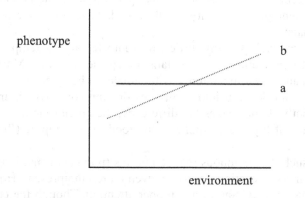

Figure 1. The reaction norm concept: two (arbitrary shaped) continuous norms of reaction for genotypes a and b.

Norms of reaction of this smooth form are found when some aspect of the phenotype, say the length of the organism or one of its parts, varies continuously under the influence of some environmental factor. A discontinuous reaction norm, which makes sudden jumps, is found in the case of *polyphenisms*. This is to say that two clearly different phenotypes such as two sexes (e.g. in reptiles) or two different wing patterns (in butterflies) are the outcome of organism-environment interaction.

Studying norms of reaction meets with huge difficulties, because it requires developmental conditions to be under control. It is therefore not surprising, if for ethical reasons alone, that little is known about them in humans. Growing plants does not present these dilemmas and indeed one of the first norms of reaction that became well known concerned a plant: it turned out to make a lot of phenotypic difference whether milfoil was raised at sea level or in the mountains.

Apart from controlled experiments, simply looking at natural phenomena also yields information: the introduction of the concept in 1909 by Woltereck was based on the observation of curious variations in animals. Woltereck observed water fleas that developed a "helmet" in the presence of the larvae of a certain species of midge, which is a predator of water fleas. The helmet was absent when the larvae were absent. Woltereck's (justified) idea was that the water fleas developed the helmet under the influence of a substance associated with the larvae. The helmet appeared to protect the fleas to a certain extent from predation. The example shows that the development of a phenotypic character can be influenced by a different species. This phenomenon is also common in humans. Our main predators are micro-organisms and we react to them by developing an antigen-specific immune system. In other words, our immune system develops in interaction with other species.

As more has become known about organism-environment interaction, it has turned out that phenotypic plasticity is all around. Here are some arbitrary examples in animals:

- Crocodiles, as well as many turtles, turn into females at temperature x and into males at temperature y. The details are species specific. Males may also result at high and low temperatures, with females in between.

- Some frogs can develop into carnivores or omnivores which are built somewhat differently. Food makes the difference; the crucial factor turns out to be the quantities of thyroid hormone in the food of the tadpole (Pigliucci, 2001).

Examples of such discontinuous reaction norms (polyphenisms) in biology abound[5] and continuous reaction norms are even more omnipresent. However, it has taken biology a long time to pay proper attention. Though the concept, stemming from 1909, is not exactly new, for most of the twentieth century it hardly played any role in biology. It slowly began to come to life as late as the sixties. Richard Lewontin's efforts were important in this (re-)vitalization; he drew figures such as the above one in order to explain the concept of reaction norm and to make it clear how important it is for understanding the true nature of heritability, and the relation between nature and nurture. If genotype a does "better" in environment x than genotype b, it is not al all self evident that the same will be true in environment y. The assumption that reaction norms will seldom cross and that the shape of reaction norms thus hardly contains interesting information, is a central tenet of genetic determinism (see, for example, Lewontin, 1974).

Ecological developmental biology

The revitalization of the concept did not take place overnight. For example, because of the ongoing underrating of environmental causes in development, developmental disruption from environmental causes could go unnoticed until the nineties. Only then was it recognized that various chemical substances have estrogen-like effects. Since estrogen plays an important role in sexual

---

[5]For more examples, see Van der Weele (1995/99), Gilbert (1997, 2001), Pigliucci (2001).

development these chemical substances, such as PCBs which have been released into the environment for decades, cause great damage, notably sexual damage such as infertility. By the time these effects began to be studied, the effects on some animal populations was already dramatic. As long as development is supposed to be fully controlled by genes, such phenomena cannot be given the attention they need. An ecological approach to development is necessary (Van der Weele, 1995/99).

The subsistence of the reaction norm concept at the margins of biology for the larger part of the twentieth century is consistent with its being out of tune with the dominant images of biological causation, in which the genome is the central player. This misfit with traditional images perhaps also explains why even many biologists, or at least biology students, find it difficult to grasp the concept of reaction norm.[6] But in recent years the biological scene seems to be definitely changing and reaction norms are now regularly placed at the center of the stage. Developmental biologist Scott Gilbert, in a programmatic review entitled "Ecological developmental biology: Development meets the real world" (2001), has defined some lines of research for an ecological approach to development in living beings, including humans. In the play called *Development* he proposes a central role for reaction norms and almost randomly lists some phenomena in humans that qualify for reaction norm description: sunlight activates vitamin D production, in the mountains our bodies produce more red blood cells than at sea level, exercise stimulates the production of muscle tissue, our immune system develops in reaction to micro-organisms, brains develop through environmental signals. He could have gone on almost endlessly: our intelligence develops under the influence of upbringing and education, body length is influenced by food, behaviour develops through experience... In another paper Gilbert stresses a fascinating aspect of our ecological embeddedness: our identities and our bodily boundaries should be put into perspective. People symbiotically live with billions of micro-organisms, which inhabit our bodies, especially our intestines, surpassing in numbers our own body cells by far and influencing our gene expression. Our bodies are adjusted to and dependent on these guests. Gilbert (2002) concludes that each of us is a "we".

Massimo Pigliucci's (2001) book on phenotypic plasticity likewise proposes a leading role for reaction norms in the drama of development. The book's subtitle "beyond nature and nurture" emphasizes an important implication of taking them seriously. Reaction norms inevitably make you realize that development without genes and development without an environment are equally nonsensical ideas and that both genes and the environment make all kinds of causal difference for the phenotype, whether or not the details are known. Thinking in terms of reaction norms raises certain

---

[6]Rolf Hoekstra, personal communication.

questions, such as the shape of the reaction norms, and makes others disappear, such as the question how many percent of our characteristics is determined by either genes or environment. That question makes no sense for individuals; if one insists on percentages it is perhaps better to say that we are one hundred percent caused by genetics and one hundred percent by other causes.[7]

2.3 EPIGENESIS AND THE EVOLUTION OF IMAGES IN BIOLOGY

Has the changing climate in biology with regard to reaction norms generated new images of causation, which make reaction norms more easy to grasp for biologists as well as non-biologists? Before answering this question it will be helpful to obtain a wider perspective on this changing climate because the growing recognition of reaction norms cannot be separated from new views of the genome; with genomics research as the driving force.

The mechanisms leading to phenotypic plasticity have long been largely mysterious; they belonged or still belong to the "black box" of development. Genomics is opening this black box. The concept of reaction norm so far does not figure in genomics. But epigenetics does and it is through epigenetics that genomics can deal with a diversity of causes. Epigenetics is defined in different ways but a usual view, since "epigenesis" has turned into "epigenetics", is that it is the field that studies the regulation of gene expression.[8]

As the understanding of gene structure and function is increasing, so is the appreciation of the complexities of gene regulation.

Even without (or before) explicitly taking genes into account, many things can be found out about the mechanisms of phenotypic plasticity. Hormones, the long distance messengers in organisms, often play a role as Pigliucci (2001) strongly emphasizes. One of the examples he refers to concerns temperature dependent sex determination in turtles. In a certain species, males are produced between 23 and 27^0C and females when the temperature is over 29.5^0C; in between, males as well as females come out of the eggs. In order to

---

[7]What makes sense with regard to percentages concerns *local differences within populations*; heritability is a measure for genetically determined *variation* in populations, in a certain environment. For an almost imperceptible transition from statements about percentages of the heritability of *variation* in a character to percentages of the heritability of the character *itself* ("accidentally" IQ) see Fukuyama (2002).

[8]In former times epigenesis was contrasted with preformationism. "Epigenetics" is a blending of epigenesis and genetics. For a thorough historical and conceptual overview, see Van Speybroeck *et al.* (2002).

test the hypothesis that a specific hormone causes the differences, turtle eggs were treated with this hormone. This indeed produced males at "female" temperatures. Subsequent studies have revealed how the normal process, in which temperature causes the switch, works. Substance A and B can both bind to the same substrate, testosterone. A binds more effectively at higher temperatures and converts testosterone into estradiol which leads to females. B wins out at lower temperatures; it turns testosterone into dihydrotestosterone, which turns the organism into a male. Because of their large roles in the formation of organisms and as messengers of long distance signals, Pigliucci sees hormones as the unsung heroes of the nature-nurture field.

But closer to the skin of genes, almost literally, many more epigenetic phenomena are to be found such as chromatin and DNA methylation. Chromatin is the DNA/protein complex that is sometimes described as the phenotype of chromosomes. It can be in different states. When it is in a compact state, DNA cannot be transcribed; only in the looser state in which it is called euchromatin is transcription possible. Chromatin is thus an important factor in the regulation of gene expression. Within genomics it is now a centrally staged character in the study of epigenetic phenomena. In one play genes are depicted as puppets which are controlled by puppet players. A whole bunch of enzymes, hormones, and RNA's play the roles of the puppet players that pull the genes' strings and make them dance. Chromatin modifying enzymes are the master puppet players or so we read in a *Science* special devoted to epigenetics (Pennisi, 2001). Those enzymes, in collaboration with other players such as the enzymes for the methylation and demethylation of DNA, cause the changing patterns of gene expression during embryonic development as well as in the rest of life. They allow the cell to respond to environmental signals brought in by hormones, growth factors and other regulative molecules.

In the development of cancer, too, patterns of DNA methylation and chromatin structure are fundamental. A review article (Jones and Baylin, 2002) emphasizes promoter hypermethylation as a mechanism that "silences" genes and disrupts the normal functioning of cells. Such epigenetic phenomena play a role in almost every step in the development of cancer and it is expected that many more regulating mechanisms will be found. This research takes place in the context of functional genomics and includes techniques to screen the activity of large amounts of genes at once. Time and again findings in this field are unexpected. They inspire increasingly complicated models of genes and gene expression.

Master? Servant? Rebel?

The surprising findings concerning the genome are not restricted to gene expression. The structure of the genome has also given rise to wonder and new images. In the eighties it came as a surprise that large parts of the genome appeared to be useless. Apart from stretches of DNA that were clearly important because they code for protein (the exons), much larger pieces of DNA were found (the introns) that consisted of endless repeats and had no obvious function. Immediately a whole series of social metaphors came into being which pictured this DNA in an unfavourable light: it was called "junk" or "parasitic" DNA. The genome emerged as a state that is hard to govern. Well meaning "good citizen genes" do their best to make this state work in proper democratic fashion with the help of a "genetic parliament", a task that is frustrated by the actions not only of "immigrants" and "nomad genes" but also by large amounts of profiteers, scroungers, vagrants and junks who threaten to disrupt the proper functioning of the state.

John Avise (2001), writing about these metaphors and their evolution notes that images such as blueprint which suggest that DNA is a kind of ordered and efficient plan, used to prevent biologists being receptive to such chaotic scenes. He also notes that the evolution of images continues: biologists gradually revise their negative evaluations of the uncomprehended DNA. Just like anthropologists, who see old dumping grounds as valuable sites to gather new knowledge about life in former days, biologists increasingly come to see the DNA outside the exons as containing important information about our evolutionary past rather than merely a useless mess or a dangerous social wilderness.

Not only ethics and public debate have trouble keeping up with the speed and direction of biological findings; the same is true for the language of biology itself. Yet, increased recognition of the complexities of both gene organization and gene expression do lead to an ongoing innovation of images in biology. Basic tenets of the play that is being performed have become uncertain and central metaphors, plots, and leading roles are being rewritten. DNA is now silenced, awakened, altered, renovated, attacked and protected... It has become less powerful and well organized, more passive, chaotic, vulnerable and reactive. Instead of being the master of the cell, the genome has sometimes come to look like a servant, serving the proteins by being their memory (Morange, 2001; Shapiro, 2002). The cell is seen as having a hard time in keeping this memory sufficiently well-organized and up to date, and to making it useful in the right place at the right time. Thus, a new hierarchy sometimes seems to take shape, in which proteins are the main players in biology, and genes "just" help them reproduce.

Since research in genomics has in fact only just begun, it is only to be expected that these are not final images and that the evolution of metaphors continues. A reversion of cell hierarchy may not be the most satisfactory way to deal with complexity but images suggesting more complex causality are already increasingly popular. For example, when I conducted a modest inventory among biologists and philosophers of biology, asking for images in connection with reaction norms[9], many of the answers were images that referred to decentralized and potentially chaotic processes involving heterogeneous actors: a meeting, a jazz orchestra, a village, a parliament, etc. Such images picture developmental processes as social processes and suggest that DNA is one player among many.

The special characteristics of biological reality also stimulate more specific dramatic images and plots. For example hormones, honoured by Pigliucci (2001) as the unsung heroes of gene environment interaction, may inspire new plays in which they (the hormones) jump on their horses like medieval knights in order to deliver their emergency messages to the kings of neighbouring countries who can then take measures to call some rebellious genes to order.

2.4 METAPHORS AND THEIR INTERPRETATION: TEXT AND CONTEXT

How may we expect the newer metaphors such as DNA as the cell's memory, or the cell as a parliament, or hormones as knights, to be helpful in public debates on nature and nurture?

Through Lakoff and Johnson's *Metaphors We Live By* (1980), it has become a widespread assumption that metaphors guide our thought. Lakoff and Johnson argued convincingly that metaphors are not linguistic ornaments but comprehensive conceptual tools that characterize large parts of our thought, including our understanding of abstract concepts. They help us to understand phenomena of a new and complex domain through (more familiar) images from other domains and in so doing direct attention to specific aspects of a subject, while hiding others from sight: metaphors generate searchlights as well as blind spots.

The blind spots of the old images of DNA have extensively been criticized by biologists as well as others. They overrate the power of DNA and in the context of public debates they lead to a one-sided and reductionist consideration of social problems and their possible solutions. Dorothy Nelkin and M. Susan Lindee, for example, are very outspoken in their book *The DNA Mystique* (1995) in which they explore popular images of genetics. These deterministic images of DNA convey a message they call genetic essentialism which equates human beings with their genes and which they find socially dangerous.

However, there is reason for some caution because much is unknown about the precise role of metaphors in thought. A background assumption in the fears just mentioned is that metaphors have a clear interpretation and so give clear

---

[9]Among philosophical/biological colleagues in the US and the Netherlands: Scott Gilbert, Peter Taylor, Kelly Smith, Lenny Moss, Fred Nijhout, Rasmus Winther, Rolf Hoekstra, Susan Oyama, Jason Scott Robert, Ron Amundson and Arno Wouters.

and unambiguous conceptual guidance. Communication scientist Celeste Condit has challenged this assumption, arguing that in reality the interpretation of metaphor is far from univocal. Her arguments are theoretical as well as empirical. The theoretical reasons stem from Josef Stern's approach to metaphor presented in his book *Metaphor in Context* (2000). According to Stern, Lakoff and Johnson have rightly awakened us to the omnipresence and conceptual importance of metaphor but their approach overlooks the importance of the context in which metaphors are used; they appear to assume that meanings from the source domain are simply transferred to the target domain. Stern proposes instead that the interpretation of metaphor is a two-step process. First, each metaphor has a large amount of possible interpretations in connection with the connotations of words in the different domains that are brought together in a metaphor. Second, by using the metaphor in a specific context a selection or filtering process takes place in which one of these potential meanings becomes the dominant one.

Condit has devoted several papers to the interpretation of DNA metaphors. On the basis of audience studies among students she suggests that the blueprint metaphor is not necessarily interpreted in a deterministic way. One usual but hardly deterministic interpretation is that a blueprint is a kind of plan which can always be changed in response to new situations (Condit, 1999). Building on this finding as well as on Stern's approach, Condit *et al.*, in their paper "Recipes or blueprints for our genes?" (2002) compare the interpretations of two DNA metaphors. The recipe metaphor was advanced in the eighties by critics who rejected the blueprint metaphor for the usual reasons: too rigid, too gene centric, too deterministic. "Recipe" allegedly would call attention to procedural aspects of development and create room for what is contingent, unexpected or otherwise not predetermined. Flexibility thus was an important consideration; besides, the critics reasoned, images such as baking a cake do not presuppose a one-to-one relation between the ingredients and the final result.

As they expected, Condit *et al.* found, through interviews and surveys, that both metaphors can be interpreted in various ways. They found relatively open interpretations of the blueprint metaphor, such as a plan or sketch, as well as more deterministic ones, such as a map. Recipe was also interpreted in different ways. It struck the authors that this metaphor was often interpreted rather statically, as a list of ingredients rather than as a procedure, while the latter was the interpretation intended by those who actively favoured the metaphor. Condit *et al.* also notice that, contrary to Stern's assumption, there often appears to be a favoured interpretation of metaphor even prior to a specific context.[10]

---

[10]An alternative hypothesis might be that each individual automatically or implicitly has some specific context in mind.

The introduction of an explicit context did have clear effects, however. This part of the study was conducted through focus groups in which students guided by a supervisor talked about the two metaphors. The open discussions with which the groups started brought out the same diversity of interpretation as was found in the surveys. Subsequently the supervisor introduced biotechnology, in particular genetic modification of humans, to which all the students expressed opposition. When the students were asked to think about the two metaphors in this context, new and surprisingly uniform interpretations emerged. "Recipe" was now interpreted as an invitation for manipulation, associated with ideas such as "cooking up humans". A blueprint was no longer interpreted as an open plan, but as a map, suitable to study and consult but not to alter. Because of the repulsive implications the recipe metaphor seemed to have, as opposed to the safety of the blueprint, students now clearly preferred the blueprint metaphor in a deterministic interpretation. Moral worries about genetic modification in humans apparently formed a selective context for the interpretation of the metaphors.

Context thus appears to play an important role in the interpretation of a metaphor. But naturally this can be true only insofar as a metaphor is interpreted at all. The analysis of popular science texts suggests, according to further findings reported by Condit, that this is not necessarily the case. In such popular texts metaphors are often used in a rather sloppy and perhaps simply decorative way. They are hardly elaborated and seem to have very limited didactic use. Briefly, in such cases, it does not appear to make much difference which metaphor is used.

This finding deserves further reflection. If authors in some contexts use metaphors in thoughtless ways, the same might well apply to readers. More generally metaphors about DNA might sometimes, or even often, be used in passive and thoughtless ways without a clear interpretation. It is not hard to imagine that, if not pressed by an interviewer, many people are not very precise about what is really meant when they read or say that DNA is a blueprint for organisms.

This finding fits in with the experience of Peter Taylor[11] who asked his students to play around with various metaphors. He noticed that, once students became actively engaged in exploring the images and their implications, they became much more interested in what exactly it was they wanted to say as well as in biological details. This in turn corresponds with one of the more elementary insights in how people learn best; active interest stimulates learning capacities. In the context of the theatre of genetics it implies that playwrights who want to stimulate people to think for themselves about genetics might consider ways to actively engage the public, perhaps even through courses in creative playwriting. Active engagement not only

---

[11]Personal communication.

stimulates the imagination but might also stimulate a deeper interest in biological phenomena which in turn is probably the best antidote against rigid, simplistic or empty ideas about biological mechanisms. In this line of thought, biological researchers would be the most active and innovative metaphor-interpreters of all since they have reason to explore the implications of metaphors in biology more deeply and systematically than most other people.

I will return to these implications for science at the end of the paper. Let me first spend some time on further thoughts on the social contexts of genetics and on the importance of a rich ethics.

2.5. POSSIBLE CONTEXTS: ETHICS AND THE IMAGINATION

The theatre of genetics takes place in a cultural context and this implies, as José van Dijck (1998) recognizes, that changes of meaning in the theatre do not only originate in the plays themselves but also in social, political and cultural developments in the world at large. Like Liakopoulos (2002) she mentions dominant ecological worries that prominently influenced the approach of biotechnology in the seventies as an example.

Genetic manipulation is an important context for images on the role of genes.[12] It does not take much reflection to see that it is not one context, but a multitude, within which we can detect further contexts in so far as we are interested in emphasizing further differences. The ecological risks of genetically modified organisms, golden rice, prenatal diagnosis, individualized medicine, new forms of eugenics: these are all different subjects. What many biotechnology issues do have in common however, is that the practical choices are often puzzling and difficult. Because the discovery of the structure of DNA in 1953 not only led to knowledge but also to technologies to alter the genome, this discovery was followed by a golden age for ethicists as well as geneticists. In the possible world in which we, inspired perhaps by Peter Taylor, all become playwrights for the theatre of genetics, we are also more or less forced to become philosophers and ethicists to a certain extent. The powerful tools and implications of genetics may make us feel uncomfortable and hesitant but I think that Peter Sloterdijk (1999) is right when he says we cut a poor figure if we refuse to think about them.

It is not only in a distant future that the choices confronting us will be awesome. Some choices have already been awesome for quite a while, as many women can testify who faced the seemingly simple choice of prenatal

---

[12]It may be questioned what is text and what is context (or figure and ground). The best answer, it seems to me, is a pragmatic and context dependent one. Genetics has to do with gathering knowledge as well as tools for action. This paper deals primarily with knowledge and sees the biotechnology aspects as context. In other situations it may be the other way around.

diagnosis during pregnancy. Do you want to know whether your future child will have Down's syndrome, and if so, do you choose to have an abortion? Many women, confronted with these questions, feel aversion to having to make this choice: they feel it is really too big for humans, they feel forced to "play god" against their will (Van Berkel and Van der Weele, 1999). Finding honest answers to such choices implies that you cannot avoid judging the value of human lives, wondering what "the best for the child" means in this case, thinking about the extra burden of a handicapped child and the role you allow that thought to play, etc. And such considerations are not free speculations; what is at stake is a possible abortion at 18, 19, or 20 weeks of pregnancy. Yet prenatal diagnosis has been "routine" for over thirty years now and genetic diagnosis has much more to offer these days. Even without genetic modification, there is much real life drama in the theatre of genetics.

Because the moral questions on genetics and genomics, in combination also with developments in neurology and informatics, may become huge and in part already are huge, we shall have to put into action everything that can help us to deal wisely with them. We cannot afford to think about them in terms of simple black/white or yes/no schemes: the choices these schemes offer are simply too poor and too rigid. We need an ethics that does not ask questions in the form of simplified dilemmas and old dichotomies but that invites us to rethink the questions, think of new solutions and use everything that may help us in these searches. Such an ethics certainly needs rich knowledge about biological complexity, and also rich imagination, which can help to devise various plays in the theatre of genetics, a process which the philosopher Dewey called "dramatic rehearsal". Additional helpful areas could easily be listed; it is a typical situation of all hands on deck.

The strategy of the students in Condit's focus groups, who fell back on blueprint metaphors in their uneasiness about genetic modification in humans, will not do; its comfort will only be briefly effective. At the same time, the students' uneasiness is not hard to understand: non-deterministic images of DNA do indeed seem to suggest the possibility of genetic manipulation more readily than deterministic ones. There are more examples than Condit's focus groups of this association, such as a recent paper by Tim Lewens (2002). Lewens, starting from an interactionist view of development, with norms of reaction as the frame of reference, notes that in this approach genes and environment should be considered as developmental resources of equal standing. Does this imply that thoughts about a just distribution of resources are also relevant for genes? Although there are many caveats, in principle the answer is yes, he argues. In other words, the question of whether we are obliged to provide our children with good genes, in the same way as we feel obliged to provide them with good education, is almost on the table. Genetic solutions for unequal chances in life may as yet still be risky, inefficient and impractical, but that does not change the general idea. And since such lines of reasoning are so near at hand, they deserve very serious consideration.

From the perspective of "all hands on deck" it is only too welcome that writers and visual artists, such as Eduardo Kac, increasingly try their fantasy and imagination on genetics, creating "dramatic rehearsals" in the genetics theatre. José van Dijck discusses quite a few of the science fiction books that have been dealing with genetics. For example, Amy Thomson's (1993) *Virtual Girl* describes the adventures of a girl who started her life as a computer program and was subsequently downloaded into a body. In this possible world, in which genetics and informatics blend, the idea that the essence of humans is to be found in their genes is rejected and Frankenstein-like elements receive a new twist: the monster develops into a self-conscious being who functions well. But the story is not innovative on all fronts. The world still consists of two sexes, men and women, and their relationship is very traditional, Van Dijck notices. More generally, apart from the technical possibilities, which are often highly original, such books often have very traditional plots from a social perspective, along the lines of old dichotomies. Why not put the imagination to work in this respect, too, in order to replace old roles and scripts by experimental ones? This gradually begins to happen as well; Van Dijck refers to a novel in which not two sexes figure but five. Further examples could be added; Houellebecq's (1998/2001) *The Elementary Particles* ends with a world in which sexual differences have disappeared completely.

It is interesting to think about possible worlds in which only one sex exists, or three, four, five, or in which people easily change their sexual identity, and how such a world could free us from the dichotomies in which thinking about men and women is often imprisoned. In our ordinary world, meanwhile, in some species of fish individuals have been changing their sex in the course of their lives for millions of years. Many species have no sex differentiation at all (mushrooms, unicellulars), hermaphroditic snails are all around, all kinds of remarkable relations between the sexes exist[13] and mechanisms of sex determination are often stranger than we could dream. For example, in the sea worm *Bonellia,* a larva develops into a female if it lands on a rock, but if it lands on the proboscis of a female, it migrates into the female and develops into a tiny male (orders of magnitude smaller than the female), which remains inside the female for his whole life, fertilizing her eggs. In short, we hardly need science fiction as a starting point for new metaphors and plots on sex and sexual relations. Real world biology is a rich source of unusual stories.

---

[13]Patterns that tend to be surprising for humans include "female eats male after copulation" , "male takes care of the children" etc.

Frankenstein and diarrhoea

There are many more social contexts in which we can think about genomics and genetics. The context of hunger, for instance, or the world health situation: as discomforting as Frankenstein or the boys from Brazil, less sexy.

How can genomics and biotechnology contribute to global health equity? This is the question in a paper by Peter Singer and Abdallah Daar (2001). They refer to the "10/90 gap": 90% of research capacity is spent on the health problems of 10% of the world population. But they also point to initiatives to put genomics and biotechnology to work for health problems in developing countries. Thus, PCR technology has improved the diagnosis of leishmaniasis and dengue fever in some Latin American countries while in Cuba, a country that invests much in biotechnology, a vaccine against meningitis B has been developed.

The "10/90 gap" applies to more domains, each time with slightly different figures: income, food, energy use. When we are deeply occupied with Frankenstein, virtual girls and fluorescent rabbits, while each year millions of children die from elementary problems like diarrhoea, we may wonder about the selectivity of our attention.[14] In terms of the theatre of genetics the challenge is how to write plays on global issues that can compete with Frankenstein in capturing our attention. Here again, norms of reaction are relevant. Whatever may be the unknown shape of reaction norms for human characteristics, we may safely assume that a better environment improves the achievements of almost any genotype. Good food, good schools, clean drinking water, public transport, vaccines, a peaceful society and loving parents are all factors that lead to more healthy and more intelligent human phenotypes. The call for sexy plays about complex causation is again on the table.

Ethical issues associated with genetics and genomics tend to have an urgency that is often annoying, because we have not asked for such complex choices. But because the issues can not be avoided, we would do well to mobilize all the sources of wisdom we can think of. In my view, biology and the imagination are two important sources of enrichment for ethics.

---

[14]In *Images of Development* I have been arguing, in connection with the problems of selective attention, for an "ethics of attention". However, an ethical evaluation of attention makes no sense without a psychological, sociological and biological analysis of attention.

2.6 BIOLOGY AND EMBODIED THOUGHT

The place of metaphor in biology was not the central subject of this essay, but let it at least briefly be the final one. The presence of metaphors in science is now widely acknowledged, but still also generates uncomfortable feelings. In popular thought, and for heuristic uses, metaphors may be needed, but is not the real search for truth a different matter... should scientific theory not as far as possible be free from them?

A call for a metaphor free science presupposes that such a thing is conceivable in the first place and this idea has been losing credibility through empirical studies of scientific explanation as well as through the development of cognitive science in recent decades. The result of the quiet revolution that has taken place following Lakoff and Johnson's *Metaphors We Live By* is that metaphors are now seen as inevitable tools of thought. In more recent work, Lakoff and Johnson (1999) claim that conceptual thought is inherently metaphorical and they start to explore the neurobiological reasons of the pervasiveness of metaphor. The embodiment of thought is their main subject in this book. They see the origin of metaphorical thought in the sensorimotor relations through which children learn to know the world. For example, for young children knowing in many instances is synonymous with seeing. This later results in the omnipresent metaphor "knowing is seeing", as expressed in "I see what you mean", "this is an interesting point of view" etc. Our interactions with the world yield a large number of such "primary metaphors"[15], as Lakoff and Johnson call them, which are subsequently elaborated and combined into more complex metaphorical systems. Their function is generally to approach complex, abstract or unstructured domains with the help of concepts from more familiar, concrete and well-known domains. Abstract concepts, according to Lakoff and Johnson, often do have a literal skeleton, but this is never rich enough to express what we want to express. Our concepts about time, causation, the mind, force, the self, morality or other philosophical and scientific subjects are largely metaphorical. Metaphor should not be seen as a hindrance to serious knowledge but as a precious intellectual gift.

This view, fascinating though it is, leaves much to be studied, and discussions about the role of metaphor in science are certainly not over. But the idea that metaphors are essential ingredients of science has been gaining ground. It can be found, for example, in Michael Bradie's (1999) brief review of science and metaphor (with special reference to biology and Lewontin in particular). Bradie distinguishes three overlapping roles of metaphor in science: rhetorical, heuristic and cognitive (or theoretical) and argues that each of these roles is indispensable.

---

[15]Knowing is seeing, affection is warmth, importance is bigness, intimacy is closeness, etc.

In relation to the heuristic and cognitive roles of metaphor, let me briefly return to the role of context. Condit found that in popularized texts the role and interpretation of metaphor is often superficial. When we think of scientific contexts, the situation is very different: implications of models and views are explored much more actively and systematically. The limitations of existing images and the need for new ones will also be felt with more urgency in the context of research. It is thus plausible that the heuristic and cognitive role of metaphors is stronger and more explicit in research contexts than in most popularized contexts. Therefore, once we are used to the presence of metaphor in science, reflections on context may lead to a reversal of the traditional picture: metaphors might well be *more* important, heuristically and cognitively, in real science than in popularization. And if science explores images more actively, it can also be expected to be a more active source of metaphorical innovation than popularized science. That the images on DNA we have been encountering all seem to have their origins within biology is in line with this thought.

Now that metaphors, and the imagination in general, are increasingly acknowledged as essential elements of conceptual thought, the study of metaphors has gradually found a place in the philosophy of biology. But through neurobiology and cognitive science, the relationship between metaphors, biology and philosophy is becoming even more complex and fascinating. A biology of thought and philosophy emerges from those disciplines, which is beginning to explain the role of metaphor in thought. In his most recent book, neuroscientist Antonio Damasio (2003) warmly embraces Lakoff and Johnson's view that the embodiment of the mind shows up in the bodily metaphors that pervade our thoughts about happiness, health, and many other concepts. His suggestion is that body mappings in the brain are the basis of thought and imagination. Brain scientist Ramachandran (see the 2003 Reith lectures at www.bbc.uk, also Ramachandran and Hubbard, 2003) meanwhile calls attention to synesthesia, the phenomenon that two different senses are blended, as when people "see" numbers or music in colours or in shapes. Everyone is a synesthete to some extent and in some form or another and Ramachandran speculates not only about the brain mechanisms explaining synesthesia but also about implications for the origins of metaphor, language, and abstract thought.

In a biological approach to thought, at least in Lakoff and Johnson's, bodily metaphors stemming from basic and universal phenomena such as seeing and feeling are cognitively fundamental. On that basis it tends to call attention to universal and cognitive aspects of metaphor. Social and contextual approaches to metaphor, on the other hand, tend to emphasize variability and, in connection with this, rhetorical, ideological and heuristic aspects of metaphors.

The tension between these approaches defines a vast field of further questions about attention, interpretation, cognition and meaning. In this field, developments in neurobiology as well as the use of metaphors in various contexts are empirically important.

In the process, borders between disciplines are constantly crossed and fields merged. The theatre of genetics may become a theatre of genetics, neurobiology, nanotechnology, ethics, art…

2.7 CONCLUSION

This essay started with a search for images in connection with reaction norms. Reaction norms have long occupied a marginal position in biology, since in the dominant images of biological causation DNA was central. However, this situation is changing. Genomics research is a great driving force behind the emergence of new metaphors that depict a more complex causal situation.

Metaphors are now widely thought to play a conceptual role but the idea that metaphors determine thought in a straightforward way is too simplistic. Condit's work shows that metaphors can be interpreted in different ways and that the context plays a large role. Some contexts may not invite any interpretation at all. I suggest that an important variable between contexts consists in the kind and amount of active interest they invite in the details and implications of specific images.

Subsequently, I paid separate attention to images of DNA in social and scientific contexts. In social contexts, genetics and genomics are associated with ethical choices that are often hard and puzzling. I propose that drawing on a biology that does justice to complexity can be very helpful for enriching the moral imagination and the quality of ethical debates.

As to science itself: the metaphorical richness of biology is not just an empirical finding. Reflection on the importance of context with regard to metaphor suggests that it is only to be expected that biology is a more active source of new imagery on genetics than popular thought: in general, researchers will have more reason and opportunity to explore the implications and limitations of genetic metaphors thoroughly.

Much remains to be found out about the use of metaphor. Neuroscientists increasingly present intriguing findings and speculations suggesting that in the coming years brain research will shed a new light on the origin of metaphors and their role in language and thought.

ACKNOWLEDGEMENTS

The writing of this essay was subsidized by the Netherlands Organization for Scientific Research (NWO) in the context of the program "Social components of genomics research" and, in addition, by the Agricultural Economics Research Institute (LEI) of Wageningen University and Research Centres (Wageningen UR).

I thank Scott Gilbert, Peter Taylor, Kelly Smith, Lenny Moss, Fred Nijhout, Rasmus Winther, Rolf Hoekstra, Susan Oyama, Jason Scott Robert, Ron Amundson and Arno Wouters for their responses to my question about metaphors and reaction norms. Further, I thank Jozef Keulartz, Volkert Beekman, Jan Vorstenbosch, Rolf Hoekstra, Bas Kooijman, Joël van der Weele, Gerda van der Weele and two anonymous referees for helpful suggestions and/or comments.

REFERENCES

Avise, J. C. (2001). Evolving Genomic Metaphors: A new look at the language of DNA. Science 294: 86-87.

Berkel, D. van, and C. van der Weele (1999). Norms and prenorms on prenatal diagnosis: New ways to deal with morality in counseling. Patient Education and Counseling 37: 153-163.

Bradie, M. (1999). Science and metaphor. Biology and Philosophy 14: 159-166.

Condit, C. M. (1999). How the public understands genetics: Non-deterministic and non-discriminatory interpretations of the "blueprint" metaphor. Public Understanding of Science 8: 169-180.

Condit, C. M., B. R. Bates, R. Galloway, S. Brown Givens, C. K. Haynie, J. W. Jordan, G. Stables and H. M. West (2002). Recipes or blueprints for our genes? How contexts selectively activate the multiple meanings of metaphor. Quarterly Journal of Speech 88: 303-325.

Damasio, A. (2003). Looking for Spinoza. Harcourt, New York.

Fukuyama, F. (2002). Our Posthuman Future: Consequences of the Biotechnology Revolution. Farrar, Straus & Giroux, New York.

Gilbert, S. F. (1997). Developmental Biology. Fifth Edition. Sinauer Press, Sunderland.

Gilbert, S. F. (2001). Ecological developmental biology: Developmental biology meets the real world (Review). Developmental Biology 233: 1-12.

Gilbert, S. F. (2002). The genome in its ecological context: Philosophical Perspectives on interspecies epigenesis. Annals of the New York Academy of Science 981: 202-218.

Griffiths, P. E. (2001). Genetic information: A metaphor in search of a theory. Philosophy of Science 68: 394-412.

Griffiths, P. E. (in press). The Fearless Vampire Conservator: Philip Kitcher and Genetic Determinism. In: Neumann-Held, E. M. and C. Rehmann-Sutter (Eds). Genes in Development: Rereading the Molecular Paradigm. Duke University Press, Durham.

Houellebecq, M. (2001). The Elementary Particles. Vintage Books, London. (Originally: Les Particules Elémentaires, 1998).

Jones, P. A. and S. B. Baylin (2002). The fundamental role of epigenetic events in cancer. Nature Reviews/Genetics 3: 415-428.

Keller, E. F. (1995). Refiguring Life; Metaphors of Twentieth-Century Biology. Columbia University Press, New York.

Keller, E. F. (2000). The Century of the Gene. Harvard University Press, Cambridge, Mass.

Lakoff, G., and M. Johnson (1980). Metaphors We Live By. University of Chicago Press, Chicago.

Lakoff, G., and M. Johnson (1999). Philosophy in the Flesh; The Embodied Mind and its Challenge to Western Thought. Basic Books, New York.

Lewens, T. (2002). Development aid: On ontogeny and ethics. Studies in the History and Philosophy of Biology and Biomedical Science 33: 195-217.

Lewontin, R. C. (1974). The Analysis of Variance and the Analysis of Causes. American Journal of Human Genetics 26: 400-411. Reprinted in: Levins, R. and R. C. Lewontin (1985). The Dialectical Biologist. Cambridge University Press, Cambridge.

Lewontin, R. C. (2001). In the beginning was the word. Science 291: 1263-1264.

Liakopoulos, M. (2002). Pandora's Box or Panacea? Using metaphors to create the public representations of biotechnology. Public Understanding of Science 11: 5-32.

Madoff, S. H. (2002). The Wonders of Genetics Breed a New Art. New York Times, May 26.

Morange, M. (2001). The Misunderstood Gene. Harvard University Press, Cambridge, Mass.

Nelkin, D. and M. S. Lindee (1995). The DNA Mystique; The Gene as a Cultural Icon. Freeman, New York.

Oyama, S. (2000). The Ontogeny of Information; Developmental Systems and Evolution. Duke University Press, Durham (originally 1985), Cambridge University Press, Cambridge, Mass.

Pennisi, E. (2001). Behind the scenes of gene expression. Science 293: 1064-1068.

Pigliucci, M. (2001). Phenotypic Plasticity; Beyond Nature and Nurture. Johns Hopkins University Press, Baltimore.

Ramachandran, V. S. and E. M. Hubbard (2003). Hearing Colors, Tasting Shapes. Scientific American, April 15.

Shapiro, J. A. (2002). Genome organization and Reorganization in Evolution: Formatting for Computation and Function. Annals of the New York Academy of Sciences 981: 111-134.

Singer, P. A. and A. S. Daar (2001). Harnessing genomics and biotechnology to improve global health equity. Science 294: 87-89.

Sloterdijk, P. (1999). Regeln für den Menschenpark. Suhrkamp Verlag, Frankfurt am Main.

Stern, J. (2000). Metaphor in Context. MIT Press, Cambridge, Mass.

Thomson, A. (1993). Virtual Girl. Ace, New York.

van Dijck, J. (1998). ImagEnation; Popular Images of Genetics. New York University Press, New York.

Van Speybroeck, L., G. van de Vijver and D. de Waele (Eds) (2002). From Epigenesis to Epigenetics; The Genome in Context. Annals of the New York Academy of Sciences 981.

Weele, C. van der (1999). Images of Development; Environmental Causes in Ontogeny. SUNY-Press, Albany. (Originally 1995, PhD Thesis, Free University Amsterdam.)

Cor van der Weele
Department of Applied Philosophy, Wageningen University

3

The Functional Perspective of Organismal Biology

Arno Wouters

ABSTRACT

Following Mayr (1961) evolutionary biologists often maintain that the hallmark of biology is its evolutionary perspective. In this view, biologists distinguish themselves from other natural scientists by their emphasis on why-questions. Why-questions are legitimate in biology but not in other natural sciences because of the selective character of the process by means of which living objects acquire their characteristics. For that reason, why-questions should be answered in terms of natural selection. Functional biology is seen as a reductionist science that applies physics and chemistry to answer how-questions but lacks a biological point of view of its own. In this paper I dispute this image of functional biology. A close look at the kinds of issues studied in biology and at the way in which these issues are studied shows that functional biology employs a distinctive biological perspective that is not rooted in selection. This functional perspective is characterized by its concern with the requirements of the life-state and the way in which these are met.

3.1 INTRODUCTION

In the wake of Mayr (1961) many evolutionary biologists and many philosophers of biology maintain that biology distinguishes itself from other natural sciences by its evolutionary perspective. This perspective is introduced by asking why-questions. Such questions can be legitimately asked (it is said) in biology but not in the physical sciences. This is because of the special character of the process by means of which organisms get their characteristics: evolution by natural selection. As selection operates on certain effects it makes sense to ask why (i.e. for which effects) a certain trait/item/activity was selected. Because mere physical objects do not acquire their characteristics through a selection history, it makes no sense to ask why-questions with regard to mere physical objects (see especially Mayr, 1997: 115-116).

Within biology how-questions and why-questions are said to be the subject of "two largely separate fields which differ greatly in method, *Fragestellung*, and basic concepts" (Mayr, 1961: 1501), functional biology and evolutionary biology. Functional biologists address how-questions, and answer them by describing the operation of mechanisms. They orient themselves towards the

T.A.C. Reydon and L. Hemerik.,(eds.), Current Themes in Theoretical Biology,
33-69.
© 2005 *Springer. Printed in the Netherlands.*

physical sciences, reach their conclusions by means of experimentation and favour a reductionist approach. Evolutionary biologists, on the other hand, address why-questions and answer them by recounting an evolutionary history. Both kinds of questions are equally legitimate and complementary to each other, but the distinguishing biological perspective is provided by the why-questions of evolutionary biology. Functional biology in isolation reduces living objects to their physical and chemical characteristics and lacks a biological point of view of its own.

In this paper I argue against this reductionistic image of functional biology. This image is grounded on the mistaken view that the main or only source of biology's autonomy is the process of natural selection. I shall draw attention to another source of autonomy, namely the organized way in which the life-state (the ability of organisms to maintain themselves, to grow, to develop and to produce offspring) is obtained. Functional biology attempts to understand this organization by depicting the parts and behaviours of an organism as solutions to certain design problems. Functional biologists ask questions such as 'what problem does it solve?', 'why is this problem a problem?', 'how is this problem solved?' and 'why is this solution better than another?'. This way of understanding organisms is commonly called the functional perspective. This perspective pervades the whole of organismal biology: it is not just a matter of asking an additional question in the manner of Mayr but it determines the way in which all issues are tackled. I shall describe this perspective in some detail and argue that it is less reductionistic than most evolutionary biologists suggest: it understands the structure and behaviour of the parts in the context of the whole. I shall also argue that the legitimacy of the functional perspective has nothing to do with the evolutionary origin of the objects under study, but rather with the difficulties in maintaining the life-state. Although functional biologists make ample use of physics and chemistry their focus on understanding the life-state makes their work thoroughly biological.

My strategy of argument is as follows. I start with an inventory of kinds of issues pursued by organismal biologists: description, biological role, causes and underlying mechanisms, biological value, development, and evolutionary history (Section 3.2). My inventory is a modification of Tinbergen's (1963) four-fold classification of issues in biology: causation, survival value, ontogeny and evolution. The main difference is this: I have added the category 'biological role'. As a corollary, I make an explicit distinction between the question 'what is the function?' and 'why is it useful?' and between function attributions and functional explanations. I explain these differences and their importance in Section 3.3. Then, I describe the functional perspective and illustrate it with an example (Section 3.4). In Section 3.5 I argue that this perspective is independent of selection. In Section 3.6 I discuss the ontological relations with which the different kinds of explanations are concerned. Finally, in Section 3.7 I draw some conclusions with regard to the autonomy of biology.

3.2 SIX ISSUES IN ORGANISMAL BIOLOGY

In order to understand the functional perspective in organismal biology, it is useful to take a brief look at the kinds of issues studied by organismal biologists. Many textbooks in behavioural biology introduce their subject matter by distinguishing four kinds of explanatory issues (see for example Alcock, 1998: 5; Raven and Johnson, 1999: 1214; McFarland, 1999: 1; Goodenough *et al.*, 2001: 1). This inventory originates from Tinbergen's now famous "On Aims and Methods in Ethology" (1963). Tinbergen (one of the founders of behavioural biology) states (p. 411) that Julian Huxley (one of the founders of the modern synthesis) liked to speak of "the three major problems of biology": "causation, survival value and evolution",[1] to which Tinbergen himself "should like to add a fourth, that of ontogeny". Tinbergen had already outlined these four problems in his *The Study of Instinct* (1951: 1-2) (the first major comprehensive textbook on behavioural biology in English), but his 1963 text provides a more extensive treatment. Tinbergen's 'four problems of biology' became the guiding principles of research in behavioural biology.[2]

---

[1]Actually, Huxley speaks of three "aspects of biological fact": "the mechanical-physiological aspect, the adaptive-functional aspect, and the historical aspect" (e.g. Huxley, 1942: 40).

[2]In a recent anniversary essay, Alcock (2003) contends that "Tinbergen's four-part scheme is still highly relevant and useful" (p. 4). Important discussions of Tinbergen's classification can be found in Sherman (1988), Armstrong (1991) and Dewsbury (1992). Nice illustrations can be found in Holekamp and Sherman (1989) and in the proceedings of the symposium on *Animal Behavior: Past, Present and Future* at the 1989 meeting of the American Society of Zoologists (*American Zoologist* 31: 283-348 (1991)). It should be noted however, that some authors did not only change the labels of the problems, but also their content (see Dewsbury (1992) for an excellent review). For example, McFarland (1999: 1) makes a distinction between the following kinds of questions:

(1) Why do animals respond to environmental stimuli in a particular way?
(2) Why do animals respond to internal stimuli in a particular way?
(3) Why do some animals respond in one way and others in another way to the same situation?
(4) Why do animals of a particular species characteristically behave in a particular way in a particular situation?

McFarland attributes this classification to Tinbergen (1963), but the connection is far from obvious. According to McFarland, the first and second questions both deal with "immediate causes and mechanisms" (Tinbergen's 'causation'). Question (4) was according to McFarland introduced by Darwin; it concerns survival value. This suggests either a confusion of Tinbergen's 'survival value' and 'evolution', or a neglect of 'evolution'. More importantly, it seems to me that McFarland's questions

His classification also applies to other disciplines of organismal biology, such as animal morphology. For example, Dullemeijer (1974) (an extensive treatment of methodological and conceptual issues in functional animal morphology by a veteran in the field) distinguishes four issues concerning the relation between form and function (activity): "how is the relation to be explained in terms of underlying mechanisms or factors, how in terms of the biological role or meaning, and how has it evolved in the ontogeny and evolution?" (p. 95). Indeed, although many authors speak about "Tinbergen's four problems[3] in *ethology*", Tinbergen himself saw his classification as a general classification of problems of *biology* (just as Huxley did).

The present paper is concerned with the legitimacy of the explanations in the problem area that Tinbergen called 'survival value'. The term 'survival value' is somewhat confusing because it is not only survival but also reproduction that is at stake. Following Mahner and Bunge (1997) I shall use the term 'biological value'. Tinbergen defined the study of survival value as the study of effects as opposed to causes. He used the term 'function' as a synonym of 'survival value'. Because other biologists use the term 'function' in a different way (see Section 3.3 of this paper) this use of the term 'function' should be avoided, if the classification is to serve as a general classification of problems in biology.

Although Tinbergen emphasizes the importance of observation and description, he did not include this issue in his taxonomy of problems. This is understandable as his problems are explanatory problems. However, if the taxonomy is meant as an overview of the questions that guide research, the issue of description should be mentioned explicitly. Tinbergen's classification of research questions can be further improved by adding an extra issue namely the search for biological roles. The biological role of an item or activity is the

---

fail to distinguish between different interpretations of the pronoun 'why'. For example, the question why a certain organism responds in a particular way to a certain (external or internal) stimulus can be answered by appeal to the mechanisms underlying the connection between stimulus and response (Tinbergen's 'causation)', the utility of responding to a certain stimulus in a certain way (Tinbergen's 'survival value'), the way in which the connection between stimulus and response was brought about in the ontogeny (Tinbergen's 'ontogeny'), and the way in which stimulus and response became associated in the course of evolution (Tinbergen's 'evolution'). Similarly, the behavioural differences between two individuals of different species (McFarland's third type of question) can be explained by appeal to the differences in mechanisms ('causation'), the different demands on those mechanism's ('survival value') and the different histories of those mechanisms ('evolution').

[3]Other phrases used to refer to what Tinbergen called "problems" are: "issues", "questions", "whys", "why-questions", "areas of study", "behavioural determinants", "causes", "levels of analysis" and "classes of models".

way in which it contributes to an activity or capacity of a larger system.[4] As I will discuss, biological roles are central to all explanatory issues in organismal biology.

Problem area	Typical questions	Type of answer
(1) Character	What does it look like? How is it built? What is its structure?	Description of the form of an item or behaviour
	What does it do? What is it capable of doing?	Description of the activity characteristics of an item or behaviour
(2) Biological role	How is it used?	Attribution of one or more biological roles
(3) Causes and underlying mechanisms	How does it work?	Physiological explanation
(4) Biological value	Why does the organism have an item/behaviour that performs this role?	Design explanation (of the need to perform a certain role)
	Why does it have the form it has? Why does it work the way it does?	Design explanation (of the character of an item or behaviour)
(5) Ontogeny	How did it develop in the course of the ontogeny? What are the mechanisms that bring about this process? How is this process regulated?	Developmental explanation
(6) Evolution	How did it evolve? Why did it do so?	Evolutionary explanation

Figure 3.1. Different issues concerning the form and function of a certain item or behaviour.

In summary, organismal biology is guided by six types of questions about an item or behaviour in study (see Figure 3.1). These questions concern:

---

[4]I took the term 'biological role' from Bock and Von Wahlert (1965), but they define it in a slightly different way. Most biologists use the term 'function' to refer to this issue, but this is confusing as this term is also used to refer to the activity of an item and to the biological value of an item or behaviour having a certain character.

(1) the form and activity of that item or behaviour (description);

(2) its biological roles;[5]

(3) the causes and underlying mechanisms resulting in the performance of those roles (Tinbergen's 'causation');[6]

(4) the biological value of that item or behaviour having the character it has and of the performance of that role (Tinbergen's 'survival value');[7]

(5) the development of that item or behaviour in the course of the ontogeny (Tinbergen's 'ontogeny');[8]

(6) the origin and modification of that item or behaviour in the course of the evolution (Tinbergen's 'evolution').[9]

Functional morphologists ask these questions typically about an item (such as the heart), behavioural biologists ask these questions typically about a behaviour (such as a bird's song).

The first type of question concerns the form and activity of the item or behaviour under study. What does the item look like? How is it built? What does it do? What is the structure of the behaviour? An example of a question of this kind in morphology is the question 'how is the heart built?'; an example from behavioural biology is 'what is the structure of a bird's song?'. Research into this kind of question aims for accurate descriptions of the item or behaviour under study.

The second type of question concerns the way in which the item or behaviour under study is used by the organism. Examples of questions of this type are 'what is the biological role of the heart?' and 'what functions does a bird's song have?' These questions are answered by means of one or more attributions of a biological role. Examples are the attribution of the role to pump the blood around to the heart and of the role to claim a territory to that bird's songs. Attributions of biological roles describe how a certain item or activity contributes to the emergence of a complex capacity of an organism. The most important complex capacities at the level of the organism are the organism's capacities to maintain itself, to grow, to develop, and to produce offspring.

---

[5] Other phrases used to refer to this issue: "function", "role" and "meaning".

[6] Other phrases used to refer to this issue: "immediate causation", "mechanisms", "underlying mechanisms" and "control".

[7] Other phrases used to refer to this issue: "function", "adaptive function", "functional consequence", "functional significance", "reproductive value", "fitness value", "selective advantage", "adaptive value", "adapative significance", "adaptiveness", "adaptation" and "evolutionary significance".

[8] Other authors use "development", "developmental history", "ontogenetic development" or "ontogenetic processes".

[9] Other authors use "phylogeny", "history" or "evolutionary history".

Questions of the third type ask 'how does the item or behaviour in question work?'. That is, how is that item or behaviour able to produce the activities that allow it to perform the biological roles attributed to it in answer to a type (2) question. Examples are 'how is the heart able to propel blood?' and 'how are bird songs produced?' These questions concern the causes and underlying mechanisms of the activity of organisms and their parts. The answers to such questions are usually called 'causal explanations' by biologists. To avoid confusion with other uses of the term 'causal explanation', I shall use the term 'physiological explanations' to refer to explanations in this area of research. Physiological explanations come in two different kinds: explanations that specify a cause and explanations that describe an underlying mechanism.[10] The first kind of physiological explanation (explanations that specify a cause) explains certain changes in the state of an organism (such as changes in the frequency of the heartbeat, or changes in a bird's readiness to sing) as the effect of changes that happened before the changes to be explained in the organism or its environment. For example, changes in the frequency of the heartbeat are explained by changes in the activity of the nerves that innervate the heart which in turn are explained by, say, the fact that the organism hears the alarming call of another organism. Similarly, changes in a bird's readiness to sing are explained by changes in the level of certain hormones in the blood in response to changes in day length. Physiological explanations of the second kind (explanations that describe an underlying mechanism) explain how the properties (including dispositions) and activities of a part of an organism (or of the organism as a whole) result from the properties and activities of that item's parts and the way they are organized. For example an explanation of an organism's capacity to circulate oxygen would point out that oxygen circulation is brought about by a system of vessels which contain blood. The blood carries the oxygen and is pumped around by a heart. The two kinds of physiological explanations are related in the following way: explanations of the second type are concerned with the mechanisms that connect the causes and effects mentioned in explanations of the first type. For example, an explanation of the second type might concern the mechanism that brings about changes in the frequency of the heartbeat in response to changes in the activity of the nerves that innervate the heart.

The fourth kind of question is concerned with biological value (utility), that is with the effect of a certain trait on the life chances of an organism in comparison with other traits that could replace the trait. Once the character and biological role of a certain item or behaviour are known, two types of question about the biological value of that item/behaviour are raised. One is the question 'why is it useful to the organism to have an item or behaviour that

---

[10]My classification of physiological explanations is inspired by Cummins (1983: Chapter 1, 2). See also Glennan (1996, 2000) and Craver (2001).

performs the biological roles attributed to that item or behaviour?' Examples of such questions are 'why is it useful to circulate the blood?' and 'why is it useful to defend a territory?'. The other is the question 'why is it more useful to the organism to perform the relevant biological role in the way it is performed than in some other way?'. Examples are 'why is it more useful to mammals and birds to pump the blood around by means of a four chambered pump than by means of a single chambered one?' and 'why is it useful to birds to mark a territory by singing (rather than by e.g. dung)?'. An answer to a question of this kind is usually called a 'functional explanation' by biologists. To avoid confusion with other uses of the term 'functional explanation', I have introduced the term 'design explanation' to refer to answers of this kind (Wouters, 1999).

Questions of the fifth kind concern the ontogenetic development of the item or behaviour under study. How did this item or behaviour develop in the course of the ontogeny and how is this development controlled? Examples of such questions are 'how does the heart develop and how is this development regulated?', 'how do bird songs develop?', 'is the song pattern innate or learned from parents?'. The explanations proposed in answer to questions of this kind are usually called 'developmental explanations'. Developmental explanations relate how a certain trait arises in the course of the ontogeny. In the example of the circulatory system, a developmental explanation would (among other things) point out that the initial differentiation of blood vessels is probably caused by a process of induction. The first blood vessels develop before circulation starts. If the heart rudiment is removed before it starts to beat, the large blood vessels continue to develop for some time. Further development depends on the direction and amount of the blood flow through these vessels. Developmental explanations and physiological explanations shade into each other. The main difference is that developmental explanations are concerned with transitions that usually occur only once in the lifetime of an organism and physiological explanations with transitions that may occur repeatedly.

The last kind of question consists of questions concerning the evolution of the item or behaviour under study. How and why did this item or behaviour evolve and how and why did it acquire the character it has? Examples of such questions are 'how did the heart acquire four chambers?' and 'how did bird songs became complex?' Explanations that answer questions of this kind are called 'evolutionary explanations'. Evolutionary explanations explain how a certain trait developed in the course of the history of the lineage. Evolutionary processes include mutation, gene flow, recombination, selection and genetic drift. I shall call evolutionary explanations that focus on evolution by natural selection 'evolutionary selection explanations'.[11] Evolutionary selection explanations explain the presence or character of a certain item or behaviour by recounting how and why natural selection modified that item or behaviour in the course of history.

3.3 DIFFERENT NOTIONS OF 'FUNCTION'

A conspicuous difference between biology and other natural sciences such as physics and chemistry is its appeal to functions in explanations.[12] As several authors (e.g., Bock and Von Wahlert, 1965; Hinde, 1975) have noted, biologists use the term 'function' in a number of different ways. Wouters (2003) distinguishes four kinds of function: (1) function as activity (function$_1$); (2) function as biological role (function$_2$); (3) function as biological advantage (function$_3$); and (4) function as selected effect (function$_4$). Function$_1$ (activity) refers to what an item does by itself; function$_2$ (biological role) refers to the contribution of an item or activity to a complex activity or capacity of an organism; function$_3$ (biological advantage) refers to the value for the organism of an item having a certain character rather than another; function$_4$ (function as selected effect) refers to the way in which a trait acquired and maintained its current share in the population. For the purpose of this paper two of these especially deserve attention, namely function as biological role (function$_2$) and function as biological advantage (function$_3$).

Function as biological role

The biological role (function$_2$) of an item/activity is the manner in which that item/activity contributes to the activity of a complex system. For example, the biological role of the lung is the manner in which this item contributes to the system that transports oxygen and carbon dioxide from the environment to the inner organs and back. It provides the site where these gases are exchanged between the ambient medium and the blood. Other examples of attributions of biological roles are: 'some functions of the circulatory system are to transport oxygen, carbon dioxide, nutrients and heat' and 'the heart is the source of energy of blood movement'. As Cummins (1975) points out, attributions of biological roles are closely connected with the mechanistic strategy of explanation (see also Craver, 2001). A central theme in organismal biology is

---

[11]A second kind of selection explanation is equilibrium selection explanation. Evolutionary selection explanations explain trait modification as the result of the operation of natural selection. Equilibrium selection explanations appeal to natural selection to explain how the frequencies of certain traits in the population remain unchanged.

[12]This is a feature functional biology shares with the engineering sciences.

the explanation of what might be called 'the life-state': the capacity of an organism to maintain itself, to grow, to develop and to produce offspring. Organismal biologists explain these capacities by analyzing the organism into a number of systems (such as the circulatory system, the digestive system and the musculoskeletal system), each of which has one or more roles in the maintenance of the life-state. Each of these systems in turn is split up into a number of subsystems. Each subsystem has a specific role in bringing about the capacity of the system of which it is a part to perform the role of that encompassing system. For example, one of the main roles of the circulatory system in vertebrates is the transport of oxygen, carbon dioxide, nutrients and heat. The capacity to perform this task is the result of the co-ordinated action of the parts of that system (say, heart, blood-vessels and blood), which each have specific roles in bringing about that capacity. The heart pumps the blood around, the blood carries the gasses, nutrients and heat, and the blood vessels contain and direct the blood. To explain the capacity of a subsystem to perform its role in the system each subsystem is analyzed into a number of subsubsystems which have a role in bringing about the capacity of the subsystem to perform its role in the system. For example, the capacity of the heart to pump the blood around is explained in terms of its internal structure, its ability to contract, its rhythmicity and the nervous control. And so on, until a level is reached at which the capacities, which an item needs to perform its role, are sufficiently simple to explain them in terms of the physical and chemical characteristics of the components of that item (see, for example, Robinson (1986) for an elaborate discussion of the explanation of the heart's capacity to contract).

Function as biological advantage

The notion of function as biological advantage (function3) is concerned with the biological value (utility) of a trait, that is with the way in which that trait influences the life chances of an organism as compared to other traits that might replace it. The biological value of a certain trait (as compared to another trait) can be positive, negative or neutral. An advantage of a trait is an ability resulting from that trait due to which the life chances of organisms that have that trait are higher than the life chances would be of organisms in which that trait is replaced. Advantage articulations tell us how a trait effects that the life chances of its bearers are higher than those of hypothetical organisms in which this trait is replaced by another one. An example of an advantage articulation is 'the increased surface area in specialized respiratory organs such as lungs and gills increases the oxygen uptake'. This statement expresses the hypothesis that the life chances of organisms that have specialized respiratory organs with a large respiratory surface are greater than the life chances would be of those organisms if their respiratory surface were smaller, due to the fact that a large area increases the oxygen intake in a certain amount of time.

Note that reports about biological value, such as advantage articulations, are comparative: they compare an organism with a certain trait with a hypothetical organism in which that trait is replaced by another one (or removed). Advantages are therefore relative to the trait(s) chosen for comparison. The organisms with which the real organisms are compared need not be real. In evolutionary biology one usually compares an existing variant with other variants that regularly turn up in the population or with other plausible variants (i.e. variants that are only a few mutations away from the existing variants). In morphology one usually compares the real organisms with hypothetical organisms that are highly implausible. For example, one might compare land vertebrates that have lungs with land vertebrates in which the lungs are replaced by gills to determine what advantages lungs have to land vertebrates over gills. Behavioural biologists make both kinds of comparisons.

The biological value of an item/activity having a certain character rather than another is typically assessed in relation to the biological role of that item/activity. For example, the biological value of the large surface area in respiratory organs is determined in relation to the role of that item (respiration): that role is performed better if the respiratory surface is large rather than if it is small. This means that advantage articulations typically depend on a preceding attribution of a biological role. It should be noted however, that the biological value of an item/activity having a certain character rather than another is not intrinsic to the role of that item/activity: what counts as better is ultimately dependent on the effect on the life chances of the organism and not on the efficiency (or some other technical criterion) with which the role is performed. Such criteria are indications of the effect on the life chances, not determinants of what is better.

The distinction between biological role and biological advantage

Basically the distinction between biological role and biological advantage is a distinction between 'how it is used' (biological role) and 'how it is useful' (biological advantage). The main differences between biological roles and biological advantages are discussed in this section.

First, roles are attributed to items or activities, advantages to the character of those items/activities. For example, it is an item (the respiratory surface) that has the respiratory role and it is the character of that item (having a large surface area) that has the advantage that it increases the oxygen intake.

Second, advantage articulations are essentially comparative and attributions of a biological role are not. This means that advantages are relative to an alternative chosen for comparison, and roles are not. When biologists attribute a respiratory role to the tetrapod lung they say something about this item in certain organisms that is independent of what that item does in other organisms (whether hypothetical or real). However, when biologists say that an advantage of an increased surface area is an increased oxygen intake they compare one trait (a certain area of respiratory surface) with another (a lesser area of respiratory surface). This advantage is relative to the trait with which the surface area is compared: the verdict may differ if the respiratory surface is compared to one with a greater surface area.

Third, attributions of biological roles tell us how an item/activity fits into an organism's machinery, whereas advantage articulations evaluate the effect of a trait on the life chances. A certain organ can be classified as a respiratory organ even it is detrimental to survival, for instance because the inhaled air contains carbon monoxide. However, in order that a certain area of respiratory surface is more advantageous than another it must have a positive effect on the life chances. In an environment that contains carbon monoxide a large surface area is definitely not more advantageous than a small one.

Fourth, advantage articulations are relative to a certain environment, and attributions of a biological role are not. Although it depends on the environment whether an item/activity is capable of performing its biological role, it does not depend on the environment whether it *has* that role. Whether an item has a role depends on the organism's organization (see Craver, 2001, for an account of this notion of organization). For example, in fishes gills have a respiratory role. This role depends on the way in which fishes are organized, but not on the environment. The gills can perform that biological role only if the fish is submerged in water that contains enough oxygen, but they have that role in other environments too. If a fish falls on land, it dies precisely because in that environment the gills can not perform their role. On the other hand, the advantage of the gills having minutely divided and thin filaments turns into a disadvantage on land: the filaments collapse against each other and, as a result, the area available for respiration is reduced to such an extent that the fish cannot get enough oxygen.

Fifth, attributions of a biological role are empirical generalizations, whereas advantage articulations are projectable (i.e. lawlike) generalizations. Biological roles are in the first place attributed to individuals ('the lung of this organism is the site of respiration'). General claims about biological roles such as 'the lungs of tetrapod vertebrates have a respiratory function' are generalizations about the way in which homologous items are used by the members of a vaguely defined class (here: tetrapod vertebrates). Such generalizations do not allow conclusions about yet unknown species. The mere discovery of individuals in which the lung has no respiratory function would not affect our judgement about the role of the lung in the currently known individuals. Advantage articulations on the other hand are projectable generalizations (laws)[13] about what is needed or useful under certain conditions. Such generalizations allow conclusions about yet unknown species. If air breathing vertebrates were found that respire by means of gills, this would cast doubt on the claim that lungs are more advantageous than gills for air breathing vertebrates.

Why these two notions of function should be kept apart

In line with the distinction between function as biological role and function as biological advantage a distinction should be made between two readings of the question 'what is the function of x of s-organisms?'. The first reading is 'what is, in s-organisms, the biological role of item/activity i?', the second is 'why is trait t useful to s-organisms?'. The first question is answered by means of an attribution of a biological role (a function$_2$ attribution), the second by means of a design explanation (i.e. the kind of reasoning that biologists call 'functional explanation').

The failure to distinguish between these two questions is one of the sources of the misunderstanding (widespread among philosophers)[14] that biologists explain the presence of an item or behaviour by merely citing its function. In fact, when biologists ask 'what is the function of x?' and are satisfied with the answer 'the function of x is to do f', they are asking for the biological role (function$_2$) of an item/activity. A functional explanation (design explanation), as will become clear in the course of my argument, is more complicated: it relates the utility of a certain trait to the other traits of the organism and the characteristics of the environment in which it lives, in terms of projectable generalizations (see Wouters, forthcoming).

More pertinent to the present paper is another reason to distinguish biological roles from biological advantages. In Tinbergen's classification it is not clear whether the search for biological roles belongs to the problem of causation or to the problem of survival value (or to both). This obscures the central role of attributions of biological roles in biological enquiry. As I will discuss in the next section, the answer to the question 'what is the biological role of item/activity i?' is the key to understanding biological organization. That answer is important to all subsequent explanatory issues (causation, biological value, development and evolution). Because of this key role of attributions of biological roles (function$_2$ attributions), the overview of

---

[13]Projectable generalizations are the kind of things the average scientist would call 'law'. I avoid that term because philosophers use the term 'law' in a much narrower sense (meaning something like exceptionless universal generalizations not bound to any place or time).

[14] See for example Canfield (1964), Wright (1973), Millikan (1989), Neander (1991), Kuipers and Wisniewski (1994), Woodfield (2000).

problems in biology gains clarity if the study of biological roles is explicitly mentioned.

Even more important, Tinbergen's association of the term 'function' with 'survival value' easily gives rise to the misunderstanding that the functional perspective is a matter of studying utility in addition to causation. In reality the functional perspective pervades all issues in organismal biology. As I will discuss in the next section causal explanation in biology is functional through and through.

3.4 THE FUNCTIONAL PERSPECTIVE

We can now see that the functional perspective has two aspects: (1) the appeal to functions as biological roles (function$_2$) in explanations that describe an underlying mechanism (i.e. in reasonings of the kind that biologists call 'causal explanation'); and (2) the application of a special kind of reasoning, design explanation (i.e. the kind of reasoning that biologists call 'functional explanation') which appeals to biological value (needs, demands and advantages) to explain certain traits of an organism (such as the presence of an item that has a certain structure or performs a certain activity). I shall discuss these issues in that order.

The appeal to biological roles in explanations that describe an underlying mechanism

Attributions of biological roles are central to physiological explanations that describe an underlying mechanism. As I discussed in Sections 3.2 and 3.3 mechanistic explanations explain a property of a system by describing the mechanism that produces that property. A mechanism consists of parts that together produce that property. The explanation specifies the parts and explains how the activity of those parts produces that property. The standard example of a mechanistic explanation in physics is the explanation of the behaviour of a certain amount of gas as described by the general gas law. This explanation details how this law results from the activity of the molecules of which that gas consists. In this example, the mechanism is relatively simple. It consists of identical parts. The explanation "merely" aggregates the behaviour of those identical parts. It does not appeal to functions. Living mechanisms (as well as artificial mechanisms) are usually more complex. Their parts differ in character and their properties depend critically on the spatial arrangements of those parts and the timing of their activities (see Craver, 2001). The properties of the whole are explained as the result of the activity of the parts and the way in which they are organized. Attributions of functions as biological roles are crucial in this kind of explanation. Such attributions situate the different items and activities in hierarchical organized mechanisms. They thereby provide the handle by means of which to understand those mechanisms. As I discussed in Section 3.2 an attribution of a biological role describes how an item contributes to the emergence of the properties of an encompassing system. This serves to explain that system (see Section 3.3). Furthermore, the attribution of a biological role to a part singles out which properties of that part are important in the life of the organism. These are the properties to be explained by describing an underlying mechanism. Without an idea of that role, one would not have any clue as to which causal connections are important. The functional perspective also assures that what is done at the lower level is relevant to the higher level. In addition, the biological role provides clues about the kind of underlying mechanism: that must be such that it can perform the role. Finally, the functional perspective provides a unifying framework that enables organismal biologists to see different causal explanations as parts of the larger project to understand how organisms work. Thanks to this perspective they can see different explanations as parts of one big puzzle. For example, it allows them to understand how the blood is circulated without knowing how the heart contracts. It allows them to see the explanation of muscle contraction and the explanation of the blood circulation as different subprojects and to connect them after both are elaborated. In short: without the functional perspective it would be impossible to understand how organisms work.

The appeal to biological value in design explanations

When the biological role of an item or behaviour (e.g. a lung or a gill) is known, two kinds of design questions are raised. The first is 'why is it useful that the organism has an item that performs this biological role?' (e.g. 'why do most animals need a respiratory organ?'). The other is 'why is it useful that this item/behaviour has the character it has?' (e.g. 'why is it useful that the gills and lungs have a large surface area?'). The answers to questions of these kinds are design explanations. Design explanations compare real organisms with hypothetical organisms in which this trait is lacking or replaced by another one. For example, to study the utility of a large surface area in gills and lungs the real organisms are compared with hypothetical organisms in which these organs have a smaller surface area. The aim of the comparison is to explain how and why the trait to be explained is more useful than the alternatives. This can be done by summing up the advantages and disadvantages of the trait in comparison with the alternatives considered. Quite often it is shown that the real trait is the only viable one among those alternatives. In such cases it is said that the trait is 'needed' or 'required'.

For example, given the biological role of the lungs and the gills (respiration) two kinds of questions about biological value arise. The first is concerned with the biological value of the respiratory system as a whole. The second with the specific character of the organs that perform the respiratory task. The need for fulfilling a respiratory task is related to the activity and size of the organism. In a moderately active organism in which the distance between the outside and the inner cells is larger than about 0.5 mm, diffusion does not suffice to bring enough oxygen to the inner parts of the organism. This problem is solved by means of a circulatory system that actively transports oxygen from the outside to the inner organs and carbon dioxide the other way round. The need to perform the respiratory role (exchange of oxygen and carbon dioxide between the circulatory system and the ambient medium) emerges as a corollary of this solution. Hence, this design explanation compares the real organisms with hypothetical organisms that lack a circulatory system. It shows that the latter are not viable if the distance between the outside and the inner cells is too large (see Wouters, 1995, for a more elaborate discussion of this example).

A further question is why this role is concentrated in specialized organs. The answer is that there are many disadvantages to the use of the entire body surface for respiratory exchange. To maintain a diffusion rate large enough to fulfil the organism's needs: (i) the distance across which the gas diffuses must be small; (ii) the surface available for diffusion must be large; and (iii) the material across which the gas diffuses must be readily permeable to that gas. In other words, the respiratory surface must be thin, large and permeable. This precludes the animal from using the entire body surface: a thin skin is easily damaged; enlargements of the outer surface would disturb the streamline; and a skin that is easily permeable to oxygen and carbon dioxide is also easily permeable to water (which is a severe disadvantage on land and in aquatic environments with an osmotic pressure that differs from the organism). Specialized respiratory organs provide the means to solve this problem. Hence, this explanation compares the real organisms (with a specialized respiratory organ) with hypothetical organisms that use their skin for respiration. It sums up the disadvantages which the latter would suffer.

An example of a question about the specific character of the respiratory organs is the question of why respiration in fishes is typically performed by means of gills and in land vertebrates by means of lungs. By definition, gills are evaginated organs and lungs are invaginated. The gills of fish typically consist of a minutely subdivided surface area, composed of numerous ultra fine lamellae, across which water is pumped in one direction (from the mouth over the gills to the outside). The lamellae are richly supplied with blood vessels; the distance from the water to the blood can be less than $1\,\mu$. The tetrapod lung is essentially an ingrowth of the foregut. In its simplest form it is just a sac with ridges and edges that increase the surface area. More advanced lungs consist of a system of conducing and distributing tubes which divide further and further until they end in blind, thin-walled sacs (alveoli) in which gas exchange takes place. Lungs are ventilated by pumping air in and out of the lungs. The flow of air in the lungs has a tidal character: there is an inhalation and an expiration phase.[15] In contrast with gills, there is no active transport of gases over the respiratory surface of lungs: the alveoli contain so-called dead air in which oxygen and carbon dioxide are transported by means of diffusion alone.

To answer the question of why fishes respire by means of gills and land vertebrates by means of lungs, biologists compare the physical qualities of the medium of respiration: water for fishes, air for land vertebrates. The concentration of oxygen in air-saturated water is about 1/30 of that in air and the rate of oxygen diffusion in air is about 30,000 times higher than in water. As a result, aquatic animals need to ventilate a vastly larger volume than land animals to extract the same amount of oxygen. Air breathing faces other difficulties: there is the continuous risk of desiccation and there are the problems caused by gravitation. As Archimedes testifies these problems do not occur in water. The main differences between gills and lungs are explained by pointing to the different biological advantages given the physical differences. The unidirectional flow of water across the gills increases the efficiency of ventilation. Compared to lungs, gills have a much larger respiratory surface with a much thinner membrane. This compensates for the smaller difference of the concentration in and outside the membrane. Such a structure would not fit for air breathing. Due to the problems of gravitation the immense increase of the surface area in gills would not be possible on land: the finely divided and thin filaments would collapse against each other. Lungs are internal, which reduces the risk of desiccation and provides the means for structural support to counteract gravitational effects. The tidal flow in lungs is much less efficient than the continuous, unidirectional flow in gills (recall that air breathing imposes lesser demands on ventilation) but it reduces the loss of water. The pumping mechanism in tetrapods has much less power than in fishes. For these two reasons, lungs would not work in an aquatic environment. Hence, this explanation compares the real organisms (water breathing fishes respectively air breathing tetrapods) with hypothetical organisms that breathe in a different environment (air breathing fishes respectively water breathing tetrapods). It sums up the disadvantages which the latter would suffer. It draws the conclusion that the latter would not be viable.

Understanding organization

Explanations that describe an underlying mechanism (causal explanations) and design explanations (functional explanations) are complementary to each

---

[15] In birds the flow in the lungs is virtually unidirectional. This meets the increased demand for oxygen imposed on them by their flying lifestyle. The flow in the upper respiratory track is tidal.

other and often integrated into one account. Together they provide an understanding of biological organization. They do so by viewing organization as a solution to a design problem. A design problem occurs if an organism would not be viable or less fit, if a certain trait were replaced by another one. A design problem is thus defined by comparison with a hypothetical organism and relative to this hypothetical organism. It is not a problem that is experienced by the real organism, but a problem that would arise if a trait of the real organism were replaced by another one.

Given a certain item (e.g. lungs or gills) or activity the first question to be answered in order to understand the organism's organization is 'what biological role (function$_2$) does it fulfil? The answer to this question (an attribution of a biological role) situates the item or behaviour in the organism's organization. Given the answer to this question (e.g. respiration) one may look upwards or downwards in the organization. Upwards one may ask what problem is solved by performing this role and why that problem is a problem (e.g. Why do fishes need to respire?). The answer to this question is a design explanation. One may also look downwards and ask how this role is fulfilled. The answer to that question is an explanation by specification of an underlying mechanism. Another question is the question why the problem is solved in the way it is solved (e.g. why is it useful for fishes to respire by means of gills rather than lungs). The answer to this question is a design explanation. In sum, explanations that describe an underlying mechanism explain how a certain design problem is solved, design explanations explain why that problem is a problem and why a certain solution is better than another.

For example: my first example of a design explanation in this section explains why oxygen transport is a problem: the distance between the inner organs and the outside is too large to provide the required amount of oxygen by means of diffusion alone. The related mechanistic explanation explains that this problem is solved by transporting oxygen actively by means of a circulatory system that is loaded from the outside. The design explanation of the presence of a special organ for respiration explains why using a special organ for respiration is a better solution to this problem than using the entire outside. Then we have more specific mechanistic explanations that describe two different solutions: gills in fishes, lungs in (most) tetrapod vertebrates. The associated design explanation explains why the first solution is better in fishes and the second in tetrapods. The investigation of mechanistic and design questions is, thus, intimately connected.

The above discussion of the functional perspective shows that the dismissal, by Mayr and many other evolutionary biologists, of functional biology as reductionistic is mistaken. It is true that mechanistic explanations explain the life-state of the organism as the result of the operation of systems of subsystems that are ultimately physical and chemical in nature. However, the way in which this is done (by appeal to roles in the organization) means that the parts are understood in the context of the whole. The problems that figure in the explanations are problems of the organism as a whole and the existence of these problems is explained by the traits of the organism as a whole and the environment in which it lives. This is a holistic point of view. A point of view that is, moreover, thoroughly biological in its concern to understand the life-state.

3.5 SELECTION AND THE LEGITIMACY OF THE FUNCTIONAL PERSPECTIVE

As stated in Section 3.2 Tinbergen thought of function (survival value) as an effect of behaviour rather than a cause. However, many behavioural ecologists present Tinbergen's 'function' as an ultimate cause in Mayr's sense (e.g. Krebs and Davies, 1984; Sherman, 1988; Holekamp and Sherman, 1989; Alcock, 1998, 2003). These authors see Tinbergen's (1963) four-fold taxonomy as a subdivision of Mayr's (1961) distinction between proximate and ultimate causes. They distinguish two kinds of proximate causes (mechanisms and development) and two kinds of ultimate causes (current utility and evolutionary origin). The first kind of ultimate cause is said to correspond to Tinbergen's 'survival value', the second to Tinbergen's 'evolution'.[16] Francis (1990), Armstrong (1991) and Dewsbury (1999) rightly protest that this use of the term 'cause' is confusing. A cause is something that brings about (produces) the behaviour to be explained. As behavioural biologists agree that a behaviour cannot be brought about by its current utility, they should not talk of current utility as a cause.[17]

---

[16]Sherman (1988) and Holekamp and Sherman (1989) even say that Tinbergen (1963) *meant* his four-fold classification as a subdivision of Mayr's (1961) two-fold distinction and that Tinbergen's classification was meant to avoid semantic confusion in the nature/nurture debate. There is no evidence at all for this position. Tinbergen does not cite Mayr; Tinbergen does not speak of his four problems as different kinds of causes; he does not group causation together with development and survival value with evolution; and he mentions the nature/nurture debate under the heading 'ontogeny' without relating this to a confusion of one or more of his four problems. Furthermore, Tinbergen presented his four-fold classification for the first time in his "The Study of Instinct" (1951), ten years before Mayr's (1961) appeared.

[17]This is not the only problem with this attempt to amalgamate Tinbergen (1963) with Mayr (1961) (see Armstrong, 1991; Dewsbury, 1992; Alcock and Sherman, 1994; Dewsbury, 1994, 1999). First, biological value includes past, current and hypothetical utility. It is clear from Tinbergen's examples that his 'survival value' is to a large extent hypothetical (useful as compared to hypothetical variants that do not necessarily exist or have existed). The amalgamation restricts Tinbergen's 'survival value' to current utility and identifies this, in turn, with the causes of current maintenance in the population. This practice seems to originate from Klopfer and Hailman (1967) who

The appeal to biological value in design explanations raises an important problem with regard to the legitimacy of this kind of explanation. Tinbergen (1963) points out that the study of survival value is concerned with cause-effect relations, and that it is, hence, as legitimate as the study of causation. He is right, of course, but this observation does not solve the problem of how appeals to effects can be explanatory. Intuitively, explanations show how the phenomenon to be explained is brought about by the explanatory facts. This intuition is worked out in philosophy as the causal theory of explanation (Salmon, 1984). Functional explanations however seem to appeal to consequences. How can such an appeal be explanatory? This problem became known in philosophy as 'the problem of functional explanation'. A solution is sought by grounding the legitimacy of appeals to function in the process of natural selection. There are two ways to do this: one is to view functional explanations as appeals to past selection, the other is to view functional explanations as concerned with the current maintenance of the trait.

Salmon (1989: 111-116), Neander (1991) and Mitchell (1993), three philosophers who accept the causal theory of explanation, have argued that the problem of functional explanation can be solved by defining 'function' in historical terms. Their 'selected effect' theories define the functions of a trait as the effects for which ancestral occurrences of that trait were maintained by the process of natural selection. For example, in their view it is the function of the heart to pump the blood around, because pumping blood is what hearts in the past did that explains their current presence. Past effects can be causes of present traits, of course, and functional explanations explain the presence of a certain trait by specifying the past effects that were, as a matter of fact, in the past, causally effective in maintaining the trait to be explained. Unfortunately these philosophers did not realize that the term 'function' is used in different

---

framed Tinbergen's problem of survival value as the question 'how is the behaviour maintained in the population?' Second, as Dewsbury (1992) points out, the evolution of an item or behaviour has several aspects: its origin, its subsequent modification (if any) and its current maintenance/elimination/modification (if any). Tinbergen did not distinguish between these aspects but it is clear from his discussion that his notion of evolution includes all three. Those who synthesize Tinbergen's classification with Mayr's restrict Tinbergen's 'evolution' to 'evolutionary origin'. Third, Mayr's notion of ultimate cause, on the other hand, is explicitly historical and includes origin and subsequent modification but not maintenance. The synthesis replaces this historical notion of ultimate cause (i.e. the cause of a past change) by a broader one that is more or less equivalent to selection (past and/or present). In other words: the amalgamation of Mayr's (1961) and Tinbergen's (1963) classification as it is presented by many behavioural ecologists restricts Tinbergen's notion of survival value to current utility, identifies current utility with maintenance by selection, restricts Tinbergen's notion of evolution to evolutionary origin and stretches Mayr's notion of ultimate cause so as to include current maintenance. This is done in passing, without providing any argument.

ways. They also did not provide real examples of the kind of explanations they call 'functional explanation'. As a result, it remains unclear whether or not they meant their accounts to apply to design explanations. Anyway, in the next section I argue that design explanations are not historical in nature and, hence, that their legitimacy is not rooted in past selection.

Another kind of solution is suggested by the accounts of the philosophers Bigelow and Pargetter (1987), Kitcher (1993) and Walsh (1996), and that of the behavioural biologists Reeve and Sherman (1993). These authors suggest that explanations that appeal to current utility to explain a trait explain how that trait is maintained by means of natural selection. I am very sympathetic to this kind of account. However, as I will argue in the section after the next section, this type of account does not account for design explanations that appeal to merely hypothetical utility. Hence, the legitimacy of this kind of explanation is not rooted in selective maintenance.

Why design explanations are not historical in character

Lungs and gills have a long selection history, but the examples of design explanations I presented in Section 3.4 do not discuss that history. Rather, they explain why the organism cannot survive if the trait to be explained (such as a specialized organ for respiration, lungs or gills) would be replaced by an alternative. They do so by: (1) specifying a biological role for those items; (2) specifying conditions that: (i) apply to the relevant organisms, and (ii) make it necessary to perform that role in the way it actually is performed (e.g. by means of lungs or gills) rather than in the alternative ways; (3) explaining (2ii) by appeal to physical laws (such as Fick's law of diffusion). Such explanations do not even presuppose that the lungs and gills have a selection history. They would not need modification if it were discovered that the lungs and gills evolved as the result of self-organization rather than selection. Even a creationist could accept them. This is because the needs of an organism are completely determined by the conditions at the moment that the need comes up. The study of the history of the lungs and the gills can yield information about why and how the gills and lungs originated and changed in the course of the history, but to determine that they are needed and why they are needed we need experiments, physics and chemistry rather than historical studies. This is analogous to the situation in a game of chess: although the state after a certain number of moves depends on what moves were made, it depends only on the state which move is the best one. Knowledge of the course of the game neither helps to determine which move is the best nor to understand why that move is the best.

Why design explanations are not concerned with selective maintenance

It might be replied that whereas design explanations are not historical in character, they are evolutionary in a broader sense: they explain why the population is immune to invasion by the alternatives and, hence, why a trait is maintained in the population. The problem with this response is this. Design explanations often compare the real organism with hypothetical organisms that are highly implausible (for instance they compare real vertebrates which hypothetical vertebrates that respire by means of their skin, real fishes with hypothetical fishes in which the gills are replaced by lungs and real tetrapods with hypothetical tetrapods in which the lungs are replaced by gills). If certain alternatives do not plausibly turn up in the population the maintenance of the real variants is not the result of the elimination of those alternatives but of the fact that those alternatives did not turn up. Hence, while a design explanation that compares the real organisms with implausible variants implies that the population would be immune to invasion by those alternatives, it says nothing about the actual dynamics of the population. Furthermore, a design explanation is not concerned with the dynamics of the population, at least not in the first place. A design explanation is primarily concerned with what kinds of organisms can viably exist. More precisely, it is concerned with how matter must be organized to obtain an organism that is able to maintain itself, to grow, to develop and to produce offspring (that is with what Cuvier, the founding father of the discipline of functional animal morphology at the end of the 18th century, would call "the conditions of existence"). In the examples above, the design explanations show that an organism of a certain size and activity that attempts to respire by means of its entire skin cannot viably exist, that a lung breather would not be viable if it breathed water and that a gill breather would not be able to live on land. By doing so these explanations exhibit dependencies between the different parts and activities of the organism and between those parts and the environment of the organism. For example, they show that having a certain size and activity is dependent on having a specialized organ for respiration, breathing water is dependent on having gills and living on land is dependent on using lungs rather than gills for respiration. So the main insight provided by a design explanation is an insight into individual level dependency relations. From this one might conclude that the population would be immune to invasion by certain variants but this kind of conclusion is not essential to design explanations. No more than the equally valid conclusion that even an almighty creator could not create these variants.

Design explanations are concerned with the requirements for being alive

The arguments in the preceding sections show that the functional perspective is not rooted in the selective character of the evolutionary process, but rather in that other special character of life: the ability of an organism to maintain itself, to grow, to develop and to produce offspring. It is the difficulty to maintain this ability (the life-state) that motivates the use of the functional perspective. The appeal to biological roles is legitimate because the life-state is brought about by organized mechanisms. Design explanations are legitimate because the presence of one combination of characters (e.g. living an active life on land) functionally depends on the presence of other characters (e.g. the presence of lungs to respire).[18] It is, in other words, not the (selective) character of the production process but the (organized) character of the products that bestows the functional perspective its legitimacy and utility.

Opposing views

In Mayr's view functional biology and evolutionary biology are two independent, equally legitimate fields of study albeit that the first merely applies physical sciences whereas the latter provides the distinctive biological perspective. There are others who have argued for a more intimate connection between the two. Dobzhansky (1973), one of the founding fathers of the modern synthesis, expressed this idea in a paper with the telling title "Nothing in Biology Makes Sense Except in the Light of Evolution":

Seen in the light of evolution, biology is, perhaps, intellectually the most satisfying and inspiring science. Without that light it becomes a pile of sundry facts some of them interesting or curious but making no meaningful picture as a whole (Dobzhansky, 1973).

Dobzhansky's paper is directed against anti-evolutionist creationism. Like Darwin before him, Dobzhansky presents many examples of biological phenomena that would not make sense if there was no evolution: the diversity of living beings, the universality of the genetic code, metabolic uniformities, variations in the amino acid sequences of specific proteins, homologies, similarities in ontogenetic development, adaptive radiation and so on. Such examples establish the scientific credentials of the theory of evolution. It is however, a large exaggeration to conclude from this that *nothing* in biology makes sense except in the light of evolution. This conclusion and the title of the paper are of a rhetorical nature and should not be given too much weight. As my examples above show, many design explanations are concerned with the conditions of being alive and such talk makes sense independent of the process by means of which living beings come into being. Functional biology without evolution is incomplete in the sense that it ignores many important questions about life, but not in the sense that no aspect of life can be understood without invoking evolution.

---

[18]"Functionally depends" means, roughly speaking, that an organism with the dependent characters (*A*) cannot be viable if the characters on which *A* depends are replaced. See Wouters (1999: Section 8.3.4) for an elaborate discussion of this notion of functional dependence.

An interesting discussion of the relation between the functional perspective and the study of evolution can be found in Daniel Dennett's *Darwin's Dangerous Idea* (1995). An anonymous reviewer of the present paper suggested to me to discuss this book, as it would oppose my view that design explanations are not evolutionary in character. As I read Dennett, this is not what he argues for or what he intends to argue for, but I can see why one would think so. For example, the part of the book discussing the role of the engineering perspective (Dennett's term for what I call the functional perspective) in evolutionary biology is called "Darwinian Thinking in Biology" and this part starts by quoting the title of Dobzhansky's (1973) paper (p. 147). What Dennett does show and aims to show is that the engineering perspective is essential to the study of evolution but he slips repeatedly from 'engineering perspective' to 'Darwinian thinking' and from 'evolutionary biology' to 'biology', giving the impression that he is concerned with showing that the engineering perspective is evolutionary in nature and central to all of biology.

A central theme in Dennett's book is the legitimacy of why-questions. According to Dennett, "some biologists and philosophers" (p. 213) maintain that Darwin did away with this kind of question. Dennett aims to show that, in fact, it is one of Darwin's most fundamental contributions to biology that he showed a new way of making sense of such questions (p. 25) and that this new way of answering why-questions (represented by the engineering perspective) is central to evolutionary biology.

> "I want to make out the case that the engineering perspective on biology is not merely occasionally useful, not merely a valuable option, but the obligatory organizer of all Darwinian thinking and the primary source of its power." (Dennett, 1995: 187)

The engineering perspective treats evolution as a problem solving process and organisms as the products of that process. This perspective helps evolutionary biologists to reconstruct the past, to find yet undiscovered features of the present and to predict the course of evolution. Examples of the first kind of use include the behaviour of Archaeopterix: a design analysis that shows the feathers of this creature are well designed for flight, supports the conclusion that Archaeopterix almost certainly flew (p. 233). An example of the second is Von Frisch's discovery of colour vision in fish and honeybees, driven by the belief that colours of fish and flowers are there for some reason (p. 233). Predictions concern on the one hand things that are technically speaking needed, in certain conditions, to stay alive (Dennett calls them 'forced moves') and on the other hand things that will be beneficial in many circumstances (Dennett calls them 'good tricks'). Examples of the first one are the prediction that all life forms will have autonomous metabolism and definite boundaries (p. 128). Examples of the second one are the prediction of the evolution of streamlining, vision and intelligence.

The reason why the engineering perspective is essential is that it allows evolutionary biologists to ignore a mass of (possibly) messy details. This is achieved by means of, what Dennett calls, the intentional stance. The intentional stance is the approach taken by engineers when they try to reconstruct the reasons why a certain design was chosen:

"When Raytheon wants to make an electronic widget to compete with General Electric's widget, they buy several GE's widgets and proceed to analyze them: that's reverse engineering. They run them, benchmark them, X-ray them, take them apart, and subject every part of them to an interpretive analysis: Why did GE make these wires so heavy? What are these extra ROM registers for? Is this a double layer of insulation, and, if so, why did they bother with it?" (Dennett, 1995: 212)

More generally:

"They treat the artifact under examination as a product of a process of *reasoned* design development, a series of *choices* among alternatives, in which the *decisions* reached were those *deemed best* by the designer. Thinking about postulated functions of the part is making assumptions about the *reasons* for their presence, and this often permits one to make giant leaps of inference that finesse one's ignorance of the underlying physics, or the lower-level design elements of the object." (Dennett, 1995: 230)

In evolutionary biology, the intentional stance is known as 'adaptationism'(p. 238). It is concerned with what Dennett calls "the reasons of Mother Nature":

"Darwin's revolution does not discard the idea of reverse engineering but, rather, permits it to be reformulated. Instead of trying to figure out what God intended, we try to figure out what reasons, if any, "Mother Nature"—the process of evolution by natural selection itself—"discerned" or "discriminated" for doing things one way rather than another." (Dennett, 1995: 213)

The ability to see the forest through the trees is, according to Dennett, not the end of the story. There is a deeper reason for the usefulness of the engineering perspective, namely the fact that evolution proceeds in many ways like an engineer.

"In chapters 7 and 8, we saw how the engineering perspective informs research at every level from the molecules on up, and how this perspective *always* involves distinguishing the better from the worse, and the reasons Mother Nature has found for the distinction." (Dennett, 1995: 233)

According to Dennett evolution really is a process of reasoned design. This process operates by producing variants and seeing how they fare. The reasons of mother nature are the reasons why one form is more successful than the other. It is, in the end, this reasoned character of the process by which the different forms are created that explains why the intentional stance works:

"If there weren't design in the biosphere, how come the intentional stance *works*?" (Dennett, 1995: 237)

I do not dispute that evolutionary biologists view organisms as the products of a process of reasoned design (in Dennett's sense). Neither do I dispute that the utility of this way of looking at organisms gives reason to say that evolution really is a process of reasoned design. However, I do want to make clear that the application of the functional perspective in organismal biology as I sketched it above is to a large extent independent of assumptions about the way in which organisms come into being.

Even in the case of artifact reverse engineering it is not clear that the reverse engineer is concerned with reconstructing reasons. It is, for example, not really important to the reverse engineers of Raytheon to reconstruct the reasons why General Electric's engineers made certain wires so heavy. What they really want to know is why General Electric's new widget is better than theirs (or better than the old one). Whether or not the engineers made the changes they made for good reasons, for bad reasons or just by accident is irrelevant. In the same way, in functional biology, the way in which an item or behaviour came into being is irrelevant if one wants to determine how a certain item works or why it is better than another. (Of course, the reverse is not true: if one wants to know why a certain design evolved one should know why this design is better than its predecessors.)

Talk of biological roles (which is, as I have argued, the central notion of function in organismal biology) is rooted in the fact that organisms are organized beings: their being alive critically depends on the spatial arrangement of their parts and the timing of the activities of those parts (see Craver, 2001). It is this organized character that makes it useful to talk about role functions not the way in which those organisms came into being. This is indicated by the fact that physicists, earth scientists and ecologists do talk of role functions of parts of wholes that are not the product of natural or artificial selection. Physicists for instance say that in elementary particles gluons glue the quarks together and that neutrons hold the atomic nucleus together; earth scientists talk of the function of rivers in the water cycle; and ecologists talk about the functions of different kinds of prey and predators in maintaining a certain community.

In the case of talk of biological values a distinction must be made between talk of needs and talk of optimalities. Talk of needs is talk of what kinds of designs can viably exist in certain circumstances. This kind of talk does not make assumptions about the process by which those designs come into being. To say that something is needed is just to say that this kind of organism cannot viably exist if that something is replaced by another thing. The choice of a criterion for optimality, on the other hand, depends on the process by means of which organic form is modified: 'better' means better in the context of the process of evolution by natural selection; the process creating and modifying those designs. However, given the choice of a fitness criterion, the answer to the question of whether a certain design is better than another depends only on the designs and the circumstances in which they are compared, not on the history of those designs or of those circumstances.

Biological value (the reasons why in certain circumstances a certain trait is more useful than another one) must be clearly distinguished from evolutionary reasons (the effects for which a trait was selected in the (recent or distant) past). The design explanation discussed above says that to fishes gills are more useful than lungs because of the physical characteristics of the medium in which they live. This is not the same as saying that fishes have gills rather than lungs because in the past fishes with gills were favoured over variants that had lungs instead of gills, due to physical characteristics of water. Neither does the latter explanation follow from the first. The design explanation in this example is well-founded, the corresponding evolutionary explanation is, as far as we know, plainly false. Statements of biological value say something about *what would be useful* to the organism in certain circumstances. Evolutionary reasons concern *what actually happened* in the past. Biological value is often determined relative to hypothetical organisms that cannot exist, that did not exist, or did not compete with the organisms in study (as when comparing fishes with gills and fishes with lungs). In those cases the reasons why that trait is useful need not be the evolutionary reasons. To determine evolutionary reasons one must always take the actually existing competitors in the (recent or distant) past into account. But even advantages relative to actually existing competitors in the actual historical circumstances are not always evolutionary reasons: the competitors might have died for other reasons before the disadvantage came into play.

The distinction between evolutionary reasons and biological value is especially clear in the case of lungs. As I have discussed, lungs are useful to tetrapod vertebrates because, due to their internal character, they enable the tetrapods that have them to live on land. However, living on land was not the reason why lungs evolved. Lungs evolved in the early Silurian, more than 420 million years ago, long before the first vertebrates went on land (in the early Devonian, about 400 million years ago) as an adaptation to very low oxygen concentrations in large tropical fresh-water basins.[19]

Dennett would probably call moving to the land a good trick and the development of lungs a forced move. He introduces these terms to emphasize that the evolution of structures and processes (such as the development of autonomous metabolism and of streamlining in moving organisms) needed to

---

[19]The species that evolved lungs are ancestral both to tetrapods and to modern teleost fish. In the latter the lung was later modified into the swim bladder.

solve a certain design problem is not dictated by the laws of physics. According to Dennett, biological necessities are like forced moves in chess (see pp. 127–129). A forced move is called 'forced' not because it is dictated by the rules of chess, but because it is obvious to anybody who knows the rules that it is the only sensible move in a certain situation. That forced moves are not dictated by the laws of physics, does not mean that physics has nothing to do with it. The laws of physics do not dictate (or force) forced moves, but they can provide reasons to execute such a move. Biological necessities are in Dennett's words "necessities of reason" and the reasons can be physical.

I completely agree with this point, provided that 'reasons' is read as 'reasons for being useful'. Giving reasons why a certain trait is useful is usually called functional explanation by biologists and design explanation by me. I have repeatedly emphasized that design explanations explain why a certain trait is useful, not why it evolved (see above and my 1995 paper) and that seems to be what Dennett's point amounts to. In Dennett's terms: design explanations explain why a move is forced, evolutionary explanation explain why a forced move happened.

Confusion can arise because Dennett does not make an explicit distinction between the reasons why a trait is needed (or useful) and the reasons why that trait is there.[20] If one fails to see this distinction design explanations (explanations that explain why a trait is useful) are easily taken for explanations that explain why a trait is there. However, as my examples make clear these two kinds of reasons do not always coincide: the reason that the first land vertebrates *needed* lungs lies in the physical conditions on land, but the reason that the first land vertebrates *had* lungs lies in the fact that lungs facilitated life in water with a low oxygen content. The explanation of why land vertebrates need lungs points to a projectable synchronic connection between conditions and utility, the explanation of why land vertebrates had lungs points to past events.

Note that a similar distinction (similar to the distinction between reasons why a trait is useful and reasons why that trait is there) applies to the chess example: the reasons why a certain player made the best move can be very different from the reasons why that move is the best move. The answer to the question why the best move is the best one depends only on the state of the game (and the rules of chess). The answer to the question why a certain player made a certain move depends, on the other hand, on what that player actually thought if it is a human player and on the actual program if it is a machine. Given a certain state and the rules of chess, the explanation of why a certain player made a certain move will differ from player to player, even if both

---

[20]Another problem is the amalgamation of 'forced' in the sense of 'the only solution' with 'obvious to anyone'. Many forced move are not obvious and design explanations are often difficult to find.

players made the same move. In addition, the explanation why a certain brute force machine made that move will be very different in kind from the explanation why a certain human player made it. Yet, the explanation why that move is the best one will be the same.

In summary, the notion of function as biological role makes sense and is useful because organisms are organized beings and the meaning and utility of that notion is independent of the historical process by means of which organized beings are created and/or modified. In the case of biological value a distinction must be made between necessities and optimalities. The meaning and utility of talk of necessities (needs and demands) is independent of evolution. The choice of a criterion for optimality, on the other hand, is determined by the character of the process that generates and/or modifies living beings. However, given the choice of a criterion, what is optimal depends not on the history but on the relevant circumstances. The utility of talk of optimalities is of course determined by the extent to which the evolutionary process really is a process of optimization (as Dennett rightly notes). Finally there is nothing in Dennett's argument that opposes my view that design explanations explain the utility of a certain trait on the basis of laws and conditions, rather than the presence of that trait as the result of historical processes.

3.6 KINDS OF EXPLANATORY RELATIONS

In order to understand how the different kinds of explanation in organismal biology fit together, I propose classifying the relations to which explanations in biology appeal along two dimensions: (1) the level of organization (individual/population); and (2) the nature of the relation (causes/functional dependencies). The individual level/population level distinction replaces Mayr's proximate/ultimate distinction. The causes/functional dependencies distinction is added. I discuss these distinctions in that order.

Individual level/population level

According to Mayr the existence of two kinds of causes (proximate and ultimate) is one of the things in which the living world fundamentally differs from the non-living. He talks about these different kinds of causes as if they explain the same phenomenon. For example, in *This is Biology* he remarks:

"Every phenomenon or process in living organisms is the result of two separated causations, usually referred to as proximate (functional) causations and ultimate (evolutionary) causations." (Mayr, 1997: 67)

Proximate and ultimate explanations address different questions about the same phenomenon. Both questions are equally legitimate and the answers are complementary to each other. As an example, Mayr (1997) mentions sexual dimorphism. The proximate explanation of this phenomenon appeals to hormones and sex-controlling genes; the ultimate explanation to sexual selection and predator thwarting. Mayr emphasizes again and again that many controversies in the history of biology could have been avoided if the apparent opponents had been aware that rather than offering competing explanations, one party was talking about proximate causes and the other about ultimate ones.

This raises the question of how the two causes are related. How is it possible that two sets of causes explain the same phenomenon without being competitors? Mayr does not present a clear answer to this question. A possible answer is that each set is partial in the sense that both are needed to bring about the phenomenon to be explained (in the same way that both the presence of oxygen and an initial spark are causes of a certain fire). However, this is not what Mayr seems to mean. In this way biological phenomena would not differ from physical ones. More importantly, it is certainly not the relation between appeal to hormones and appeal to sexual selection to explain sexual dimorphism: both the proximate set and the ultimate set are by themselves sufficient to bring about the phenomenon they explain. A better solution seems to view the ultimate causes as the causes of the proximate causes. This is what Mayr repeatedly suggests. For instance when he says that proximate explanations are about the decoding of the genetic program and ultimate explanations about the history of that program. Note, however, that physical causes too might be related in this way. However, if you would follow the line from the materials and events that caused a certain bird individual to be male, back via the zygote out of which it grew to its very distant sexless ancestors you will find modifications caused by mutation and recombination but you will not find modifications caused by sexual selection, predator thwarting and so on. So this too seems not to be the relation between the proximate and ultimate causes.

The reason why tracing back the modifications of a trait (or of the genes for that trait) from ancestor to ancestor does not give insight in selection is, of course, that natural selection is essentially a population phenomenon (see Mayr, 1959; Lewontin, 1983; Sober, 1984; Matthen and Ariew, 2002; Walsh et al., 2002).[21] Sexual selection and predator thwarting do not cause modifications of individual traits, they cause modification of the share of certain traits in the population. For that reason, the answer to the question of the relation between proximate causes and ultimate causes is that these causes explain different phenomena. Proximate causes are causes that operate at the individual level, ultimate causes are causes that bring about changes at the

---

[21]Note that the term 'population' is used here in a broader sense than usual in biology. A population does not necessarily consist of interbreeding organisms but applies to collections of entities that originate as inaccurate copies from other entities (that is as entities among which there is heredity and variation).

population level.[22] Proximate causal explanations are concerned with the emergence of traits, abilities or whatever in certain individuals. Ultimate causal explanations (evolutionary selection explanations) are concerned with the dynamics of populations.[23]

Causes/utilities

Both physiological explanations, developmental explanations and design explanations are concerned with individual level relations. The difference between on the one hand design explanations and on the other hand physiological and developmental explanations has to do with the nature of the explanation. Both evolutionary and physiological/developmental explanations are causal in character. They tell us how a certain event, state, trait, capacity or process is brought about. Design explanations on the other hand tell us what is useful to or needed by an organism that has certain characteristics (e.g. that lives on land, has a certain size and a certain level of activity) and why this is the case.

In the pre-Darwinian traditions of Cuvier and Von Baer (as described by Coleman, 1964; and Lenoir, 1982: 196) it was supposed that the needs a trait satisfies causally explain the emergence and maintenance of that trait at the individual level. The biologists working in these traditions were impressed by the interdependence of the different parts and processes of an organism. The harmony of the different parts of an organism was understood as the result of interaction of the needs of that organism at the individual level. In Darwin's trail it became clear that functional interdependencies should be distinguished from causal interactions.

Cuvier's principle of the conditions of existence states that the different parts and processes of an organism depend on each other and support each other.

"Since nothing can exist without the reunion of those conditions which render its existence possible, the component parts of each being must be co-ordinated in such a way as to render possible the whole being, not only in itself, but also with regard to its surrounding relations." (Cuvier, 1817, Vol. 1: 6)

The question of how the different parts and processes of an organism became geared to each other was one of the central theoretical issues in eighteenth and nineteenth century biology. As is well-known, in British Natural Theology this question was answered by appealing to the hand of a

---

[22]Note that some causes such as the presence of mutagens are both proximate and ultimate.
[23]I originally proposed replacing Mayr's proximate/ultimate distinction by the individual/population distinction in Wouters (1995). Ariew (2003) comes to a similar conclusion.

benevolent creator. In mainstream biology this doctrine never made headway. Cuvier saw the harmony between the different parts and processes of an organism as the result of the causal interaction of the interdependent parts at the individual level. This interaction was assumed to be a kind of material exchange, called "*tourbillon vitale*" ("*Stoffwechsel*" in German). Interdependent parts were thought to maintain each other by means of this *tourbillon vitale*. The same process operates in development and regeneration. Cuvier and his followers tended to confuse this assumed causal interaction between functionally interdependent parts with the relation of functional interdependency itself. In their view the fact that several organs are functionally interdependent maintains the gearing between those organs. The needs of an organism (such as the need for strong claws in an organism capable of digesting only flesh) act as efficient causes which organize the process of material exchange in such a manner that the organism's needs are satisfied. Hence, design explanations (which appeal to the need for a certain structure) were seen as explanations that causally explain how the harmony between the parts and the processes of an individual organism is maintained.

The process of material exchange was thought to explain (in principle) how the harmony of an organism is maintained. The origin of this harmony was seen as another issue. According to these biologists causal interaction in the organic world differs from causal interaction in the non-organic world. Causal interactions in the non-organic world were supposed to be linear, ($A \rightarrow B \rightarrow C \rightarrow D$), causal interactions in the organic world are "clearly" cyclic ($A \rightarrow B \rightarrow C \rightarrow A$). In the views of the late eighteenth, early nineteenth century science it is impossible to explain how such a cyclic arrangement of causes came into being.[24] What we can try however, is to explain how this arrangement is maintained (in the individual) and modified (in the course of the ontogeny) given the fact that there is such an arrangement.

Darwin's theory offers the solution to the problem of the initial organization. According to this theory the answer to the question 'how did the parts and organs of an organism become geared to each other and to the environment in which it lives?' must be sought in the evolutionary history of the lineage rather than in immediate causal interaction between the parts that are in harmony. In modern biology, the metabolic interaction (if any) between two functionally interdependent organs does not explain the gearing of those organs. Harmony between parts and processes is "pre-stabilized" in the genes. For instance, the lungs of birds have a very complicated structure which is needed to enable flight. In the view of Cuvier, Von Baer and their followers, this harmony is established and maintained by a metabolic process operating between the lungs and the wings of the individual that has both items. In the view of modern biology there is no such exchange. In the course of the

---

[24]This view is most clearly expressed in Kant's *Kritik der Urteilskraft* (1790).

ontogeny the lungs and the wings acquire their structure independently. The fact that these structures are in harmony is explained by the fact that in the course of evolution the structures of wings and lungs became tuned to each other (due to selection).

Explanatory relations in organismal biology

When the two dimensions of explanation (level and nature) are combined one has four kinds of fundamental relations: (1) causal relations at the level of the individual; (2) causal relations at the level of the population; (3) functional interdependencies at the individual level; and (4) functional interdependencies at the population level (see Figure 3.2). Relations of the fourth kind are not relevant to understanding explanations in organismal biology. Their utility in explanations in ecology is one of my current research projects, as are design explanations at the cellular and molecular level.

	Causes	Interdependencies
Individual	Immediate causes, underlying mechanisms, development	Design explanations
Population	Evolution	(requires further research)

Figure 3.2. Kinds of relations to which explanations appeal.

Why-questions in biology

We can now answer Dennett's question about Darwin's position with regard to why-questions (see Section 3.5). Did Darwin throw those questions away (as, according to Dennett, many biologists and philosophers think)? Or did he find a new way to answer them (as Dennett thinks)? Perhaps, the best answer to this question is that in the light of Darwin's theory it became clear that why-questions in biology are ambiguous. In the old days before Darwin a question of the type 'why do s organisms have trait t?' was answered by means of a teleological explanation in which the utility of t explains why t emerged in the cause of the ontogeny. This kind of teleological explanation is nowadays unacceptable and replaced by three kinds of explanations: (1) developmental explanations that specify one or more factors that explain why individuals of kind s develop t (this kind of explanation does not appeal to t's utility); (2) design explanations that specify characteristics of the organism and its environment that explain why t is more useful to organisms of kind s than some conceivable alternative (this kind of explanation does not explain why t developed); (3) evolutionary selection explanations that specify the past effects of occurrences of t in the history of the lineage of s organisms that explain t's current share in the population (current utility and mere hypothetical utility do not help to explain the current presence of t, but a design explanation of why t is more useful than the actually existing alternatives in the actual historical circumstances can be part of an evolutionary selection explanation).

3.7 THE AUTONOMY OF FUNCTIONAL BIOLOGY

As I said in the introduction, many evolutionary biologists and many philosophers of biology maintain that the autonomy of biology is rooted in the selection process and consists of asking evolutionary questions in addition to questions about the operation of mechanisms. Those biologists and philosophers depict functional biology as a reductionist science that restricts itself to the application of physics and chemistry and lacks a biological point of view.

My arguments show that explanation in functional biology differs from explanation in the physical sciences in a number of ways. First, mechanistic explanations in biology differ from mechanistic explanations in the physical sciences by their appeal to biological roles. Although such explanations explain the activities of the whole as the result of the operation of systems of subsystems that are ultimately physical and chemical in nature, they provide an understanding of the parts in the context of the organism as a whole. Furthermore, functional biologists explain the traits in which they are interested not simply by specifying underlying mechanisms, they also explain those traits by pointing to other characteristics of the organism that functionally depend on the trait to be explained, that is by appeal to the utility of that trait.

My arguments show, furthermore, that this distinctive character of explanation in biology is not rooted in the selective character of the process by means of which life gets shape but rather in that other special character of life: the ability of an organism to maintain itself, to grow, to develop and to reproduce. Appealing to biological roles in mechanistic explanations is legitimate because the properties of the whole are not a simple aggregate of the properties of the parts, but the result of the way in which the parts are organized. If the life-state could be brought about by a simple aggregate of parts, appeal to biological roles would not be necessary. Appealing to utility is legitimate because the presence of one combination of characters (e.g. living an active life on land) puts constraints on the way in which the life-state can be maintained. Not all combinations of characters are viable.

In other words: biology distinguishes itself from the physical sciences not only by the evolutionary perspective but also by its functional perspective, that is by its concern with the requirements of the life-state and the way in which these are met. Although the life-state is the product of evolution, the difficulty of acquiring and maintaining that state is a fundamental characteristic of the world, that is independent of evolution. Functional biology is a real biological science with its own point of view: the study of the life-state.

REFERENCES

Alcock, J. (1998). Animal Behavior. An Evolutionary Approach. 6th edition. Sinauer, Sunderland.

Alcock, J. (2003). A textbook history of animal behavior. Animal Behavior 65: 3-10.

Alcock, J. and P. W. Sherman (1994). The utility of the proximate-ultimate dichotomy in biology. Ethology 96: 58-62.

Ariew, A. (2003). Ernst Mayr's ultimate/proximate distinction reconsidered and reconstructed. Biology and Philosophy 18: 553-565.

Armstrong, D. P. (1991). Levels of cause and effect as organizing principles for research in animal behaviour. Canadian Journal of Zoology 69: 823-829.

Bigelow, J. and R. Pargetter (1987). Functions. Journal of Philosophy 84: 181-196.

Bock, W. J. and G. von Wahlert (1965). Adaptation and the form-function complex. Evolution 19: 269-299.

Canfield, J. (1964). Teleological explanation in biology. British Journal for the Philosophy of Science 14: 285-295.

Coleman, W. (1964). Georges Cuvier, Zoologist: A Study in the History of Evolution Theory. Harvard University Press, Cambridge, Mass.

Craver, C. F. (2001). Role functions, mechanisms, and hierarchy. Philosophy of Science 68: 53-74.

Cummins, R. (1975). Functional analysis. Journal of Philosophy 72: 741-765.

Cummins, R. (1983). The Nature of Psychological Explanation. The MIT Press, Cambridge, Mass.

Cuvier, G. (1817). Le Règne Animal Distribué D'apres Son Organisation. Deteuille, Paris.

Dennett, D. C. (1995). Darwin's Dangerous Idea. Evolution and the Meanings of Life. Simon and Schuster, New York.

Dewsbury, D. A. (1992). On the problems studied in ethology, comparative psychology, and animal behavior. Ethology 92: 89-107.

Dewsbury, D. A. (1994). On the Utility of the Proximate-Ultimate Distinction in the Study of Animal Behavior. Ethology 96: 63-68.

Dewsbury, D. A. (1999). The proximate and the ultimate: past, present, and future. Behavioural Processes 46: 89-199.

Dobzhansky, T. (1973). Nothing in biology makes sense except in the light of evolution. American Biology Teacher 35: 125-129.

Dullemeijer, P. (1974). Concepts and Approaches in Animal Morphology. Van Gorcum, Assen.

Francis, R. C. (1990). Causes, proximate and ultimate. Biology and Philosophy 5: 401-415.

Glennan, S. S. (1996). Mechanisms and the nature of causation. Erkenntnis 44: 49-71.

Glennan, S. S. (2000). Rethinking mechanistic explanation. Philosophy of Science 69: S342-353.

Goodenough, J., B. McGuire and R. A. Wallace (2001). Perspectives on Animal Behavior. John Wiley and Sons, New York.

Hinde, R. A. (1975). The Concept of Function. In: Baerends, G. R., C. Beer and A. Manning (Eds). Function and Evolution in Behaviour. Clarendon, Oxford. p. 3-15.

Holekamp, K. E. and P. W. Sherman (1989). Why male ground squirrels disperse. American Scientist 77: 232-239.

Huxley, J. L. (1942). Evolution: The Modern Synthesis. Allen and Unwin, London.

Kant, I. (1790). Kritik der Urteilskraft. Lagarde und Friederich, Berlin.

Kitcher, P. (1993). Function and Design. In: French, P. A., T. E. Uehling and H. K. Wettstein (Eds). Philosophy of Science. University of Notre Dame Press, Notre Dame. p. 379-397.

Klopfer, P. H. and J. P. Hailman (1967). An Introduction to Animal Behavior. Prentice-Hall, Englewood Cliffs.

Krebs, J. R. and N. B. Davies (Eds) (1984). Behavioral Ecology: An Evolutionary Approach. Blackwell, Oxford.

Kuipers, T. A. F. and A. Wisniewski (1994). An erotetic approach to explanation by specification. Erkenntnis 40: 377-402.

Lenoir, T. (1982). The Strategy of Life. Reidel, Dordrecht.

Lewontin, R. (1983). Darwin's revolution. New York Review of Books 30: 21-72.

Mahner, M. and M. Bunge (1997). Foundations of Biophilosophy. Springer, Berlin.

Matthen, M. and A. Ariew (2002). Two ways of thinking about fitness and natural selection. Journal of Philosophy 99: 55-83.

Mayr, E. (1959). Typological Vs. Population Thinking. In: Evolution and Anthropology: A Centennial Appraisal. The Anthropological Society of Washington, Washington. p. 409-412.

Mayr, E. (1961). Cause and effect in biology. Science 134: 1501-1506.

Mayr, E. (1997). This Is Biology: The Science of the Living World. Harvard University Press, Cambridge, Mass.

McFarland, D. (1999). Animal Behavior. Longman, London.

Millikan, R. G. (1989). An ambiguity in the notion "function". Biology and Philosophy 4: 172-176.

Mitchell, S. D. (1993). Dispositions or etiologies? A comment on Bigelow and Pargetter. Journal of Philosophy 90: 249-259.

Neander, K. (1991). The teleological notion of 'function'. Australian Journal of Philosophy 69: 454-468.

Raven, P. H. and G. B. Johnson (1999). Biology. McGraw-Hill, Boston.

Reeve, H. K. and P. W. Sherman (1993). Adaptation and the goals of evolutionary research. Quarterly Review of Biology 68: 1-32.

Robinson, J. D. (1986). Reduction, explanation, and the quests of biological research. Philosophy of Science 53: 33-353.

Salmon, W. C. (1984). Scientific Explanation and the Causal Structure of the World. Princeton University Press, Princeton.

Salmon, W. C. (1989). Four Decades of Scientific Explanation. University of Minnesota Press, Minneapolis.

Sherman, P. W. (1988). The levels of analysis. Animal Behavior 36: 616-619.

Sober, E. (1984). The Nature of Selection. The MIT Press, Cambridge, Mass.

Tinbergen, N. (1951). The Study of Instinct. Clarendon, Oxford.

Tinbergen, N. (1963). On aims and methods of ethology. Zeitschrift für Tierpsychologie 20: 410-433.

Walsh, D. M. (1996). Fitness and function. British Journal for the Philosophy of Science 47: 553-574.

Walsh, D. M., T. Lewens and A. Ariew (2002). The trials of life: Natural selection and random drift. Philosophy of Science 69: 452-473.

Woodfield, A. (2000). Teleological Explanation. In: Newton-Smith, W. H. (Ed.). A Companion to the Philosophy of Science. Blackwell, Oxford. p. 492-494.

Wouters, A. G. (1995). Viability explanation. Biology and Philosophy 10: 435-457.

Wouters, A. G. (1999). Explanation without a Cause. Utrecht University. Ph.D. thesis. Available at <http://www.knoware.nl/users/arnow/diss/>.

Wouters, A. G. (2003). Four notions of biological function. Studies in History and Philosophy of Biology and Biomedical Science 34: 633-668.

Wouters, A. G. (forthcoming). Functional Explanation in Biology. In: Festa, R., A. Alisda and J. Peijnenburg (Eds). Cognitive Structures in Scientific Inquiry. Rodopi, Amsterdam.

Wright, L. (1973). Functions. Philosophical Review 82: 139-168.

Arno Wouters
Department of Philosophy, University of Nijmegen

4

Infectious Biology: Curse or Blessing? Reflections on Biology in Other Disciplines, with a Case Study of Migraine

Wim J. van der Steen

ABSTRACT

Biology has come to play important roles in many other disciplines. Some applications of evolutionary thought outside biology are disappointing, but promising approaches are feasible in medicine. Biology has a rich store of valuable knowledge extending to evolution, which is often disregarded in medicine. For example, the role of omega-3 fatty acids in the genesis of disease deserves much more attention. Biomedical research should pay more attention to higher levels of organization and to functional explanation. A case study of migraine illustrates all this. Biology is an infectious discipline in that parts of it are incorporated in many disciplines. But in the process of incorporation it is often distorted by transformations and omissions. Nobody would deny that applications of biology to medicine, for example, often amount to a blessing. At the same time, some distortions of biology may tend to transform blessings into a curse.

Keywords: aura, dietary deficiency, endothelium, evolutionary psychiatry, evolutionary psychology, evolutionary medicine, fatty acid, functional explanation, level of organization, migraine, platelet, serotonin, vasopressin.

4.1 INTRODUCTION

If we drown in details, we miss general patterns. This is a mundane truth, one that applies with increasing force to biology and its applications. Biology has become molecularized and geneticized. Details of ongoing research concerning low levels of organization are being produced in huge quantities, and higher levels of organization no longer receive the attention they deserve. If we take care of all the details in our own specialties, we are unable to see what is happening in other significant disciplines. The present chapter is an attempt to indicate how we may begin to manage the existing overloads of information, and redress imbalances among specialties.

T.A.C. Reydon and L. Hemerik.,(eds.), Current Themes in Theoretical Biology,
71-94.

My main concern here is with the roles played by biology in other disciplines. These roles are becoming more and more pronounced. Biology appears to be an infectious discipline. At times, its being infectious is a blessing, in many areas of medicine for example. But examples also exist of biological subject matter mutating into something else while infecting other disciplines. Also, essential biological information may be deleted in the process. This represents a curse rather than a blessing.

Considering evolution, I note that evolutionary biology is often distorted in other disciplines (Section 4.2). In medicine, valuable extensions of evolutionary biology are feasible. Evolutionary thinking applied to diets has far-reaching implications for medicine that are often disregarded (Section 4.3). The main case study in this chapter concerns migraine (Section 4.4). I use a generalist approach of migraine to uncover lacunae in existing research. The approach illustrates how systematic generalist work results in new hypotheses. The themes of evolution and diet recur in the case study. Apart from this, I argue that existing research on the role of hormones and neurotransmitters in migraine is one-sided. The emphasis is overmuch on serotonin. However important this neurotransmitter may be in migraine, other neurotransmitters and hormones should be equally important. Thus, researchers have disregarded vasopressin, which presumably has a major role in migraine. On the positive side of migraine research we may note that the study of the role of serotonin has yielded powerful medications, the triptans. Unfortunately these drugs have negative side effects. If biological research on migraine were less one-sided, a richer variety of treatments would presumably ensue. Hence this case study illustrates both the positive and the negative aspects of biology infecting medicine.

4.2 EVOLUTIONARY THINKING IN MEDICINE, PSYCHOLOGY, AND PSYCHIATRY

The study of evolution belongs to the core of biology. Indeed, evolutionary biology has become popular in many other disciplines, and we now have new interdisciplines such as evolutionary anthropology, evolutionary theology, and many more. In some instances, the biology found in these interdisciplines may deviate considerably from the genuine article (for a survey, see Van der Steen, 2000). I briefly consider the fate of biology in evolutionary medicine, evolutionary psychology, and evolutionary psychiatry, in this order. In Section 4.3, I continue the analysis with a focus on evolutionary aspects of diet that have far-reaching implications for medicine.

As biology is a foundation for medicine, a natural expectation would be that evolutionary thinking is common in medicine. It is not. But the tides are turning. Thanks to foundational work of Nesse and Williams (1994, 1998), evolutionary (or Darwinian) medicine is developing into a promising discipline. The same goes for evolutionary epidemiology, a recent offshoot of evolutionary medicine (Ewald, 1994; Frank, 2002; Greenblatt and Spigelman, 2003). Nesse and Williams distinguish five categories of explanation concerning features of diseases.

First, some features of diseases are evolved defences that help us make pathogens harmless. Responses such as fever and vomiting may help us get rid of pathogens, for example. Less obvious examples also exist. Bacteria need iron. Hence the reduction of blood iron upon invasion by pathogens appears to be an evolved defence response. Physicians not trained in evolutionary thinking may not be aware of this and help the pathogen rather than the patient by prescribing iron as a remedy. Second, some diseases represent enduring conflicts with other organisms, especially pathogens. Our evolved defence mechanisms do not make all pathogens harmless for us, since pathogens evolve much faster than we can, owing to short generation times. Third, time lags of natural selection may have left us with old features that are no longer adaptive in present-day environments. Our ancestors may have been well advised to consume as much fat, salt and sugar as they could lay their hands on, since these things were scarce. We still crave for these elements of diet, but now they are readily available. The ensuing over-consumption often impairs our health. Fourth, some diseases represent evolutionary trade-offs as some genes code for disadvantageous and advantageous phenotype features at the same time. This explains, for example, the persistence of sickle-cell anaemia in some regions. In homozygous form, this disease is lethal. But persons who are heterozygous for the sickle cell gene do not get the disease and they have an enhanced resistance against malaria. This explains the persistence of the anaemia in regions where malaria is common. Fifth, constraints on selection may preclude the elimination of maladaptive features. Natural selection has not eliminated the appendix, a vestigial organ that may cause death upon infection. A decrease in size of the organ would decrease blood flow and thereby increase the probability of infection. This has apparently precluded the selective elimination of the appendix.

By and large, explanations in evolutionary medicine appear to be reasonable as they are backed up by accessible evidence. I am less positive about the young disciplines of evolutionary psychology and evolutionary psychiatry, which are rife with speculation. I illustrate this with a sample of critical comments from a previous analysis (Looren de Jong and Van der Steen, 1998; for details and references, see this article) and some additional sources.

Evolutionary psychologists, particularly Cosmides and Tooby, regard the mind as an organ of the body with hard-wired mechanisms representing a universal human nature. According to them, "mental modules" serving specific functions are hard-wired. They oppose what they call the Standard

Social Science Model (SSSM), which reputedly explains human behaviour as an outcome of cultural influences. Evolutionary psychologists would not deny that cultural influences produce variations in behaviour, but they regard this as a secondary manifestation of shared universal mechanisms.

Evolutionary psychologists aim to explain psychological processes as biological adaptations. They assume that human traits such as personality, sexual preferences, love, and many more, derive from psychological mechanisms that enhance survival or have done so in our evolutionary past. Prominent representatives such as Cosmides and Tooby (1994) and Buss (1995, 1998) have proposed that the new discipline should replace outmoded forms of social science, and that its focus on man's evolutionary history of natural selection is the best way to unify psychology with hard science (see also the earlier volume edited by Barkow et al., 1992). The implementation of this unification should be difficult since psychological mechanisms and behaviours do not fossilize.

In evolutionary biology, explanations that portray histories of natural selection and adaptation, should meet several conditions. The philosopher of biology Brandon (1990), in a well-known book, distinguishes five conditions of adequacy for "adaptive explanations." First, we have to show that selection has acted on types of organisms that differ in traits affecting reproductive success in the relevant environment. Second, we must identify ecological factors implicated in selection. Third, we must show that the trait investigated is heritable. Fourth, we need information about population structure regarding gene flow, and about the structure of the selective environment. Finally, we have to know about phylogenetically primitive traits, and derived traits.

Richardson (1996) has applied Brandon's five criteria to proposals in evolutionary psychology that rationality, language, and capacities for social exchange have been selected as adaptations. He demonstrates that all proposals fail dismally to provide evidence concerning an evolutionary history of natural selection for these features. True, we may reasonably maintain, for example, that language is adaptive since it facilitates communication, that rationality has to represent an adaptation to changing environments, and that social skills adaptively foster group cohesion. But as Richardson rightly notes, such considerations amount to facile, general, vacuous explanations (for additional critical comments from a variety of perspectives, see Davies, 1996; Griffiths, 1996; Lloyd, 1999; Bjorklund and Pellegrini, 2002; Laland and Brown, 2002; Wilson, 2002).

Some well-known explanations of human behaviours found in evolutionary psychology appear to fly in the face of plain logic. I illustrate this by an example concerning murder and infanticide. Daly and Wilson (1988a and b, 1998), whose work has been approvingly assimilated by evolutionary psychology (see Rose, 2000), have done extensive research on murder and infanticide (for details and critical comments, see Rose, 2000; Van der Steen and Ho, 2001a: 88-90). Daly and Wilson provide evolutionary explanations, for example, for the finding that children are murdered or abused more often by stepfathers than by biological fathers. This is taken to be a result of our evolutionary heritage, since stepfathers, unlike biological fathers, may enhance their own reproductive success in this way. Daly and Wilson stick to this evolutionary explanation even though they have to acknowledge that the murders represent very rare events. In almost all cases, stepfathers refrain from killing their stepchildren just like biological fathers refrain from killing their own children. If the explanation made sense, the murders should be more common.

Evolutionary psychology has a close cousin called evolutionary psychiatry, which shares many of its features (for reference, see Van der Steen, 2000: Chapter 10). Like evolutionary psychologists, researchers in evolutionary psychiatry are concerned about showing how natural selection can be responsible for particular psychological processes and behaviours. For the evolutionary psychiatrist, the most important problem to be explained is the commonness of psychopathology. Why has natural selection failed to decrease mental illnesses toward lower prevalence? The most common answer given boils down to the assumption that genetic factors implicated in psychopathology must in some way have benefits associated with them. To the extent that this is not so in our culture, we may be dealing with leftovers from our ancestors. Stevens and Price (1996: 143), for example, hypothesize that schizoid personalities have played a positive role as groups in ancestral human populations had to split up when resources became inadequate:

"At this point, the issue of leadership becomes crucial for survival, because the leader has to inspire the departing group with its sense of mission and purpose, its need to unite against all odds, its belief that it can win through and find its own "promised land." Such a leader needs the sort of charisma traditionally granted by divine will and maintained through direct communion with the gods. It is when called upon to fulfil this exalted role that the schizoid genotype comes into its own." (Stevens and Price 1996: 143)

Disordered language of the schizoid leader could promote group identity by new linguistic forms, and delusional and hallucinatory originality of the leader could make group splitting a more easy process (Idem: 149). This is a typical example of a just so story that can hardly be tested. Other examples of such stories are to be found in Van der Steen (2000) and Van der Steen and Ho (2001a).

Speculative explanations in evolutionary psychology and evolutionary psychiatry presuppose that some mental features of human beings are geared to ancestral environments rather than present-day environments. I would conjecture that food is one of the most important environmental factors to be considered, with implications for the enterprises of evolutionary psychology and evolutionary psychiatry. In the next section, I review evidence that ancestral diets deviate considerably from modern diets. Our ancestors used to eat very different foodstuffs than we do, and our current Western diets are maladaptive because, by and large, we still have the genetic make-up of our ancestors. This has resulted in all sorts of somatic and psychiatric disorders, and it does markedly affect our behaviour. The assumption that our diets are maladaptive can be backed up by evidence from many disciplines.

In the scheme of Nesse and Williams, this consideration belongs to their third category of evolutionary explanations of disease. Time lags of natural selection have left us with old features, namely the need of particular diets, which are no longer adaptive in present-day environments, to wit environments in which agribusiness and the food processing industry provide us with unhealthy foods.

For evolutionary psychology and evolutionary psychiatry, this appears to imply that much of their subject matter becomes irrelevant. Many of the behaviours seen by evolutionary psychologists as remnants of a remote behavioural past may actually result from more recent dietary changes. The evolutionary interpretations of psychopathology provided by Stevens and Price (1996) become problematic as well. They explain the commonness of psychiatric disorders by speculating that these disorders must have carried advantages with them for our ancestors. This evolutionary speculation may have to yield to a simple causal explanation: recent evidence indicates that many persons may develop a psychiatric disorder as they eat the wrong sort of food.

4.3 FATTY ACIDS, HEALTH, DISEASE, AND EVOLUTION

Diet can markedly affect health and disease. I introduce examples of this with the main emphasis on omega-3 PUFAs (polyunsaturated fatty acids). Known biological roles of these acids in health and disease are often ignored in many areas of medicine. This omission is a deplorable example of biased biology in medicine.

Erasmus (1993) has, since the 1980s, campaigned for changes in the fatty acid composition of our diet. He argued that existing methods of food production result in highly unnatural diets causing all sorts of diseases. Stoll (2001) makes the same point. In particular, modern diets lack adequate amounts of the omega-3 PUFAs EPA (eicosapentaenoic acid) and DHA (docosahexaenoic acid), and they have a high ratio of omega-6 to omega-3 PUFAs. Evidence presented hereafter demonstrates that this does impair our health.

Omega-3 and omega-6 PUFAs are converted in the body to other compounds along two pathways. First, they are incorporated into cell membranes and, second, they are converted into eicosanoids, including prostaglandins. Omega-3 PUFAs make membranes more fluid and thereby influence our biochemistry and physiology in many ways. This explains in part why shortages of these acids may result in many different diseases. In addition to this, the shortages affect the eicosanoid pathway so as to foster disease. The eicosanoid pathway is complex (for details, see Serhan and Oliw, 2001). Omega-6 PUFAs produce inflammatory eicosanoids, whereas omega-3 PUFAs have the opposite effect. We need omega-6 PUFAs, but they have to be kept in check by the omega-3 PUFAs. Omega-3 PUFAs play particularly important roles in the brain, which needs some of them in large quantities. As shortages also impair the blood-brain barrier, the brain is particularly vulnerable under deficiencies (Banks *et al.*, 1997). The medical community is now slowly recognizing all these things, although they have been known for years outside medicine proper (see, for example, Clandinin and Jumpsen, 1997).

The existing shortages and imbalances in our diet contribute to a great variety of diseases. For example, literature reviewed in a previous article (Van der Steen and Ho, 2001b) indicates that omega-3 deficiencies are implicated in autoimmune diseases such as rheumatoid arthritis, and in ulcers of the stomach and the duodenum. Treatment with fish oil, which is rich in DHA and EPA, has beneficial effects in these diseases, though it may not cure ulcers once they have developed. The use of fish oil as a food supplement anyhow serves the cause of prevention. The analysis in Van der Steen and Ho (2001b) suggests that existing drug treatments could better be replaced by diet treatments in many situations (for additional examples, see the book by Van der Steen *et al.*, 2003). The current use of drugs anyhow represents extensive overmedication. This situation exists also in psychiatry (Healy, 2002; Van der Steen, 2003; Van der Steen *et al.*, 2003).

Some authors have performed comprehensive studies of omega-3 deficiencies that cover many diseases. Holman (1997) observed the deficiencies in neurological diseases such as multiple sclerosis (MS), Huntington's disease, anorexia nervosa, and immune diseases such as Crohn's disease and sepsis, and a survey by Katz *et al.* (2001) also indicates that omega-3 PUFAs play a role in many diseases, especially neurological diseases. Treatment with ethyl-EPA has even resulted in marked beneficial effects in a disease as serious as Huntington's chorea (Puri *et al.*, 2002). Convincing evidence is accumulating that omega-3 deficiencies are also significant in the aetiology of psychiatric disorders, and that treatment with EPA and/or DHA benefits or even cures patients suffering from such disorders. Stoll (2001) has made this case for depression, and the research group of Horrobin and Peet has done the same for schizophrenia (see for example Horrobin, 1997; Peet *et al.*, 2001; Horrobin *et al.*, 2002; Peet and Horrobin, 2002). As the omega-3 PUFAs affect these disorders, we may expect them also to affect the behaviour of healthy persons. This is indeed so. For example, in some situations DHA intake reduces aggression in times of mental stress (Hamazaki *et al.*, 1996; Hamazaki *et al.*, 2002).

The most convincing evidence that inadequate diets are causing much disease in "civilized" countries concerns cardiovascular diseases and cancers. Research spanning five decades has yielded a credible reconstruction of hominid diets in the course of evolution. Ancestral diets resemble those of hunter-gatherers now living in marginal habitats. The data concerning diets together with data concerning disease prevalence demonstrate that the dietary deficiencies considered here have become a major disaster. I rely here on two excellent reviews, Eaton and Konner (1985) and Simopoulos (2001), which put the omega-3 story in a broader context; many other sources are reviewed in Van der Steen *et al.* (2003). The two reviews put together evidence from evolutionary biology, the history and prehistory of agriculture, epidemiology, experimental zoology, and the study of medical treatments with omega-3 PUFAs.

Evolution and ageing is one of the interesting themes in Eaton and Konner (1985). They oppose the view that diseases of old age are common because they do not affect reproductive survival, the target of natural selection. Eaton and Konner unearth evidence, in part dating from the 1950s, against this view. Young people in the Western world often have asymptomatic forms of cardiovascular disease, whereas old persons in technologically primitive cultures often remain free of them. In Western societies, the prevalence of cardiovascular diseases and cancers has increased over the last 100 years as dietary habits changed with modern agribusiness and food processing. Our species has much nutritional adaptability, but we have moved far beyond the range of healthy diets.

From reconstructed diets of our ancestors, and data concerning diets of existing hunter-gatherers in marginal habitats, we can infer what diets should be appropriate for ourselves. Meat from game would be excellent, but our domestic livestock is a poor source of meat, as it contains far less polyunsaturated fat, particularly omega-3 PUFAs. This is caused by unnatural food provided to our livestock. Meat and fish, together with vegetables and fruit, are the main constituents of the hunter-gatherer's diet. We have added to this much milk and milk products, and bread and cereals, which represents a highly unnatural situation. Eaton and Konner provide evidence indicating that a change in the direction of the hunter-gatherer's diet would benefit us.

Many of the suggestions put forward by Eaton and Konner have been confirmed by subsequent research as reviewed by Simopoulos (2001). Simopoulos first of all reviews the so-called Seven Countries Study, the results of which were published more than thirty years ago by Keys (1970). This study uncovered differences among countries in coronary heart disease in the order of five to 10-fold. The island of Crete had the lowest death rate and the lowest prevalence of cancer and heart disease. Later studies indicated that diets in Crete resemble the Palaeolithic diet in terms of fibre, antioxidants, fat composition, and the ratio of omega-6 to omega-3 fatty acids. Western diets deviating from this are associated with a high prevalence of cardiovascular disease, diabetes, obesity and cancer. The Lyon Heart Study, a prospective randomised trial in the 1990s, confirms that the Crete diet is healthy. The researchers doing this study compared the Crete diet with a diet recommended by the American Heart Association over a period of five years. The Crete diet resulted in a 70% decrease in death rate and a marked decline of heart disease and cancer. A rich variety of additional evidence confirms that omega-3 PUFAs play crucial roles in diseases that are common in the West, for example cardiovascular diseases, cancers, various neurological disorders, gastric ulcers, and autoimmune diseases (for more details and sources, see Simopoulos, 2001, and references in Van der Steen *et al.*, 2003).

We may reasonably generalize, and conjecture that omega-3 PUFAs should play *some* role in almost all diseases. But many diseases have simply not been investigated from the omega-3 perspective. Considering any unexplored disease, we may arrive at reasonable hypotheses concerning a possible role of omega-3 PUFAs by a search for indirect evidence in the literature. The role of omega-3 PUFAs is indeed one of my themes in the case study of migraine in the next section. Indirect evidence suggests that these PUFAs should play a role in migraine.

4.4 A CASE STUDY OF MIGRAINE

The problem of unconnected literatures

Migraine is often misunderstood. An estimated 23 million Americans suffer from disabling migraines (Cady, 1999) but only a minority are diagnosed and treated (Sheftell and Tepper, 2002). Recent scientific literature provides numerous mechanisms at the cellular and molecular level, but no integrative view of migraine. Instead of presenting a theory of my own, I locate lacunae in ongoing research, with suggestions for new approaches. The example of migraine illustrates a general problem with current research in the life sciences. As detailed knowledge accumulates, blank areas accumulate as well. Considering research on migraine in medicine, I argue that blank areas often represent missing information that could be drawn from biology. In this respect, the case of migraine resembles the case of omega-3 PUFAs reviewed in the previous section. But the story of migraine is more complicated, because the evidence from biology needed to understand the disorder is less readily available. To reach it, we have to build bridges among unconnected literatures.

I follow here a generalist approach, which resembles that of Swanson, a specialist concerned with scientific documentation in medicine. Over many years, he has developed methods including special software for locating missing links in the literature, for which he received an award (Swanson, 2001). His primary tool is search by way of key words, to find unconnected literatures dealing with similar phenomena (Swanson, 1990; Swanson and Smalheiser, 1997 and 1999). This often results in new, fruitful hypotheses. Two of his case studies happen to fit in with the subject of the present section.

In one study, Swanson identified articles showing that dietary fish oils lead to certain blood and vascular changes, and also articles with evidence that similar changes might benefit patients with Raynaud's phenomenon (reduced arterial blood flow in hands, which easily get cold as a result) (Swanson, 1986 and 1987), a phenomenon that is common in migraine patients. The two sets of articles had never been connected. Taken jointly, they generate the hypothesis that dietary fish oil may benefit Raynaud patients. This was indeed confirmed in subsequent research. Notice that this is another item confirming that omega-3 PUFAs play a role in many diseases (see the previous section). A subsequent case study concerns migraine (Swanson, 1988 and 1991). An extensive search of unconnected literatures resulted in the hypothesis that magnesium deficiency is a causal factor in migraine headache. Recent research suggests that this is an important hypothesis. Johnson (2001) reports that physiological roles of magnesium are undervalued, and that magnesium deficiency is common. Also, recent research keeps confirming that magnesium is a crucial factor in migraine (see for example Boska *et al.*, 2002; Mauskop *et al.*, 2002).

Surprisingly, Swanson did not forge connections between the case studies concerning Raynaud's phenomenon and migraine. Migraine is often associated with Raynaud's phenomenon, and it is conceivable that the two conditions share mechanisms involving the vascular system (Voulgari *et al.*, 2000; Constantinescu, 2002). This suggests that fish oil may in some cases be a promising treatment option not only in Raynaud's phenomenon, but also in migraine. I will indeed argue here for the hypothesis that deficiencies of omega-3 PUFAs may predispose towards migraine. The possible role of omega-3 PUFAs is one of the examples in this section, which indicate that potentially valuable knowledge from biology has failed to acquire a foothold in research on migraine.

Problems with classification

A recent book by Davidoff (2002) is presumably the most comprehensive biomedical text concerning migraine. Davidoff is aware that classifications of migraine phenomena are problematic. The classification published in 1988 by the Headache Classification Committee of the International Headache Society

(IHS) is presumably the best system. It recognizes seven subtypes of migrainous headaches. Most of these headaches belong to two subtypes, migraine without aura and migraine with aura. Auras are neurological phenomena that may take many forms, for example abnormalities of vision. The IHS classification stipulates that characteristic features must be present to establish a diagnosis of migraine, such as a limited duration of attacks (4 to 72 hours), intense unilateral pain, and nausea.

Davidoff notes that this classification, indeed any classification of headaches, is somewhat arbitrary, since transitions between migraine types and co-occurrences of types are common. "Migraine" is actually an abstract word for a heterogeneous category of phenomena.

Some hypotheses

Many hypotheses have been proposed to explain migraine. Wolff (1963; references and comments in Davidoff, 2002: 190) proposed that migraine is primarily a problem of the blood circulation. He argued that constriction of particular blood vessels in the brain and outside the skull, set in motion a chain of events resulting in migraine, local lack of oxygen being responsible for aura phenomena. Vasoconstriction with local sterile inflammation of blood vessel walls would be followed by reactive vasodilatation with stretched nerve endings in vessel walls generating the pain. Wolff's hypothesis cannot explain all migraine phenomena. At present, a neurogenic hypothesis postulating events in the nervous system as a cause of vascular changes has more adherents. The two hypotheses may actually supplement each other. Anyhow, both the vascular system and the nervous system play a role in migraine. Wolff's idea concerning stretched nerve endings as a cause of pain has meanwhile taken a more specific form. In migraine, vasodilatation stimulates pain receptors in blood vessel walls in a particular area of the brain, the dura. Aura phenomena are associated with a phenomenon called cortical spreading depression, which has been studied in experimental animals. Particular stimuli applied to the cortex result in a local decrease of nerve cell activity and blood flow. The effect slowly spreads over a larger area, with blood flow remaining off balance for up to an hour.

Hormones and neurotransmitters also play a role in migraine. Researchers have allotted a crucial role to serotonin in the blood, most of which is stored in platelets (Davidoff, 2002: Chapter 12). Some migraine attacks are associated with changes in platelet serotonin, which may generate pain attacks via a changed balance of vasoconstriction and vasodilatation. The popularity of the serotonin hypothesis fits with the recent emergence of triptans, drugs that are commonly used to treat migraine attacks as they affect serotonin metabolism (Davidoff, 2002: Chapter 16; see also Diamond and Wenzel, 2002, for a review of commonly used drugs). Davidoff notes that many drugs are potentially useful, but no drug benefits all patients on all occasions, and undesirable side effects are common.

The immune system is not at centre stage in Davidoff's book, but it should be important as migraine involves local sterile inflammation of the endothelium in particular blood vessels (Appenzeller, 1991). Local inflammation does result in aggregation and adhesion at the endothelium of platelets and other blood cells belonging to the immune system. Hanington *et al.* (1981) have indeed called migraine a platelet disorder (see also Mezei *et al.*, 2000).

Interlude: folk biology and common sense

Biomedical research of migraine keeps generating much sophisticated, detailed information. The details may be helpful, but they may also detract from potentially useful inputs from elementary biology. I would suggest that changing eating and drinking habits might help patients prevent or avoid migraine attacks, since it is reasonable to speculate that migraine should be affected by patterns of blood circulation in association with the water balance. For example, it may be advisable to have some exercise before breakfast. Breakfast without prior exercise delays the removal from muscles of metabolites accumulated during the night. The exercise helps with the removal of metabolites and enhances the circulation of blood throughout the body including the brain. Drink should be as important as food. The circulation of blood and the removal of waste will be enhanced if we have a few glasses of water before breakfast, for example. All in all, I would assume that experimentally altering habits of eating and drinking might provide clues for the abatement of migraine.

Next, I would suggest that the deliberate use of muscles in areas where blood circulation is limited might be helpful. For example, Raynaud's phenomenon, which often accompanies migraine, may be caused in part by restricted blood flow toward extremities. Deliberate movements of hand and feet in all possible directions, combined with local massage stimulating blood flow, could therefore be helpful. The circulation of blood is also affected by regimes of temperature exposure. Frequent exposure to alternating heat and cold has beneficial effects on blood circulation as it enhances temperature adaptation.

These considerations amount to a kind of folk biology that is lacking in all the sophisticated scientific literature on migraine. The literature focuses on the molecular and the cellular level. Higher levels of physiological integration as considered in my admittedly simplistic folk biology, are seldom considered in the literature. I am convinced that the folk biology makes sense, as I have put it into practice together with a few friends and colleagues. In some of our homegrown experiments, we did witness the disappearance of migraine attacks. This kind of anecdotal evidence does not in itself have much force. But it may set us on a profitable track of biological approaches involving higher levels of organization that are now underrepresented in medicine.

A role for ADH

At the beginning of this section, I indicated how the search for unconnected literatures could be rewarding. Swanson's methods are extremely helpful. I have used them in an informal way in my own study of migraine. This resulted in interesting new hypotheses.

Swanson found that magnesium has a pivotal role in migraine. His magnesium deficiency hypothesis is based on relations among physiological and biochemical quantities which had not been brought together in earlier research. I would accept the hypothesis, but I conjecture that numerous additional key factors at low levels of organization could be discovered with his method. To avoid getting lost in all the details, we need overarching hypotheses concerning higher levels of organization. I have uncovered two factors that may help us develop such hypotheses, to wit ADH (antidiuretic hormone, also called vasopressin) and omega-3 PUFAs.

To start with, I return to elementary folk biology to explain how I came to search for possible effects of ADH. I argued that eating and drinking habits affect the water balance in our bodies and blood circulation, with possible implications for migraine. ADH affects the water balance as it induces reabsorption of water in the kidney, and promotes thirst upon loss of water. These effects are so well known, that it is natural to think of ADH as the hormone, which has this specific function. But functions of hormones are seldom if ever that specific, and interactions among hormones are pervasive. ADH also influences the balance of vasoconstriction and vasodilatation. Thus, it affects blood circulation both via an effect on water balance, and by effects on blood vessels.

Serotonin also affects vasoconstriction and vasodilatation, in a different way. As serotonin affects migraine, we may reasonably conjecture that ADH should also affect it. A great amount of research has confirmed the serotonin hypothesis. A long chapter in Davidoff (2002) reviews all the evidence and mechanisms involved. But Davidoff does not even mention ADH. Likewise, a search of the internet yields numerous articles on serotonin and migraine, and very few articles on ADH and migraine. In addition to the scarce literature that explicitly mentions a link between ADH and migraine, I have uncovered many indirect links. Taken together, the links indicate that ADH should be a significant factor in migraine. They add up to the following picture.

Stress may provoke migraine attacks. ADH plays a crucial role in the stress response and in depression associated with chronic stress, but most researchers have disregarded this role (Scott and Dinan, 1998). Migraine is associated with local inflammation of cerebral blood vessels. This also points to a role of ADH in migraine, because ADH helps counteract inflammatory responses (Chikanza and Grossman, 1998). Buschmann *et al.* (1996) observed increased platelet ADH receptors in migraine patients. ADH subserves many functions in homeostatic adjustment to stress and pain. Therefore, ADH metabolism may have special significance in the pathophysiology of migraine (Gupta, 1997). High levels of ADH and endothelin-1 have been found in migraine patients. Endothelin-1 appears to play a role in migraine (Hasselblatt *et al.*, 1999; Flammer *et al.*, 2001, Dreier *et al.*, 2002), and ADH stimulates endothelin-1 synthesis. These results combine in suggesting that ADH is as important as serotonin in migraines.

My emphasis on ADH may serve to redress one-sidedness in existing views concerning the role of hormones and neurotransmitters in migraine. More crucially, it helps us recognize the role of higher levels of physiological integration in migraine, as exemplified by the maintenance of an appropriate water balance. Our eating and drinking habits influence ADH metabolism, and thereby may affect migraine. Considering therapies for migraine patients, my attitude would be that such habits deserve our primary interest. If a focus on such mundane matters does not solve problems with migraine, then it may be wise to resort to sophisticated biomedicine and allied medications. Elementary matters should have priority over sophistication. Unfortunately, present-day biomedicine has it the other way round.

Omega-3 fatty acids: dietary deficiency in migraine?

The subject of food brings me to the second factor to be considered here, omega-3 PUFAs. I will argue that shortages of these PUFAs in our average diet may be an important etiological factor in migraine. Research in many areas, reviewed in the previous section, has shown that these shortages explain the high prevalence of many diseases in prosperous countries, and that fish oil, which contains much EPA and DHA (both omega-3 PUFAs) often has therapeutic value. I have searched the literature to uncover links between omega-3 acids and migraine. Few direct links exist, but I have found many indirect connections.

Preliminary results obtained by Harel *et al.* (2002) suggest that fish oil and also olive oil benefit adolescent migraine patients. But Pradalier *et al.* (2001) found no significant effect in a large double-blind study. However, this finding concerns the last four weeks of a treatment period of 16 weeks. The entire treatment period yielded a significant difference suggesting that the omega-3 PUFAs are beneficial. More indirect evidence linking omega-3 PUFAs with migraine takes many forms. The summary that follows shows by way of examples that these fatty acids influence many of the processes that play a role in migraine.

Yamada *et al.* (1998) demonstrated that omega-3 PUFAs, especially DHA, suppress platelet aggregation in rats. EPA causes beneficial vasodilatation by several routes, some of which involve the endothelium (Engler *et al.*, 2000; Shimokawa, 2001). Jamin *et al.* (1999) found that EPA decreases the release of endothelin-1 in cultured bovine endothelial cells. Omega-3 acids also have anti-inflammatory properties via effects on T-cell populations of the immune system (Almallah *et al.*, 2000; Grimble, 2001; and Terada *et al.*, 2001), and via a reduced adhesion of T-cells to the endothelium (Mayer *et al.*, 2002; Grimm *et al.*, 2002; Sethi *et al.*, 2002). These findings indicate that omega-3 PUFAs could counteract inflammatory processes in migraines. Smith (1992), whose lead has not been followed up in recent research, already suggested this. The immune system is anyhow important in migraine (Martelletti *et al.*, 1989; Covelli *et al.*, 1998; Empl *et al.*, 1999).

I conjecture that comorbidity of migraine with psychiatric disorders may be due in part to omega-3 PUFA deficiencies, since the deficiencies play a role in these disorders (for references, see the previous section). Migraine is often associated with anxiety disorders and depression (see for example Guidetti *et al.*, 1998). Silberstein (2001) and Davidoff (2002: 20-22) suggest that bidirectionality of the association points to a shared aetiology, but they do not mention omega-3 PUFAs. All in all, substantial evidence supports the hypothesis that omega-3 PUFA deficiencies are an etiological factor in migraine.

Levels of organization, functions, and evolution

As I indicated, the emphasis in research on migraine is now overmuch on low levels of organization at the cost of high levels of physiological integration. Occasional exceptions are to be found in recent literature. For example, Hellstrom (1999) suggests that the homeostatic balance is disturbed in migraines. The most valuable sources are not easily located as they date from before the internet era. The book by Sacks (1970, update 1985) is a valuable source. He puts migraine into a rich historical context spanning more than two millennia. Much of what he has to say is still relevant today.

Sacks argues that migraine is at once a physical and an emotional phenomenon, and that migraine attacks may also have symbolic value. I comment here only on insights of Sacks concerning physical aspects of migraine. The passage quoted hereafter (Sacks, 1985: 7), which has not lost actuality, captures his view of ongoing research:

"The present century [the previous one by now] has been characterized both by advances and retrogressions in its approach to migraine. The advances reflect sophistications of technique and quantitation, and the retrogressions represent the splitting and fracturing of the subject, which appears inseparable from the specialisation of knowledge. By a historical irony, a real gain of knowledge and technical skill has been coupled with a real loss in general understanding."

Sacks is aware that migraine phenomena are so heterogeneous that attempts to elaborate a general, comprehensive theory covering them all are futile. Indeed, analogous situations exist in much biomedicine and biology; we have to be content with theories for limited domains of phenomena (Van der Steen, 1993; Burian *et al.*, 1996; Van der Steen, 2000). However, some open-textured generalities covering most migraine phenomena are feasible.

Sacks regards migraine as a disorder of arousal. He postulates relatively stable, essential features beneath variable and disjunctive components: alterations of conscious level, of muscular tonus, of sensory vigilance, and so forth (Sacks, 1985: 109). Under stress, arousal may result in a flight-or-fight response, in animals and in man, which is functional if it helps to escape dangerous situations. Activated arousal is associated with an altered state of the sympathetic nervous system, changes in hormone secretion, and increased energy metabolism. As Sacks was writing his book, a common assumption was that more detailed generalities concerning stress are possible. By now, it is obvious that the details are strongly context-dependent (see for example Moberg and Mench, 2000). But the general idea of stress inducing arousal and a flight-or-fight response with many physiological changes is still valid.

In animals and in man, responses to stress in the form of danger may also take the form of immobilization (freezing). This is a functional response, for example, if it prevents detection by a predator. Sacks argues that migraine belongs to the category of freezing responses that may be functional biologically, as when you have been overloaded with demands. Withdrawing somewhere with a headache where disturbance is not allowed may help a migraine patient to recuperate. However, the response may also become dysfunctional and grow out of proportion, if serious problems remain unsolved, for example.

Sacks thus puts migraine in the context of functional explanations, which characterize much biology concerned with higher levels of organization. He also puts it in an evolutionary context, rightly albeit somewhat superficially. The responses to stress occur throughout the animal kingdom, and we must have inherited them from our non-human ancestors. Hence, comparative biological research inspired by evolutionary theory (and, I would add, functional ecology) may help us better understand migraine. Such research is now sorely missing in biomedical approaches of migraine.

However, Sacks overblows the relevance of functional approaches somewhat. He often searches for unconscious motives that may provide functional explanations of severe migraine. My survey of dietary deficiencies indicates how functional explanations may miss the mark. If a person suffers from severe migraine due to a dietary deficiency, a causal explanation would be more fitting than a functional explanation, and diet therapy would then be more appropriate than psychotherapy geared to unconscious motives.

The idea that arousal may be important in migraine has recently been revived in psychophysiological research using electrophysiology (see for example Backer *et al.*, 2001; Davidoff, 2002: 153-155; De Tommaso *et al.*, 2002; Giffin and Kaube, 2002). Investigators have found that migraine patients, in between attacks, show less habituation to visual stimuli than control persons. This is consistent with the assumption that patients have a high level of arousal in normal situations. They may be punished for this by attacks.

Conclusions

My case study of migraine illustrates that resources from biology are underused in current biomedicine. Let me list what has been uncovered. The important role of ADH in migraine has virtually been disregarded. Associated with this is negligence concerning the role of the water balance in migraine. Indirect evidence suggests that shortages of omega-3 PUFAs in our diet play a role in the aetiology of migraine. Research on migraine often fails to consider higher levels of physiological integration, and it undervalues functional and evolutionary approaches. I do hope that researchers will pick up some of these themes.

4.5 CODA

Biology is infectious. It generates ideas that spread to many other disciplines. But the spread is not a balanced one. Barring illicit traffic, gems of ideas are often less infectious than germs with a habit of mutating on the way to their destination. Evolutionary thinking has mutated into ideas that are foreign to genuine biology, in evolutionary psychology and evolutionary psychiatry, for example. But medicine appears to be in a good shape to assimilate gems of evolutionary thinking. I have suggested through case studies that biology has many more gems in store to benefit medicine and thereby all of us.

The case studies in this chapter jointly indicate that we need to change common styles of research to redress the balance of germs and gems. The existing priorities are wrong. For example, if there is a grain of truth in my thesis that deficiencies of omega-3 fatty acids are a major disaster as they foster many diseases, a large share of funding should go to research on this subject. Unfortunately, scientists are seldom in a position to determine major research priorities. But they may aim to harp on existing unbalances and make the general public aware of them. The case studies also show that we need to shift the balance of specialist and generalist research in favour of generalist options. I do hope that generalist boats will keep us afloat in the ocean of information overloads.

ACKNOWLEDGEMENTS

Gratefully I thank my colleagues Huib Looren de Jong, Vincent Ho, and Agostino di Giacomo Russo for helpful comments and assistance. Comments of three anonymous referees have also been helpful.

REFERENCES

Almallah, Y. Z., S. W. Ewe, A. El-Tahir, N. A. Mowat, P. W. Brunt, T. S. Sinclair, S. D. Heys and O. Eremin (2000). Distal proctocolitis and n-3 polyunsaturated fatty acids (n-3 PUFAs): The mucosal effect in situ. Journal of Clinical Immunology 20: 68-76.

Appenzeller, O. (1991). Pathogenesis of migraine. Medical Clinics of North America 75: 763-789.

Backer, M., D. Sander, M. G. Hammes, D. Funke, M. Deppe, B. Conrad and T. R. Tolle (2001). Altered cerebrovascular response pattern in interictal migraine during visual stimulation. Cephalalgia 21: 611-616.

Banks, W. A., A. J. Kastin and S. I. Rapoport (1997). Permeability of the blood-brain barrier to circulating free fatty acids. In: Yehuda, S. and D. I. Mostofsky (Eds). Handbook of Essential Fatty Acid Biology: Biochemistry, Physiology, and Behavioural Neurobiology. Humana Press, Totowa, New Jersey. pp. 3-14.

Barkow, J., L. Cosmides and J. Tooby (Eds) (1992). The Adapted Mind: Evolutionary Psychology and the Generation of Culture. Oxford University Press, Oxford.

Bjorklund, D. F., and D. Pellegrini (2002). The Origins of Human Nature: Evolutionary Developmental Psychology. American Psychological Association, Washington.

Boska, M. D., K. M. Welch, P. B. Barker, J. A. Nelson and L. Schultz (2002). Contrasts in cortical magnesium, phospholipid and energy metabolism between migraine syndromes. Neurology 58: 1227-1233.

Brandon, R. N. (1990). Adaptation and Environment. Princeton University Press, Princeton.

Burian, R., R. C. Richardson and W. J. van der Steen (1996). Against generality: Meaning and reference in genetics and philosophy. Studies in History and Philosophy of Science 27: 1-29.

Buschmann, J., G. Leppla-Wollsiffer, N. Nemeth, K. Nelson and R. Kirsten (1996). Migraine patients show increased platelet vasopressin receptors. Headache 36: 586-588.

Buss, D. M. (1995). Evolutionary psychology: A new paradigm for social science. Psychological Inquiry 6: 1-30.

Buss, D. M. (1998). Evolutionary Psychology: The New Science of the Mind. Allyn and Bacon, Boston.

Cady, R. K. (1999). Diagnosis and treatment of migraine. Clinical Cornerstone 1: 21-32.

Chikanza, I. C. and A. S. Grossman (1998). Hypothalamic-pituitary-mediated immunomodulation: Arginine vasopressin is a neuroendocrine immune mediator. British Journal of Rheumatology 37: 131-136.

Clandinin, M. T. and J. Jumpsen (1997). Fatty acid metabolism in brain in relation to development, membrane structure, and signaling. In: Yehuda, S. and D. I. Mostofsky (Eds). Handbook of Essential Fatty Acid Biology: Biochemistry, Physiology, and Behavioural Neurobiology. Humana Press, Totowa, New Jersey. pp. 15-65.

Constantinescu, C. S. (2002). Migraine and Raynaud phenomenon: Possible late complications of Kawasaki disease. Headache 42: 227-229.

Cosmides, L. and J. Tooby (1994). Beyond intuition and instinct blindness: Toward an evolutionary rigorous cognitive science. Cognition 50: 41-77.

Covelli, V., A. B. Maffione, C. Nacci, E. Tato and E. Jirillo (1998). Stress, neuropsychiatric disorders and immunological effects exerted by benzodiazepines. Immunopharmacology and Immunotoxicology 20: 199-209.

Daly, M. and M. Wilson (1988a). Homicide. Aldine de Gruyter Hawthorne, New York.

Daly, M. and M. Wilson (1988b). Evolutionary social psychology and family homicide. Science 242: 519-524.

Daly, M. and M. Wilson (1998). The Truth about Cinderella: A Darwinian View of Parental Love. Weidenfeld & Nicolson, London.

Davidoff, R. A. (2002). Migraine: Manifestations, Pathogenesis, and Management (second edition). Oxford University Press, New York.

Davies, P. S. (1996). Discovering the functional mesh: On the methods of evolutionary psychology. Minds and Machines 6: 559-585.

De Tommaso, M., D. Murasecco, G. Libro, M. Guido, V. Sciruicchio, L. M. Specchio, V. Gallai and F. Puca (2002). Modulation of trigeminal reflex excitability in migraine: Effects of attention and habituation on the blink reflex. International Journal of Psychophysiology 44: 239-249.

Diamond, S. and R. Wenzel (2002). Practical approaches to migraine management. CNS Drugs 16: 385-340.

Dreier, J. P., J. Kleeberg, G. Petzold, J. Priller, O. Windmuller, H. D. Orzechowski, U. Lindauer, U. Heinemann, K. M. Einhaupl and U. Dirnagl (2002). Endothelin-1 potently induces Leao's cortical spreading depression in vivo in the rat: a model for an endothelial trigger of migrainous aura? Brain 125: 102-112.

Eaton, S. B. and M. Konner (1985). Palaeolithic nutrition: A consideration of its nature and current implications. New England Journal of Medicine 312: 283-289.

Empl, M., P. Sostak, M. Breckner, M. Riedel, N. Muller, R. Gruber, S. Forderreuther and A. Straube (1999). T-cell subsets and expression of integrins in peripheral blood of patients with migraine. Cephalalgia 19: 713-717.

Engler, M. B., M. M. Engler, A. Browne, Y. P. Sun and R. Sievers (2000). Mechanisms of vasorelaxation induced by eicosapentaenoic acid (20:5n-3) in WKY rat aorta. British Journal of Pharmacology 131: 1793-1799.

Erasmus, U. (1993). Fats that Heal, Fats that Kill (second edition). Alive Books, Burnaby.

Ewald, P. W. (1994). Evolution of Infectious Diseases. Oxford University Press, Oxford and New York.

Flammer, J., M. Pache and T. Resink (2001). Vasospasm, its role in the pathogenesis of diseases with particular reference to the eye. Progress in Retinal Eye Research 20: 319-349.

Frank, S. A. (2002). Immunology and Evolution of Infectious Disease. Princeton University Press, Princeton and Oxford.

Giffin, N. J. and H. Kaube (2002). The electrophysiology of migraine. Current Opinions in Neurology 15: 303-309.

Greenblatt, C. and M. Spigelman (2003). Emerging Pathogens: Archaeology, Ecology and Evolution of Infectious Disease. Oxford University Press, Oxford and New York.

Griffiths, P. E. (1996). The historical turn in the study of adaptation. British Journal for the Philosophy of Science 47: 511-532.

Grimble, R. F. (2001). Nutritional modulation of immune function. Proceedings of the Nutrition Society 60: 389-397.

Grimm, H., K. Mayer, P. Mayser and E. Eigenbrodt (2002). Regulatory potential of n-3 fatty acids in immunological and inflammatory processes. British Journal of Nutrition 87, Supplement 1: S59-S67.

Guidetti, V., F. Galli, P. Fabrizi, A. S. Giannantoni, L. Napoli, O. Bruni and S. Trillo (1998). Headache and psychiatric comorbidity: clinical aspects and outcome in an 8-year follow-up study. Cephalalgia 18: 455-462.

Gupta, V. K. (1997). A clinical review of the adaptive role of vasopressin in migraine. Cephalalgia 17: 561-569.

Hamazaki, T., S. Sawazaki, M. Itomura, E. Asaoka, Y. Nagao, N. Nishimura, K. Yasawa, T. Kuwamori and M. Kobayashi (1996). The effect of docosaheaxaenoic acid on agression in young adults. Journal of Clinical Investigation 97: 1129-1134.

Hamazaki, T., A. Thienprasert, K. Kheovichai, S. Samuhaseneetoo, T. Nagasawa and S. Watanabe (2002). The effect of docosahexaenoic acid on aggression in elderly Thai subjects—a placebo-controlled double-blind study. Nutrional Neuroscience 5: 37-41.

Hanington, E., R. J. Jones, J. A. Amess and B. Wachowicz (1981). Migraine: A platelet disorder. Lancet 2: 720-723.

Hasselblatt, M., J. Kohler., E. Volles and H. Ehrenreich (1999). Simultaneous monitoring of endothelin-1 and vasopressin plasma levels in migraine. Neuroreport 10: 423-425.

Harel, Z., G. Gascon, S. Riggs, R. Vaz, W. Brown and G. Exil (2002). Supplementation with omega-3 polyunsaturated fatty acids in the management of recurrent migraines in adolescents. Journal of Adolescent Health 31: 54-61.

Healy, D. (2002). The Creation of Psychopharmacology. Harvard University Press, Cambridge, Mass.

Hellstrom, H. R. (1999). The altered homeostatic theory: A holistic approach to multiple diseases, including atherosclerosis, ischemic diseases, and hypertension. Medical Hypotheses 53: 194-199.

Holman, R. T. (1997). Omega3 and omega6 essential fatty acid status in human health and disease. In: Yehuda, S. and D. I. Mostofsky (Eds). Handbook of Essential Fatty Acid Biology: Biochemistry, Physiology, and Behavioural Neurobiology. Humana Press, Totowa, New Jersey. pp. 139-182.

Horrobin, D. F. (1997). Fatty acids, phospholipids, and schizophrenia. In: Yehuda, S. and D. I. Mostofsky (Eds). Handbook of Essential Fatty Acid Biology: Biochemistry, Physiology, and Behavioural Neurobiology. Humana Press, Totowa, New Jersey. pp. 245-256.

Horrobin, D. F., K. Jenkins, C. N. Bennett and W. W. Christie (2002). Eicosapentaenoic acid and arachidonic acid: Collaboration and not antagonism is the key to biological understanding. Prostaglandins, Leukotrienes and Essential Fatty Acids 66: 83-90.

Jamin, S. P., M. Crabos, M. Catheline, C. Martin-Chouly, A. B. Legrand and B. Saiag (1999). Eicosapentaenoic acid reduces thrombin-evoked release of endothelin-1 in cultured bovine endothelial cells. Research Communications in Molecular Patholology and Pharmacology 105: 271-281.

Johnson, S. (2001). The multifaceted and widespread pathology of magnesium deficiency. Medical Hypotheses 56: 163-170.

Katz, R., J. A. Hamilton, A. A. Spector, S. A. Moore, H. W. Moser, M. J. Noetzel and P. A. Watkins (2001). Brain uptake and utilization of fatty acids: Recommendations for future research. Journal of Molecular Neuroscience 16: 333-335.

Keys, A. (1970). Coronary heart disease in seven countries. Circulation 41, Supplement: S1-S211.

Laland, K. N., and G. R. Brown (2002). Sense and Nonsense: Evolutionary Perspectives on Human Behaviour. Oxford University Press, Oxford and New York.

Lloyd, E. A. (1999). Evolutionary psychology: The burdens of proof. Biology and Philosophy 14: 211-233.

Looren de Jong, H. and W. J. van der Steen (1998). Biological thinking in evolutionary psychology: Rockbottom or quicksand? Philosophical Psychology 11: 183-205.

Martelletti, P., J. Sutherland, E. Anastasi, U. Di Mario and M. Giacovazzo (1989). Evidence for an immune-mediated mechanism in food-induced migraine from a study on activated T-cells, IgG4 subclass, anti-IgG antibodies and circulating immune complexes. Headache 29: 664-670.

Mauskop, A., B. T. Altura and B. M. Altura (2002). Serum ionized magnesium levels and serum ionized calcium / ionized magnesium ratios in women with menstrual migraine. Headache 42: 242-248.

Mayer, K., M. Merfels, M. Muhly-Reinholz, S. Gokorsch, S. Rosseau, J. Lohmeyer, N. Schwarzer, M. Krull, N. Suttorp, F. Grimminger and W. Seeger (2002). Omega-3 fatty acids suppress monocyte adhesion to human endothelial cells: role of endothelial PAF generation. American Journal of Physiology, Heart and Circulatory Physiology 283: H811-H818.

Mezei, Z., B. Kis, A. Gecse, J. Tajti, B. Boda, G. Telegdy and L. Vecsei (2000). Platelet arachidonate cascade of migraineurs in the interictal phase. Platelets 11: 222-225.

Moberg, G. P. and J. A. Mench (Eds) (2000). The Biology of Animal Stress: Basic Principles and Implications for Animal Welfare. CAB Publishing, New York.

Nesse, R. M. and G. C. Williams (1994). Why We Get Sick. Times Books, New York.

Nesse, R. M. and G. C. Williams (1998). Evolution and the origins of disease. Scientific American November: 58-65.

Peet, M., J. Brind, C. N. Ramchand, S. Shah and G. K. Vankar (2001). Two double-blind placebo-controlled pilot studies of eicosapentaenoic acid in the treatment of schizophrenia. Schizophrenia Research 49: 243-251.

Peet, M. and D. F. Horrobin (2002). A dose-ranging exploratory study of the effects of ethyl-eicosapentaenoate in patients with persistent schizophrenic symptoms. Journal of Psychiatric Research 36: 7-18.

Pradalier, A., P. Bakouche, G. Baudesson, A. Delage, G. Cornaille-Lafage, J. M. Launay and P. Biason (2001). Failure of omega-3 polyunsaturated fatty acids in prevention of migraine: A double-blind study versus placebo. Cephalalgia 21: 818-822.

Puri, B. K., G. M. Bydder, S. J. Counsell, B. J. Corridan, A. J. Richardson, J. V. Hajnal, C. Appel, H. M. Mckee, K. S. Vaddadi and D. F. Horrobin (2002). MRI and neuropsychological improvement in Huntington disease following ethyl-EPA treatment. Neuroreport 13: 123-126.

Richardson, R. C. (1996). Critical notice: Robert N. Brandon, adaptation and environment. Philosophy of Science 63: 122-136.

Rose, S. (2000). Colonising the social sciences? In: Rose, H. and S. Rose (Eds). Alas Poor Darwin: Arguments Against Evolutionary Psychology. Jonathan Cape, London. pp. 106-128.

Sacks, O. (1985). Migraine: Revised and Expanded. University of California Press, Berkeley and Los Angeles.

Scott, L. V. and T. G. Dinan (1998). Vasopressin and the regulation of hypothalamic-pituitary-adrenal axis function: Implications for the pathophysiology of depression. Life Sciences 62: 1985-1998.

Serhan, C. N. and E. Oliw (2001). Unorthodox routes to prostanoid formation: New twists in cyclooxygenase-initiated pathways. Journal of Clinical Investigation 107: 1481-1489.

Sethi, S., O. Ziouzenkova, H. Ni, D. D. Wagner, J. Plutzky and T. N. Mayadas (2002). Oxidized omega-3 fatty acids in fish oil inhibit leukocyte-endothelial interactions through activation of PPAR alpha. Blood 100: 1340-1346.

Sheftell, F. D. and S. J. Tepper (2002). New paradigms in the recognition and acute treatment of migraine. Headache 42: 58-69.

Shimokawa, H. (2001). Beneficial effects of eicosapentaenoic acid on endothelial vasodilator functions in animals and humans. World Review of Nutrition and Dietetics 88: 100-108.

Silberstein, S. D. (2001). Shared mechanisms and comorbidities in neurologic and psychiatric disorders. Headache 41, Supplement 1: S11-S17.

Simopoulos, A. P. (2001). The mediterranian diets: What is so special about the diet of Greece? Journal of Nutrition 131: S3065-S3073.

Smith, R. S. (1992). The cytokine theory of headache. Medical Hypotheses 39: 168-174.

Stevens, A. and J. Price (1996). Evolutionary Psychiatry: A New Beginning. Routledge, London and New York.

Stoll, A. L. (2001). The Omega-3 Connection. Simon & Schuster, New York.

Swanson, D. R. (1986). Fish oil, Raynaud's syndrome, and undiscovered public knowledge. Perspectives in Biology and Medicine 30: 7-17.

Swanson, D. R. (1987). Two medical literatures that are logically but not bibliographically connected. Journal of the American Society for Information Science 38: 228-233.

Swanson, D. R. (1988). Migraine and magnesium: Eleven neglected connections. Perspectives in Biology and Medicine 31: 527-557.

Swanson, D. R. (1990). Medical literature as a potential source of new knowledge. Bulletin of the Medical Libraries Association 78: 29-37.

Swanson, D. R. (1991). Complementary structures in disjoint science literatures. SIGIR91: Proceedings of the 14th annual international ACM. ACM Press, Chicago.

Swanson, D. R. (2001). ASIST award of merit acceptance speech on the fragmentation of knowledge, the connection explosion, and assembling other people's ideas. Bulletin of the American Society for Information Science and Technology, February/March: 12-14.

Swanson, D. R. and N. R. Smalheiser (1997). An interactive system for finding complementary literatures: A stimulus to scientific discovery. Artificial Intelligence 91: 183-203.

Swanson, D. R. and N. R. Smalheiser (1999). Implicit text linkages between Medline records: Using Arrowsmith as an aid to scientific discovery. Library Trends 48: 48-59.

Terada, S., M. Takizawa, S. Yamamoto, O. Ezaki, H. Itakura and K. S. Akagawa (2001). Suppressive mechanisms of EPA on human T cell proliferation. Microbiology and Immunology 45: 473-481.

Van der Steen, W. J. (1993). A Practical Philosophy for the Life Sciences. SUNY Press, Albany.

Van der Steen, W. J. (2000). Evolution as Natural History: A Philosophical Analysis. Praeger, Westport.

Van der Steen, W. J. (2003). Assessing overmedication: Biology, philosophy and common sense. Acta Biotheoretica 51: 151-171.

Van der Steen, W. J. and V. K. Y. Ho (2001a). Methods and Morals in the Life Sciences: A Guide for Analyzing and Writing Texts. Praeger, Westport.

Van der Steen, W. J. and V. K. Y. Ho (2001b). Drugs versus diets: Disillusions with Dutch health care. Acta Biotheoretica 49: 125-140.

Van der Steen, W. J., V. K. Y. Ho and F. J. Karmelk (2003). Beyond Boundaries of Biomedicine: Pragmatic Perspectives on Health and Disease. Rodopi, Amsterdam.

Voulgari, P. V., Y. Alamanos, D. Papazisi, K. Christou, C. Papanikolaou and A. A. Drosos (2000). Prevalence of Raynaud's phenomenon in a healthy Greek population. Annals of Rheumatic Diseases 59: 206-210.

Wilson, D. S. (2002). Darwin's Cathedral: Evolution, Religion, and the Nature of Society. University of Chicago Press, Chicago and London.

Yamada, N., J. Shimizu, M. Wada, T. Takita and S. Innami (1998). Changes in platelet aggregation and lipid metabolism in rats given dietary lipids containing different n-3 polyunsaturated fatty acids. Journal of Nutritional Science and Vitaminology (Tokyo) 44: 279-289.

Wim J. van der Steen
Maluslaan 3, 1185 KZ Amstelveen, The Netherlands

5

The Composite Species Concept: A Rigorous Basis for Cladistic Practice

D. J. Kornet and James W. McAllister

ABSTRACT

As previous work has shown, the genealogical network can be partitioned exhaustively into internodons, mutually exclusive and historically continuous entities delimited between two successive permanent splits or between a permanent split and an extinction. Internodons are not suitable candidates for the status of species, because of their short life span and the difficulty of recognizing their boundaries. However, internodons may be suitable building blocks for a viable species concept. We introduce the concept of composite species as a sequence of internodons, by qualifying only some permanent splits in the genealogical network as speciation events. The permanent splits that count as speciation events on our account are those associated with a character state fixation: this proposal ensures the recognizability of composite species. Lastly, we show how actual taxonomic practice is able to recover the phylogenetic tree of composite species from standard morphological data.

Keywords: Species concepts, genealogical network, internodons, character state, fixation, speciation.

5.1 INTRODUCTION

This paper presents a new species concept, the composite species concept, which is developed from first principles. On this concept, species are historical entities composed of parts of the genealogical network named internodons, and their recognition in nature is achievable by standard taxonomic practice. The composite species concept shows some affinity with the phylogenetic species concept, but has important advantages over it.

The main motivation for developing the composite species concept is the conviction that a good species concept ought to define species that are mutually exclusive and that exhaust the genealogical network. The composite species concept fulfils this requirement. Available species concepts, by contrast, including the phylogenetic species concept, fail to satisfy this condition. Furthermore, the composite species concept incorporates rigorous

T.A.C. Reydon and L. Hemerik.,(eds.), Current Themes in Theoretical Biology,
95-127.

definitions of taxonomic entities and phenomena, including character state and fixation, which make it more precise than available rivals.

We begin in Section 5.2 by reviewing the concept of internodon. This may be characterized informally as a part of the genealogical network delimited by two successive permanent splits, or by a permanent split and an extinction. However, a formal definition of internodon in terms of an equivalence relation is also available. Some authors, most notably Hennig, have proposed that entities approximating to internodons constitute species. In contrast, we argue that internodons are not suitable candidates for the status of species, because of their short life span and the fact that it is difficult to recognize or diagnose them with the aid of standard taxonomic data.

We use the concept of internodon in Section 5.3 to formulate the composite species concept. In informal terms, a composite species is a lineage of internodons descended from an internodon with a particular property, which we call *quality Q*. We opt initially for an abstract approach, in which we explore implications of this definition of composite species before identifying quality Q with an actual property of internodons. We argue that it is desirable that species be paraphyletic groups of lower-level entities, and point out that composite species are paraphyletic groups of internodons.

In Section 5.4, we fill in quality Q in morphological terms: on our candidate definition, an internodon has quality Q if and only if a character state reaches fixation in it. Because the concepts of character state and of fixation are not clearly defined or characterized in the theoretical systematics literature, we provide our own account of these concepts (Section 5.5). We describe various phenomena that can be expected to arise in the fixation of character states and discuss their effect on phylogeny reconstruction. We provide concise definitions of internodon and composite species for easy reference in Section 5.6.

In Section 5.7, we contrast composite species with morphological species, pointing out that—although our preferred candidate for quality Q is defined in morphological terms—possession of a certain morphological attribute is neither a necessary nor a sufficient condition for an organism to belong to a certain composite species. We then show, with the aid of a hypothesized phylogenetic tree of internodons, that the composite species concept is compatible with standard phylogeny reconstruction techniques in cladistics, and that the concept is capable of placing this practice on a more rigorous footing (Section 5.8).

In Section 5.9, we point out that composite species arise not by a symmetrical splitting up which gives rise to two descendant sibling species, but by the asymmetrical process of branching off, in which the ancestor species survives. We argue that this is a further element that weighs in favour of the concept of composite species, as it is undesirable that species be regarded as arising by dichotomous splits. We compare the composite species concept with the phylogenetic species concept in Section 5.10: we regard the latter as an imperfect approximation and operationalization of the more rigorous composite species concept. The paper concludes with some remarks on the evolutionary behaviour of composite species in Section 5.11.

5.2 INTERNODONS: BUILDING BLOCKS FOR COMPOSITE SPECIES

The genealogical network is the mapping of all actual organisms and the parental relationships holding between them. The problem of how to define species can be interpreted as the question of how to partition the genealogical network into supra-organismal entities that meet most of our pre-analytic intuitions about species. Among these intuitions is the conviction that all organisms belong to precisely one species, or, in other words, that species are mutually exclusive and exhaustive of the genealogical network.

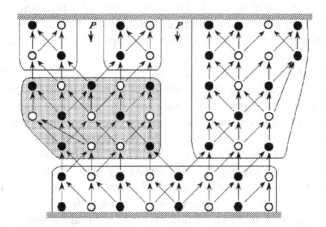

Figure 5.1. A part of the genealogical network with two permanent splits (**P**) marking the beginning and end of the life span of a historical supra-organismal entity (shaded), which we call internodon.

One way of partitioning the genealogical network into parts that are mutually exclusive and exhaustive is by dividing it into portions delimited by two successive permanent splits (Figure 5.1) or by a permanent split and an extinction. These entities will be called *internodons* in this paper. Kornet (1993) has shown that internodons can be defined formally as sets of organisms between any two of which a particular relation INT holds. The properties of relation INT guarantee that internodons have temporal continuity. Furthermore, INT is, in mathematical terms, an equivalence relation: it is reflexive, symmetric and transitive. Since an equivalence relation partitions its domain into exhaustive and mutually exclusive sets, we can be certain that internodons are exhaustive of the genealogical network and mutually exclusive (Kornet *et al.*, 1995). In this way, the concept of internodon meets two important intuitions that we have about species: that species are historically continuous and non-overlapping entities that exhaust the genealogical network.

Does this mean that internodons are suitable candidates for the status of species? This is suggested by the species concept defined by Hennig (1966) and elaborated by later authors. Hennig delimited species by reference to splits in the genealogical network, writing, "New species arise when gaps develop in the fabric of the tokogenetic relationships" (Hennig, 1966: 30). Hennig's species thus resemble internodons, though Hennig added interbreeding ability as a defining criterion for species (ibid.: 45) and stipulated a constant conjunction between splits in the genealogical network and morphological diversification (ibid.: 88). Ridley (1989) elaborates on Hennig's concept of species, coming closer to the notion of internodon by rejecting morphology and interbreeding ability as defining criteria in his "cladistic species concept" (Ridley, 1989: 5 and 11), named "internodal species concept" by Nixon and Wheeler (1990: 213).

Ridley's cladistic species could be made identical to internodons by stipulating that they arise only with permanent splits in the genealogical network. Ridley, by contrast, commits himself to the thesis that temporary splits too are speciation events, stating that hybridization followed by merging is a speciation event (Ridley, 1989: 4-5). A merging of two of his cladistic species, each consisting of a separate branch of the genealogical network, is nothing other than a closing up of a temporary split, in which Ridley must suppose that the two cladistic species which later "merge" originated.

It matters a great deal whether one considers temporary splits in the genealogical network to be speciation events. Temporary splits in the network are very frequent (Figure 5.2): they open up between, for instance, any pair of siblings that do not immediately interbreed. Because of their frequency, it is implausible to suggest that temporary splits constitute speciation events. Nor can we distinguish in a principled way between "short" and "long" temporary splits, with the intention of giving the status of speciation event only to the latter: there is a continuous gradation in the length of temporary splits, as Figure 5.2 illustrates.

Of course, when a split first appears in the network, it is impossible to tell whether it will be permanent or temporary. A permanent split is recognizable only retrospectively: the conclusive criterion for deeming a split permanent is the extinction of one of the branches in which the split has resulted. We depend on retrospective diagnosis to identify with certainty any historical entity in the genealogical network.

Notwithstanding Hennig's and Ridley's apparent belief that an entity resembling internodons should be seen as species, there are two important reasons for rejecting the concept of internodon as an acceptable concept of species: one is practical and one more fundamental.

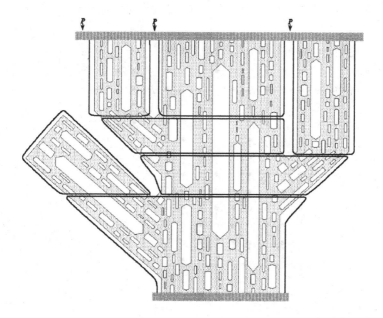

Figure 5.2. A part of the genealogical network with a split made permanent by an extinction, two further splits which we assume to be permanent, and many temporary splits. The life span of internodons is delimited only by permanent splits and extinctions.

The practical objection is the following. Whether a given organism belongs to a given internodon depends solely on the structure of the genealogical network and the organism's position in it. It does not depend on, for instance, the morphological characteristics of the organisms, either macromorphological or genetic. Since our knowledge of the tokogenetic relationships among organisms (which constitute the genealogical network) is typically scarce, and reference to morphological characteristics is in principal irrelevant to determining to which internodon an organism belongs, internodons have low recognizability. The internodal concept would therefore have very limited practical value as a species concept.

The more fundamental objection against interpreting internodons as species lies in their short life span. Both Hennig and Ridley seem to have overestimated the typical life span of their species, assuming that they extend over many generations of organisms. Hennig writes: "Species are relatively stable complexes that persist over long periods of time, but they are not absolutely permanent" (Hennig, 1966: 19; see also 30). Ridley reproduces and endorses Hennig's diagram showing species living long enough to accumulate several new character states (Ridley, 1989: 3, Figure 1; see also ibid.: 13, Figure 3).

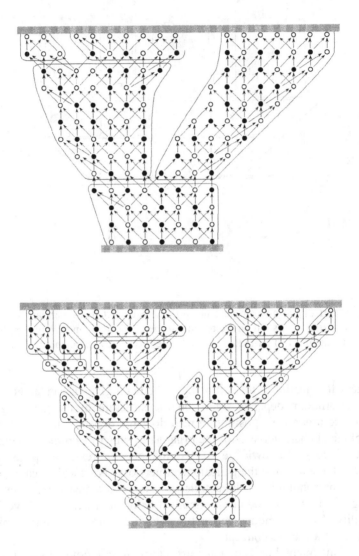

Figure 5.3. The life span of internodons should not be thought of as extended (above), but as relatively short due to the frequent extinction of small groups (below).

Clearly, they believe speciation to be a relatively infrequent event in the genealogical network. But splits, which are the speciation events that they envisage, are much more frequent than they seem to assume. As we have shown, temporary splits are so abundant that, if they were taken seriously as speciation events, they would give species only fleeting life spans. But even if we restrict speciation events to permanent splits, the life spans of the resulting species would be too short to meet our intuitions. After all, the extinction of any isolated branch of the genealogical network, no matter how small its membership, retrospectively renders a split in the network permanent (Figure 5.3). If we accepted every permanent split as a speciation event, humankind would be fragmented into two further species by every road accident in which a couple and all of its children perish. In many realistic scenarios, the life span of an internodon is shorter even than a generation. Far from needing Hennig's warning that species are not absolutely permanent, we require reassurance that they endure to any appreciable degree.

This constitutes a fundamental shortcoming of interpreting the internodon as species: the concept does not approximate closely to our intuitions about the life span of species. Together with the scarce degree of recognizability of members of internodons, this shortcoming weighs against identifying species with internodons.

Instead, we envisage a different role for internodons in the definition of a satisfactory species concept. We will show how a more inclusive and longer-lived supra-organismal entity can be defined by reference to internodons. We will identify this entity, which preserves the properties of historicity and mutual exclusivity, as species. The definition of this more inclusive entity will, unlike that of the internodon itself, refer to criteria external to the structure of the genealogical network. These criteria can be of various kinds. In this paper, we have chosen to explore the possibility of using a morphological criterion for composite species in order to meet the practical need for species recognizability.

5.3 A GENERALIZED WAY TO BUILD SPECIES FROM INTERNODONS

The practice of defining species in terms of supra-organismal entities of some kind, rather than directly of organisms, is well established. The supra-organismal entity that is most frequently chosen for this task is the population. Species are defined as "composed of natural populations" by Mayr (1957: 13), as "systems of populations" by Dobzhansky (1970: 357), as "lineages, being ancestral–descendant sequences of populations" by Simpson (1961: 153) and Wiley (1981: 25), as "populations or groups of populations" by Rosen (1979: 277), and as "the smallest aggregation of populations (sexual) or lineages (asexual)" by Nixon and Wheeler (1990: 218). Similarly, de Queiroz and Donoghue (1988: 326) consider the basal units to be populations rather than species. Indeed, Nelson and Platnick have noted that "almost all definitions of the word 'species' that have been proposed utilize the word 'population'; species are populations, or groups of populations, that meet one or more criteria." (Nelson and Platnick, 1981: 11).

Nelson and Platnick continue with a warning, however: "But the word 'population' is itself in need of definition, and is fully as difficult to define as the word 'species'." (ibid.). Clearly, a definition of species based on a less inclusive supra-organismal entity is worthless if the latter entity is not itself precisely defined. But we have at our disposal rigorously defined supra-organismal entities, which moreover are mutually exclusive and historically continuous: the internodons. There is no other well-defined supra-organismal entity on offer in the literature, let alone one having the characteristics of mutual exclusivity and historicity. Henceforth in this paper, the concept of the internodon will be not a tentative model of a species, but rather a building block out of several of which a species will be composed.

A diagram of the internodons that have resulted from a sequence of permanent splitting events shows a phylogenetic tree of internodons. (It is legitimate to use the term "phylogenetic tree" for both the mapping of the relations among internodons and that of the relations among species, since these do not form networks, unlike relations among organisms.) Figure 5.4a depicts such a tree, and should be interpreted as representing the frequent permanent splits in the genealogical network and the internodons' consequent short life spans.

We conceive of a species as the set of the organisms belonging to several consecutive internodons in the phylogenetic succession, identified and grouped together by some procedure. We will call the species yielded by this concept *composite species*. Composite species originate with the coming into being of particular internodons: we will call each of these internodons the *originator internodon* of its species.

Let us stipulate that the originator internodons of species are identified by a particular quality Q that they possess. Each originator internodon is allocated to a species together with all internodons that are its descendants and that do not exhibit Q. Every later internodon in that internodon lineage that exhibits quality Q is the first internodon of a fresh species.

Quality Q, by which originator internodons are picked out from within the succession of all internodons, could be taken to consist of any one of several different properties. It may relate to the fixation of a new character state in the internodon, to the organisms' loss of the ability to interbreed with members of other internodons, or to some other event. If composite species are to be diagnosable in practice in the genealogical network however, it will be most useful to identify Q with a morphological property shown by the organisms that are members of originator internodons. This option will be developed in the next section.

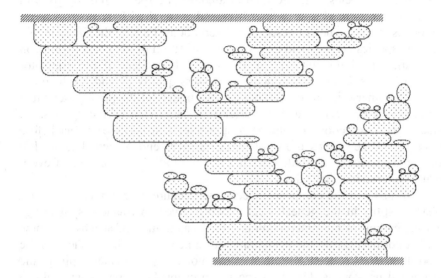

Figure 5.4a. A phylogenetic tree of internodons.

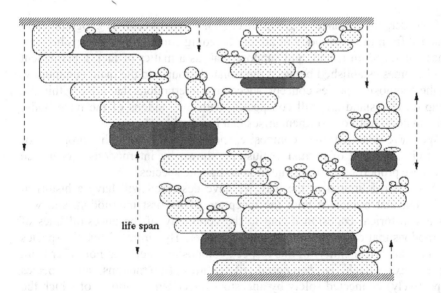

Figure 5.4b. Partitioning the phylogenetic tree into composite species. An originator internodon (shaded) is determined by some quality Q of that internodon. A composite species survives until the extinction of the last of its internodons.

Figure 5.4b, in which originator internodons (picked out arbitrarily at this stage) are shaded, shows how a phylogenetic tree of internodons is divided up into composite species. The life span of a composite species thus opens with the rise of its originator internodon. It extends over the life spans of several internodons in a sequence of descendant internodons. Finally, the life span of a composite species comes to an end with the extinction of the latest internodon that satisfies the following criteria: (a) it is a descendant of the originator internodon of the species, (b) it does not possess quality Q, and (c) there has appeared no other originator internodon in the branch of the phylogenetic tree of internodons between it and the originator internodon of the species. Of course, some internodons belonging to a species will become extinct before the life span of the species as a whole has come to an end. Nonetheless, the life span of the species as a whole does not end until all the organisms of every one of its internodons have died.

This means that the composite species will endure over, typically, several permanent splits in the genealogical network. Every permanent split brings about the ending of one internodon and the inauguration of at least two new ones, but not necessarily the inauguration of a new composite species: on the composite species concept, in other words, not every permanent split in the genealogical network is deemed a speciation event. For a permanent split to constitute a speciation event, at least one of the internodons that arise as a result of the split must have a particular property, identified arbitrarily at this stage as Q.

Consider, for instance, a group of organisms that becomes geographically detached from the main body of an interbreeding community, perhaps because it has colonized an island, and suppose that, as a matter of fact, a permanent split becomes established between the isolated group and the main community. On the composite species concept, successive internodons that arise within the group on the island are still conspecific with internodons of the main body, and remain so until one of them arises that has quality Q.

We have claimed that composite species retain both the property of historicity and that of mutual exclusivity shown by internodons. Let us see how these properties are transmitted to composite species.

A species delimited in the way we have described will have a historical beginning that coincides with the inception of its first internodon, and will possess historical cohesion owing to the continuity of the ancestral lines of internodons that originate in that first internodon. By virtue of this, the species will be a historical entity. In this respect, composite species do not differ from higher taxa. Each is composed of entities (internodons and species respectively) connected solely by ancestor–descendant relations, of which the members by definition do not interbreed and which therefore lack cohesion. (For discussion see Ereshefsky, 1991, and the references therein.)

Composite species are clearly mutually exclusive, since internodons are themselves mutually exclusive, and we allocate each internodon to only one composite species. An alternative way of establishing that composite species are mutually exclusive makes use of the concepts of monophyly and paraphyly. The objection of Nixon and Wheeler (1990: 214), among others, that the concepts of monophyly and paraphyly should not be applied to entities below the species level, because such entities form reticulate groups, does not hold for internodons, since the member-organisms of different internodons do not interbreed."

The question whether species are monophyletic (posed by de Queiroz and Donoghue, 1988: 319) makes sense only in a more specific form, as the question whether species are monophyletic groups of some specified entities. For example, we may inquire whether a species is a monophyletic group of species (consisting of just one species), of organisms (as suggested by de Queiroz and Donoghue, 1988), or of internodons. The answer will differ with the entity chosen. A composite species that has no descendant species is the smallest possible monophyletic group of species, containing just one species. Simultaneously, every composite species could (if one is willing to disregard the objection of Nixon and Wheeler mentioned above) be seen as a polyphyletic group of organisms, except if its originator internodon has arisen with a single organism.

Here, however, we are interested in considering composite species as groups of internodons. A composite species is a paraphyletic group of internodons, except if it becomes extinct without leaving a descendant species, in which case it is a monophyletic group of internodons (Figure 5.5a).

Composite species are mutually exclusive only by virtue of the fact that every ancestral composite species (i.e. every composite species that has at least one descendant species) is a paraphyletic rather than a monophyletic group of internodons. Composite species would not be mutually exclusive if they were defined in every case to be monophyletic groups of internodons, since then there would be smaller (more recent) composite species wholly included within larger (longer-established) ones (Figure 5.5b). This is a general reason why, although being a monophyletic group of species is a desirable property of all higher taxa, which are intended to form hierarchies (Figure 5.5c), it is preferable to consider species as paraphyletic groups of internodons, since species are intended to be mutually exclusive in the genealogical network.

Once the genealogical network has been divided up into internodons, the only operation that is required to unite a number of internodons into a species is to identify the internodons that are the originator internodons of each composite species: in other words, to define quality Q.

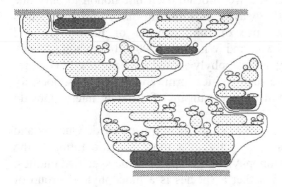

Figure 5.5a. Composite species, each arising with an originator internodon, are mutually exclusive. A composite species is a paraphyletic group of internodons, unless it becomes extinct without giving rise to a descendant species, in which case it is a monophyletic group of internodons.

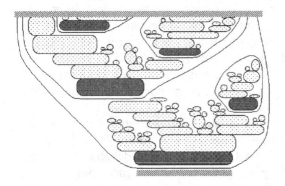

Figure 5.5b. If all species were monophyletic groups of internodons arising with originator internodons, they would not be mutually exclusive: ancestral species would include successor species.

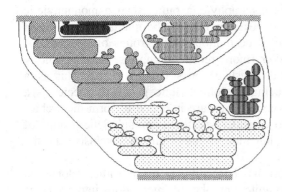

Figure 5.5c. Monophyletic groups of composite species include one another. This feature makes such groups suitable for hierarchical classification. The smallest monophyletic group of species consists of a single species.

5.4 A MORPHOLOGICAL DEFINITION OF QUALITY Q

To summarize: we posit that the originator internodon of a species is distinguished by a particular quality Q, and is allocated to a species together with all internodons that are its descendants, up to but not including the next internodons that exhibit Q. Every later internodon in that internodon lineage that exhibits quality Q is the first internodon of a fresh species.

The question remains, of course, what kind of quality Q would best serve the purpose of uniting internodons into composite species?

One possibility is to seek to define Q in terms of the logical apparatus developed by Kornet (1993). In that treatment, Kornet defines internodons (considered there as candidates for the status of species) by reference to the primitive terms of parenthood and chronological order of birth of organisms. If we took this route, Q too would ultimately be reduced to these same primitive terms. But this option cannot, for logical reasons, be successful: Kornet's logical apparatus deliberately regards internodons as equivalent to one another, while the purpose of quality Q is that it should uniquely identify certain internodons as being the first of their species.

What we need, therefore, is a defining criterion that is external to the logical apparatus used to partition the genealogical network into internodons. This means that we will construct a species concept defined jointly by two criteria: one (developed by Kornet, 1993) to group organisms into internodons, and one (under development in this paper) to unite internodons into composite species. Some previous species concepts have been flawed by their applying joint defining criteria that were incompatible, i.e. that did not always jointly apply, yielding indeterminate species boundaries (for discussion, see Kornet, 1993). However, our application of joint criteria does not introduce such flaws into our species concept. Quality Q will be applied as a criterion only after the internodons have been delimited in the genealogical network, and will therefore be only a second-stage criterion.

Quality Q, marking originator internodons of composite species, can be defined in terms of biological concepts such as morphological characteristics or interbreeding ability. Because we strive to construct a species concept with maximal practical value, the most attractive option is to define Q in terms of morphological criteria. After all, such criteria will make it possible to recognize composite species by familiar taxonomic methods. Here, we explore the possibility of defining a morphologically based quality Q by reference to the fixation of character states.

Different authors describe fixation of a character state as taking place in different supra-organismal entities. For instance, de Queiroz and Donoghue (1990: 70-71) envisage fixation as occurring in a population, while Nixon and Wheeler (1990: 217) see it occurring in "terminal lineages" and "clades".

Clearly, in our model, the supra-organismal entity in which fixation occurs is the internodon.

We now define quality Q as the property that an internodon has by virtue of the fact that a character state becomes fixed within it. On this definition, a composite species originates with an internodon in which the fixation of a character state occurs, and endures (barring its extinction) until and including the internodon before the next internodon in which the fixation of a character state occurs (that is, the next internodon that also shows quality Q).

5.5 CHARACTER STATES AND FIXATIONS

For completeness, we must specify which "character states" and "fixations" we accept for the purposes of defining quality Q. In this section, we provide a new and more rigorous account of these key taxonomic terms.

In our view, character states are to be understood as (single or multiple) genetic properties that find a phenotypic expression. (Genetic properties that have no observable expression are of little use in practical taxonomy and are therefore ignored in the present approach.) We will call the phenotypic expression of some set of genetic properties corresponding to a character state the *manifestation* of that character state. The manifestation of a character state is therefore an attribute that an organism shows by virtue of possessing that character state. Our terminology will assume that every character state can be recognized as a state of a particular character. In the symbolism that we will use, A_1 and A_2 are two states of the same character A, and have manifestations a_1 and a_2 respectively.

The finer definition of phenotypic expression depends on certain issues that we here leave open, namely what counts as a "morphological" and "observable" attribute. The notion of morphological attributes may include only macromorphological properties (such as having red petals) or also other detectable properties. If every detectable property is deemed to count as phenotypic expression, the red petals of two organisms may count as different phenotypic expressions if the chemical pathways resulting in their red petals are different.

How does a new character state come into existence? Consider two organisms x and y, each possessing character state A_1 and showing its manifestation a_1, which produce an offspring z. Suppose that the genetic material of z was affected by a mutation event, as a result of which z possesses genetic properties different from those of its parents in virtue of which the latter possessed character state A_1. If these genetic properties of z have a phenotypic expression that makes z observably different from its parents, we say that a character state A_2 has originated in z that in z has the manifestation a_2.

A character state A_2, which originated with organism z, endures as long as there are descendants of z alive that inherit it. It vanishes from the genealogical network when either there are no further descendants of z, or the character state A_2 has been replaced in all the extant descendants of z by one or more further states of the same character (A_3, A_4, etc.).

The identity of a character state is tied to its origin. If a set of genetic properties, which amounts to a character state and is already present in the genealogical network, originates afresh in the network by a separate mutation event, then what originates in this second event amounts to a new character state. In other words, a character state of one organism is non-identical to a character state of another organism if and only if there is no single ancestor organism from which they both inherited the state.

Two non-identical character states can have indistinguishable manifestations. For instance, the manifestation of a character state A_3 can be indistinguishable from those of A_1 or B_3; i.e. it may be that $a_3 = a_1$, or $a_3 = b_3$. (We interpret these phenomena as reversal and convergence respectively.) However, a new character state cannot, in the light of its definition, have a manifestation indistinguishable from that of its immediate ancestor: a new character state is said to originate only in virtue of the fact that its manifestation differs from that of its immediate ancestor.

We have chosen the fixation of a character state as the criterion for Q for its value for practical taxonomy. In order to obtain maximal recognizability for composite species, it is useful to distinguish among three senses in which a character state could be said to have become fixed. We shall call these full fixation, near fixation, and majority fixation. We will now define these, and examine which form of fixation best allows us to identify originator internodons on morphological criteria.

Of course, not every character state that arises will become fixed. Where necessary, we will distinguish a character state that becomes fixed by an asterisk (e.g., A_3^*).

The best way of judging how well a form of fixation delivers diagnosability of originator internodons is by asking to what extent the period of fixation of a character state—A_5^*, say—overlaps with the interval in which A_5^* has the highest frequency in the historical succession of states of character A. The period of fixation of a character state A_5^*, for any form of fixation, is the time interval between the fixation of A_5^* and the fixation of the next state of character A that happens to become fixed.

The *full fixation* of a character state A_5* in an internodon is the completion of the replacement of the previous state A_1* to have become fixed (Figure 5.6), or of several previous states of character A that may have been present together, by A_5* in every member organism of the internodon living at a certain time. Note that, while the replacement of some character state A_1* by A_5* is completed within some particular internodon, its frequency may well have been building up gradually in a succession of internodons; i.e. character state A_5* may well have originated in an organism belonging to a relatively distant ancestral internodon.

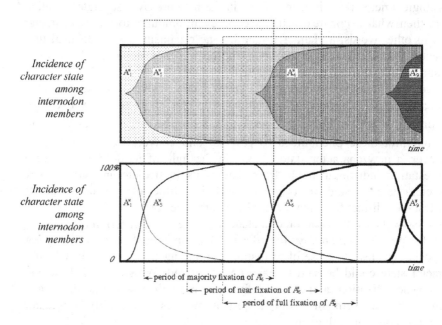

Figure 5.6. The incidence of each character state in a succession of internodons rises and falls in time. The periods of fixation associated with the three forms of fixation discussed in the text correspond to different intervals in this process.

The period of full fixation of a character state A_5* extends from the time at which character state A_5* first reaches an incidence of 100% among the then-living members of an internodon to the time at which a later state of character A reaches 100% incidence, typically in a successor internodon (Figure 5.6). A comparison of the period of full fixation of A_5* with the time during which A_5* achieves its highest frequencies among members of an internodon reveals that full fixation is an unsuitable form of fixation to which to tie property Q. The full fixation of a character state A_5^* typically occurs late in the career of the character state, after it has been present for long periods in a large proportion of the members of successive internodons. This is because the character state A_1^*, the latest predecessor character state of A_5^* that became fixed, will typically persist among members of successive internodons at low frequencies well after its heyday. This means that, before the internodon in which A_5^* achieves full fixation, there may have been many internodons of which the organisms were already characterized by A_5^*.

To remedy this shortcoming, we could relax the demand that, in order to be deemed to have reached fixation, a character state should attain 100% incidence, and be content with a specified lower incidence. A character state's first reaching this specified incidence would constitute its *near fixation* (Figure 5.6). For character state A_5^* to achieve near fixation, it is therefore not necessary for the previous character state to have become fixed, A_1^*, to disappear entirely. Near fixation still has two disadvantages for our purposes, however. The period of near fixation of A_5^* corresponds to the interval during which A_5^* achieves its greatest frequency more closely than does its period of full fixation, but it still leaves out many organisms with character state A_5^*. In addition, stipulating a precise frequency at which the near fixation of A_5^* occurs would be arbitrary.

We therefore turn to the third option, majority fixation. We define the *majority fixation* of character state A_5^* as the event in which A_5^* for the first time in a single succession of internodons in the network reaches relative majority, i.e. a frequency greater than that of any other state of the same character then represented among members of the internodon. As can be seen from Figure 5.6, the period of majority fixation of character state A_5^* coincides more closely with the interval in which the frequency of A_5^* attains its highest values. This means that the originator internodon of the species which is to be characterized by character state A_5^* will be picked out more easily in the succession of internodons: this is done by locating the internodon in which for the first time a majority of the organisms living at any one time shows A_5^*. Of course, it would be easy to mistake character state A_5^* for some other character state, if this other character state had manifestations indistinguishable from those of A_5^*; nonetheless, this other character state is non-identical to A_5^* if, as we explained earlier, it had an independent origin in the genealogical network.

If character state A_5^* achieves majority fixation in some internodon, it can happen that, after a dip in its frequency, A_5^* achieves relative majority incidence also in an internodon which is a descendant of the first one. However, this event does not constitute majority fixation, in view of our stipulation that the majority fixation of a character state occurs only when it achieves relative majority incidence *for the first time* in any succession of internodons in the genealogical network. Without this stipulation, a character state that achieved relative majority incidence on several occasions separated by periods of lower frequency would have to be judged to have become fixed, and therefore to have given rise to a new species, on all those occasions. It is undesirable to associate the origin of a new species with perhaps very small oscillations of a frequency around a boundary value. (These observations hold also for near fixation.)

Nonetheless, a character state can achieve fixation of any of our three kinds more than once in a genealogical network. For instance, a character state A_5* that has at some time achieved majority fixation in an internodon can achieve majority fixation at a later time in an internodon that is not a descendant of the first one. If this happens, each internodon in which A_5* achieves fixation is an originator internodon. We will call this phenomenon *parafixation*.

In Figure 5.6, the succession of character states is idealized in at least three ways. First, Figure 5.6 does not show the incidence of the constantly arising and declining states of the character that never attain frequencies high enough to permit fixation. These would be contained in a band at the foot of the diagrams. Second, the diagrams portray character states as becoming fixed in the same chronological order as that in which they originate and disappear. In reality, this correlation will not always hold: a state A_1* may arise earlier than A_5*, but remain for longer at low frequencies, and therefore become fixed after A_5*, or outlive it. Third, Figure 5.6 shows periods of fixation of the same duration. None of these diagrammatic simplifications invalidates our conclusion that majority fixation is a good basis for a definition of quality Q.

Majority fixation enables us also to deal adequately with a phenomenon that will here be called *scrolling*, while full and near fixation do not. In scrolling, successive new states of a character arise frequently enough that, while a particular state is still far from full or near fixation, the next state that will eventually become fixed has already appeared in some organisms (Figure 5.7). In this scenario, full or near fixation might well never occur, and yet a succession of internodons could still witness a succession of distinguishing character states. Species arising with an internodon marked by full or near fixation will then come to include many organisms lacking the character state by which the species is characterized. Majority fixation treats this phenomenon differently: every one of the character states that reaches relative majority characterizes a composite species, so species generated during a period of scrolling will not lose their diagnosability.

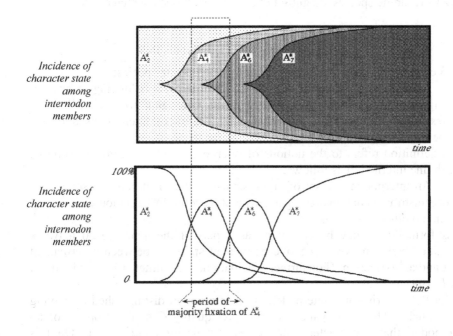

Figure 5.7. Majority fixation is the only form of fixation able to deal satisfactorily with the phenomenon of scrolling.

For practical applicability, therefore, we should indeed select as quality Q, which identifies originator internodons, the property of an internodon that a character state achieves majority fixation within it. Our assessment that majority fixation has the virtues described above does not depend on our endorsing any particular model of the evolution of novel character states, in the range stretching from gradualism to punctuated equilibria theory. Majority fixation offers good diagnosability in each of these cases. Whereas we opt for majority fixation on the strength of its diagnosability, we acknowledge that full fixation has a greater evolutionary importance than near and majority fixation, since it results in the disappearance of a previous character state.

The composite species concept stipulates that each composite species corresponds to one originator internodon. According to the composite species concept, therefore, each composite species in principle corresponds to the fixation of one character state. It may occur that more than one character state becomes fixed in a single originator internodon, though we expect this to be uncommon, in view of the short life span of internodons. In this case, we might speak of "double" speciation, resulting in "superposed" composite species. In practice, of course, we will deal with such superposed species as if they were single species, diagnosable by more than one character state.

5.6 SUMMARY OF DEFINITIONS

We are now able to state the definition of composite species:

A composite species is the set of all organisms belonging to an originator internodon, and all organisms belonging to any of its descendant internodons, excluding later originator internodons and their descendant internodons.

This definition refers to the notions of internodon and originator internodon. We define internodon as follows:

An internodon is a set of organisms such that, if it contains some organism x, it contains all organisms that have the INT relation with x, and no other organisms.

Less formally, it may be conceived as a part of the genealogical network contained between two successive permanent splits, or between a permanent split and an extinction. (For further elucidation, see Kornet, 1993, and Kornet *et al.*, 1995.)

Lastly, an originator internodon is an internodon distinguished by having some quality Q. In this paper, we interpret quality Q as the property of an internodon that a character state achieves majority fixation in it. On this interpretation, an originator internodon is an internodon in which a character state achieves majority fixation.

In the remainder of this paper, we explore some of the implications of the notion of composite species, and investigate how it may be incorporated into extant phylogenetic practice.

5.7 COMPOSITE SPECIES CONTRASTED WITH MORPHOLOGICAL SPECIES

Whereas we use morphological criteria in the delimitation of composite species in the genealogical network, ours is emphatically not a morphological species concept. This section points out the differences between these two concepts of species. Thanks to these differences, the composite species concept avoids some of the problems that affect the morphological concept, such as its lack of sharp boundaries and its ahistoricity (Mayr, 1942: 115-118; Hull, 1976; see also Kornet, 1993).

Our criterion for species membership is morphological in the sense that, in applying it, regard must be paid to morphological attributes of organisms, since it is through examination of these attributes that one detects the fixation of a character state in an internodon. On the other hand, our criterion is not morphological if by this term one means that the species membership of an individual organism can be decided by looking at nothing but its morphological attributes. The function of our morphological criterion is to allocate organisms to species in virtue of the internodons to which they belong. The criterion has regard not primarily for the morphological attributes of individual organisms, but rather for the properties of an internodon as a whole, such as its property of being an internodon in which the fixation of a character state occurs.

The most convincing way of showing the difference between our species concept and the morphological concept is by noting that, in our concept, an organism's showing a_2, the manifestation of a character state A_2, is neither necessary nor sufficient for it to be allocated to the composite species associated with the fixation of A_2. By contrast, in any pure morphological concept, an organism belongs to the species defined in terms of one or more particular attributes if and only if it possesses those attributes. We shall now demonstrate that showing a particular manifestation is neither necessary nor sufficient for an organism to be allocated to a composite species.

There are two reasons why the possession of manifestation a_2 of character state A_2 is not necessary for an organism to be a member of the composite species defined by the fixation of A_2.

First, it is unlikely that a character state should become fixed in the first generation of an originator internodon. Because of this, even if full fixation of A_2 occurs in this internodon, some of the earliest-born members of that internodon are likely to lack the character state by the fixation of which the species is identified. (For discussion of a similar phenomenon in phylogenetic species, sometimes called a "paradox", see Nelson, 1989: 286, and de Queiroz and Donoghue, 1990: 68-69.) These organisms will not show the manifestations given by character state A_2.

Second, even after character state A_2 has become fixed within an internodon, if the form of fixation to which quality Q is tied is either majority or near fixation, A_2 need not be possessed by, and therefore its manifestation a_2 need not be shown by, 100% of the member organisms of the internodon.

These are the reasons why it is not necessary for a member organism of the composite species associated with some character state A_2 to show the manifestations given by A_2. Now let us turn to consider whether an organism's showing manifestations indistinguishable from those given by A_2 is sufficient to compel its allocation to the composite species associated with A_2. There are two reasons why it is not.

First, a character state A_2, which becomes fixed in a particular internodon, can have spread also to branches of the genealogical network different from that on which this internodon is located. If it has, organisms outside the species associated with the fixation of A_2 can possess A_2 and therefore show the manifestations given by A_2.

Second, branches of the genealogical network other than the one in which A_2 has become fixed can contain organisms possessing character states that are non-identical to A_2, but that give organisms manifestations indistinguishable from those given by A_2. In this case, organisms outside the species associated with the fixation of A_2 can show the manifestations typical of A_2, despite not actually possessing A_2. This phenomenon, which we call convergence, will be discussed further in the next section.

This implies that an organism's possession of manifestations indistinguishable from those given by character state A_2 is not sufficient to allocate it to the composite species associated with the fixation of A_2, even if the organism is contemporaneous with the internodon in which the fixation takes place.

5.8 APPLICATION OF THE COMPOSITE SPECIES CONCEPT IN CLADISTIC PRACTICE

In this section, we aim to show that the composite species concept is compatible with actual cladistic practice, and that furthermore this species concept provides deep justifications of assumptions and procedures used in phylogeny reconstruction. To do this, we set up a hypothesized phylogenetic tree of internodons and investigate to what extent the taxa identified by cladistic practice correspond to composite species.

Figure 5.8a gives the phylogenetic tree of internodons that we will use for the test. This tree contains several composite species. The originator internodon of each species is one in which, as we envisage in our definition of Q, a state of a character becomes fixed. In the diagram, we deal with states of characters A to G. Each state is denoted by a subscript numeral. The time at which one of these character states becomes fixed is marked in the diagram by its name, such as $A_3{}^*$.

Each of the character states has a manifestation, represented by, e.g. a_3. As explained in Section 5.5, the manifestation of a character state is an attribute that an organism shows in virtue of possessing that character state.

Each species in Figure 5.8a is characterized by the manifestation of the character state that became fixed in its originator internodon. These morphological attributes of species are indicated in Figure 5.8a by the different shadings, distinguishing the several internodons of each composite species. These attributes are what, in practice, will be used to allocate a given organism to one of the composite species. Each of the character states $A_3{}^*$, $B_9{}^*$, $C_8{}^*$, $D_7{}^*$, and $E_4{}^*$ gives, to the organisms that possess it, manifestations that are different from those given by the other character states.

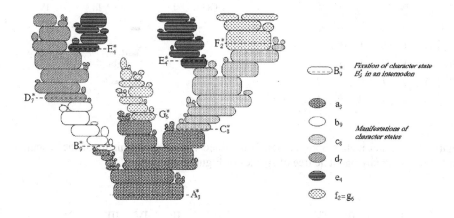

Figure 5.8a. A postulated phylogenetic tree of internodons. Internodons originate whenever a permanent split occurs in the genealogical network. In some of the internodons a character state has become fixed. The shading represents the manifestations of the character states. Character state $E_4{}^*$ has become fixed twice. Character states $G_6{}^*$ and $F_2{}^*$ have similar manifestations.

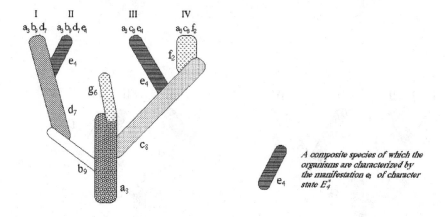

Figure 5.8b. A representation of the phylogenetic tree of species (the historical sequence of speciation events) drawn from the postulated phylogenetic tree of

internodons of Figure 5.8a. Note that composite species speciate not by splitting up, but by branching off.

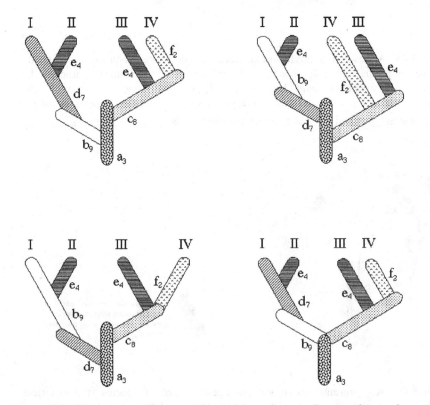

	a_3	b_9	c_8	d_7	e_4	f_2
OG	0	0	0	0	0	0
I	1	1	0	1	0	0
II	1	1	0	1	1	0
III	1	0	1	0	1	0
IV	1	0	1	0	0	1

Figure 5.8c. Data matrix and cladogram obtained from the organisms of the recent time slice of the phylogenetic tree of species of Figure 5.8b.

Figure 5.8d. Four reconstructions of the phylogenetic tree of composite species, out of the 27 that are compatible with the cladogram of Figure 5.8c.

Figure 5.8a shows two cases of homoplasy, or pairs of character states that, by the similarity of their manifestations, falsely suggest a common history of the taxa in which they are found. One case is due to the fact that, as our scenario envisages, $F_2{}^*$ and $G_6{}^*$ give to the organisms that possess them indistinguishable manifestations: $f_2 = g_6$. $F_2{}^*$ and $G_6{}^*$ are different character states, in virtue of having had different origins in mutation events in the genealogical network. This case of homoplasy is due to convergence, i.e. to the rise of character states having similar manifestations but different ancestor character states (along the lines of Wiley, 1981: 12). The second case of homoplasy is due to the fact that character state $E_4{}^*$ has become fixed twice, in two separate branches, qualifying two internodons as originator internodons and therefore giving rise to two composite species that are characterized by $E_4{}^*$ and thus have the same manifestation. This is a case of homoplasy, since the presence of $E_4{}^*$ on two different branches falsely suggests a common history of the taxa concerned. But it is not a recognized form of homoplasy: we propose to call this a case of *parafixation*. Parafixation is obtained when two organisms lying on separate branches of a phylogenetic tree show indistinguishable attributes in virtue of possessing one character state in common. For character state $E_4{}^*$ to become fixed in two internodons, as illustrated in Figure 5.8a, the mutation event in which it originated must have occurred earlier in the phylogenetic tree: more precisely, it must have taken place in an internodon no later than the last internodon that is an ancestor of both the internodons in which $E_4{}^*$ eventually becomes fixed. Clearly, parafixation is different from parallelism: while parallelism is the fixation of two character states with morphologically indistinguishable manifestations that developed from the same ancestor character state, parafixation is the fixation of one character state on two different branches.

Figure 5.8b depicts the phylogenetic tree of composite species that corresponds to the phylogenetic tree of internodons in Figure 5.8a. The attributes characteristic of each species shown in Figure 5.8a are shown here too, by the shadings as well as by the lower-case letters. Note that, as we shall discuss in the next section, composite species branch off and do not split up. Save for their latest time slice of reasonable thickness, the phylogenetic trees in Figures 5.8a and 5.8b are not accessible to the taxonomist. The numerals I to IV identify the species that are extant now, at the time of the taxonomic investigation that we here envisage.

How is this model related to taxonomic practice? To diagnose extant composite species, the following procedure suggests itself. The taxonomist, examining the latest organisms belonging to species I to IV, compiles a record of their attributes against their locations. The first task is to diagnose internodons. Groups of organisms might most plausibly be supposed to constitute separate internodons if they are isolated from one another. Then, combinations of fixed character-state manifestations should be found such that every entity believed to be an internodon shows one such combination. All internodons whose members show the same combination of fixed character-state manifestations are allocated to the same composite species. It should not be expected that every composite species has a fixed character-state manifestation that no other species has (i.e. that it has an autapomorphy of its own). For instance, in Figure 5.8b, species I, because it is ancestral, has a unique combination of fixed character-state manifestations $(a_3b_9d_7)$, but no character-state manifestation in this combination is unique to it.

The combinations of fixed character-state manifestations found for the extant composite species are recorded in a data matrix (Figure 5.8c). From the data matrix, by cladistic analysis, the taxonomist hypothesizes cladograms. (On cladograms, see e.g. Eldredge and Cracraft, 1980: 19-85.) In our model, a cladogram is a candidate reconstruction of the distribution of acquisitions of character states over the branches of the genealogical network; in other words, it is a map of the sequence of character-state fixations in branches of the genealogical network. Figure 5.8c shows the cladogram that cladistic analysis programs, such as PAUP (Swofford, 1991) and CAFCA (Zandee, 1991), indicate to be the best-supported solution admitted by the data matrix shown.

Various possible phylogenetic trees of species can be inferred from a cladogram. (Compare Cracraft, 1974; Nelson and Platnick, 1981: 169-183; and Wiley, 1981: 104-108.) Which particular trees are obtained depends on the concept of species used.

On our interpretation, the segments of a cladogram do not correspond to species, and the points at which these segments originate do not correspond to speciation events. (More about this at the end of the present section.) A segment with one character-state acquisition corresponds to a single speciation event. Empty segments of the cladogram, i.e. segments on which no character-state acquisitions are marked (such as terminal segment I), should be interpreted as indicating that an ancestral species survived one or more speciation events in which daughter species originated. Finally, segments on which more than one character-state acquisition is marked correspond in principle to an equal number of speciation events, and therefore indicate in principle the existence of an equal number of composite species that are descendants of one another. If this is so, the segment labelled with character-state manifestations b_9 and d_7 in Figure 5.8c represents a sequence of two species, one being the descendant of the other. The exception is constituted by the case of double speciation (see Section 5.5), in which more than one character state happens to become fixed in the same originator internodon, giving rise to superposed species.

On the composite species concept, a single cladogram may be compatible with several phylogenetic trees, for two reasons. First, a cladogram admits several trees if it has segments on which more than one character-state acquisition appears, since it is impossible from the data contained in the latest time slice to ascertain when these character states were acquired. Because of this, we cannot reconstruct the order in which the composite species arose, and whether any of them were superposed species. Second, the cladogram cannot indicate the order in which different species possessing the same ancestor species branched off from that ancestor. However, neither homoplasies (convergences, reversals, parallelisms, or parafixations) nor the empty segments in a cladogram increase the number of phylogenetic trees that are compatible with a cladogram.

From the morphological data available for the latest time slice, it is not possible to determine which of the candidate trees describes the historical events. To discriminate further, we would require extra data: those that may be acquired from earlier time slices, such as by palaeontology. It will also be impossible on the basis of the data matrix to decide whether the fact that taxa II and III are characterized by (i.e. have as autapomorphy) indistinguishable manifestations (e_4 in Figures 5.8c and 5.8d) is due to parafixation (as in fact our scenario stipulates in Figures 5.8a and 5.8b) or to convergence or parallelism. Since, however, we would consider taxa II and III two different species in either case, this uncertainty does not affect the construction of hypotheses of the phylogenetic tree of composite species.

By this procedure, 27 possible phylogenetic trees are obtained from the cladogram in Figure 5.8c (including nine trees in which the species characterized by b_9 and d_7 are "superposed": see Section 5.5). In Figure 5.8d, we show a sample of four of these trees. One of them, the first, reproduces the actual phylogenetic tree of species as we postulated it in Figures 5.8a and 5.8b, save for the species characterized by $G_6{}^*$, which died out before the present, and of which therefore no trace survives in the cladistic data.

The superficial similarity of cladograms and phylogenetic trees may tempt some to interpret a cladogram as a stylized phylogenetic tree, in which each segment corresponds to a species, and in which each point at which a segment originates corresponds to a speciation event. This interpretation of a cladogram however, yields species (which we shall call cladospecies) quite different from composite species.

For example, if we interpreted each of the segments of the cladogram in Figure 5.8c as corresponding to a species, the resulting phylogenetic tree of cladospecies would be that illustrated in Figure 5.9a, and the postulated phylogenetic tree of internodons would correspondingly be partitioned as in Figure 5.9b.

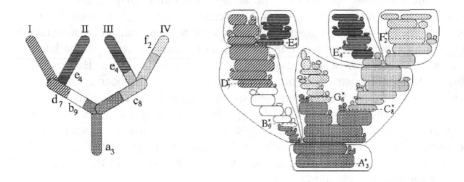

Figure 5.9. If segments in the cladogram of Figure 5.8c are interpreted as species, the phylogenetic tree of cladospecies is as shown here (left). This corresponds to a division of the phylogenetic tree of internodons into cladospecies (right). The speciation of cladospecies is necessarily dichotomous.

In the cladospecies, there is no guarantee that any character acquisitions that occur will be located in the first internodon, or even close to the beginning of species life spans. Where this does not happen, the diagnosability of these species will be lowered, as Figure 5.9b shows. Moreover, this interpretation of a cladogram condemns us to considering speciation as invariably dichotomous (as shown in the phylogenetic tree of cladospecies in Figure 5.9a), which is generally considered an artificial representation of speciation imposed by methodological principles. These are the reasons for which cladogram segments ought not to be construed as species.

5.9 COMPOSITE SPECIES DO NOT SPLIT UP, BUT BRANCH OFF

The Hennigian species concept requires the methodological principle, sometimes defended also as an empirical claim about speciation, that speciation should be seen as occurring by splitting up of branches (Hull, 1979: 425). In a splitting up, a new branch arises by the bifurcation of an extant branch of the genealogical network into two (or possibly more) new branches, which are siblings of one another. The rise of one successor branch is necessarily accompanied by the rise of a sibling of it, and the rise of these successor branches is necessarily accompanied by the ending of the ancestor branch. This is the way in which, by virtue of their definition, we envisage internodons to originate. However, if, as is generally accepted (Hull, 1979: 432), speciation may well occur through the isolation of a small interbreeding community, there is no justification for assuming that the ancestral species always becomes extinct in speciation (Figure 5.8b).

On the composite species concept, speciation occurs by branching off: a successor branch can arise without the rise of any sibling branches of it, and without the ending of the ancestor branch. Any internodon (and therefore any internodon possessing Q) arises by the splitting up of a branch of the genealogical network into two. One of these branches is occupied by the internodon with quality Q, which is the originator internodon of a new composite species. What occurs to the internodon that is its sibling? There are two possible cases.

In by far the more common case, no character state becomes fixed in the sibling internodon. This internodon cannot be an originator internodon: rather, it must belong to the same species as the latest internodon that is the ancestor of both it and its sibling (see Figure 5.8a). Therefore, the ancestor species has survived the rise of the originator internodon of the new species. This means that the emergence of a daughter species does not imply the disappearance of its ancestor species: on the contrary, the ancestor species typically persists, at least for a while, after the speciation event, so that the life spans of an ancestor and daughter composite species overlap. It follows that composite species arise by branching off rather than by splitting up.

In by far the less common case, there occurs a fixation of a character state in the second internodon, as well as in its sibling. Here both the internodons are originator internodons, of different composite species; and each of the two originator internodons and the internodon from which they arose belongs to a different composite species. While this might be viewed as a splitting up, in fact every instance of splitting up is an instance of branching off: a particular branching off in which two branches arise simultaneously. In the light of this, we are warranted in both the cases described here to speak of speciation by branching off.

5.10 THE COMPOSITE AND PHYLOGENETIC SPECIES CONCEPTS COMPARED

The question may arise how the composite species concept is related to the phylogenetic species concept. The phylogenetic species concept is defined by Cracraft (1989: 34-35) as "an irreducible (basal) cluster of organisms, diagnosably distinct from other such clusters, and within which there is a parental pattern of ancestry and descent", by Nelson and Platnick (1981: 12) as "simply the smallest detected samples of self-perpetuating organisms that have unique sets of characters", and by Nixon and Wheeler (1990: 218) as "the smallest aggregation of populations (sexual) or lineages (asexual) diagnosable by a unique combination of character states in comparable individuals (semaphoronts)."

The composite and phylogenetic species concepts show certain similarities, as follows.

First, both the composite and phylogenetic species concepts define species as collections of supra-organismal entities: these are internodons in the former concept, and populations, clusters, or samples of organisms in the latter. In both concepts, identifying a species in nature requires first that the component entities be picked out. As Nixon and Wheeler (1990: 218) state, "application of the phylogenetic species concept requires initial hypotheses of populations before relevant comparisons among individuals can be made", giving to populations the role that we give to internodons.

Second, both species concepts rely on similar diagnostic indicators to pick out these component entities in practice. Nixon and Wheeler (1990: 219) advise that populations can be "hypothesized initially on the basis of location and similarity of attributes". We likewise suggest (Section 5.8) that a group of organisms that share attributes and are isolated from other organisms should be hypothesized as constituting an internodon.

In both species concepts, component entities (internodons and populations, clusters, or samples) in the latest time slice are united into species on the basis of morphological similarity. We allocate all internodons that show the same combinations of fixed character-state manifestations to the same composite species. Likewise, Nixon and Wheeler (1990: 220) recommend that populations with the same set of character states be allocated to species, regardless of whether they interbreed.

In both species concepts, the morphological criterion used to group the component entities into species is the unique combination of character states (see Cracraft, 1983: 103; Nelson and Platnick, 1981: 12; Nixon and Wheeler, 1990: 218). A species "need not have even a single character that is unique to it", as Nelson and Platnick (1981: 12) put it and as we also allow.

Alongside these similarities, we see at least two important differences between the composite and phylogenetic species concepts.

The first is that the definitions of population, cluster, and sample of organisms used in the phylogenetic species concept are vague (see Section 5.3). It is therefore difficult to know where the boundaries of these entities lie in the genealogical network. A particular failing of these definitions is that they do not specify by what the life span of these entities is bounded. Because of this, the phylogenetic species concept meets difficulty in drawing boundaries in time between species in the genealogical network. How in the phylogenetic species concept does one demarcate an ancestor from a daughter species in the genealogical network? Because of its incapacity to answer this question, the phylogenetic species concept can be applied only to organisms in the latest time slice. In contrast, in the composite species concept the component entity of species receives a rigorous definition. The boundaries of both internodons and composite species in the genealogical network are therefore sharp, even in time.

The second important difference between the composite and phylogenetic species concepts is that, while in the composite species concept speciation is viewed as a branching off rather than a splitting up, the phylogenetic species concept is unclear on this point. In general, users of the phylogenetic species concept seem to regard it as interpreting speciation as dichotomous.

In the light of these similarities and differences, we consider the phylogenetic species concept as an approximation to the composite species concept that provides a less precisely defined theoretical framework and therefore less deep justifications of cladistic procedures and results. When it is applied to organisms in the most recent time slice, the approximation of the phylogenetic species concept to the composite species concept is quite close; but in the historical reconstruction of taxa by cladistic analysis, the approximation is loose.

5.11 CONCLUSIONS

Our attempt has been to construct a rigorously defined species concept that delivers species that are mutually exclusive, historical, and recognizable entities. To do so, we have taken a number of decisions for which there were alternatives. We have chosen, for instance, to identify quality Q with a morphological quality, rather than with, say, interbreeding ability, and to associate Q with the fixation of one character state, rather than more than one. We leave it to others to judge whether ours have been the optimal choices to reach our goal. We further discussed whether majority incidence or 100% incidence was the preferable notion to which to tie the fixation of a character state: while we have shown the merits of the former, we would not want the discussion of the composite species concept to be confined to this choice.

In conclusion, we draw attention to one peculiarity of the composite species concept. The origin of a composite species is tied to the achievement of fixation of a character state in an internodon. But a character state will achieve fixation much more frequently in small internodons than in large ones. Therefore new, small composite species will often arise with small internodons in which, by chance, a character state has high incidence. This effect is likely to manifest itself in the following two contexts.

First, if by a permanent splitting a small internodon arises of which the members happen to share a character state that is common locally but rare elsewhere, that internodon will be the originator internodon of a new, minuscule composite species. If this should happen frequently, "fringe species", defined by character states that are typically different from that defining the composite species from which they branch off, would arise continuously at the margins of interbreeding communities.

Second, if the incidence of a character state increases gradually in a branch of a genealogical network, that character state may achieve majority fixation earlier in smaller internodons than in larger ones. Then the coming into being of a larger composite species would be foreshadowed by the origination of many smaller "forerunner species" defined by the same character state. The composite species concept counts each of these as separate species, despite the fact that their member organisms are typically indistinguishable.

These features of the composite species concept might be removed by further extending the concept. One option is to stipulate that small internodons at the margins of a large species in which character states reach fixation in the ways described here do not originate new species, but are part of the species on the margins of which they develop. This stipulation could be achieved by formulating a third-stage criterion, based perhaps on interbreeding ability, to group these entities into species.

ACKNOWLEDGEMENTS

The first author presented the main conclusions of this paper at the Eleventh Meeting of the Willi Hennig Society, Paris, August 1992. We thank the participants for their constructive response. We also thank the editors of this volume and two referees for helpful suggestions. The investigations of the first author were supported by the Foundation for Biological Research (BION), which is subsidized by the Netherlands Organization for Scientific Research (NWO).

REFERENCES

Cracraft, J. (1974). Phylogenetic models and classifications. Systematic Zoology 23: 71-90.

Cracraft, J. (1983). Species concepts and speciation analysis. Reprinted in Ereshefsky, 1992, pp. 93-120.

Cracraft, J. (1989). Speciation and Its Ontology: The Empirical Consequences of Alternative Species Concepts for Understanding Patterns and Processes of Differentiation. In: Otte, D. and J. A. Endler (Eds). Speciation and Its Consequences. Sinauer Associates, Sunderland, Mass. pp. 28-59.

de Queiroz, K. and M. J. Donoghue (1988). Phylogenetic systematics and the species problem. Cladistics 4: 317-338.

de Queiroz, K. and M. J. Donoghue (1990). Phylogenetic systematics or Nelson's version of cladistics? Cladistics 6: 61-75.

Dobzhansky, T. (1970). Genetics of the Evolutionary Process. Columbia University Press, New York.

Eldredge, N. and J. Cracraft (1980). Phylogenetic Patterns and the Evolutionary Process: Method and Theory in Comparative Biology. Columbia University Press, New York.

Ereshefsky, M. (1991). Species, higher taxa, and the units of evolution. Reprinted in Ereshefsky, 1992, pp. 381-398.

Ereshefsky, M. (Ed.) (1992). The Units of Evolution: Essays on the Nature of Species. MIT Press, Cambridge, Mass.

Hennig, W. (1966). Phylogenetic Systematics. University of Illinois Press, Urbana.

Hull, D. L. (1976). Are species really individuals? Systematic Zoology 25: 174-191.

Hull, D. L. (1979). The limits of cladism. Systematic Zoology 28: 416-440.

Kornet, D. J. (1993). Permanent splits as speciation events: a formal reconstruction of the internodal species concept. Journal of Theoretical Biology 164: 407-435.

Kornet, D. J., J. A. J. Metz, and H. A. J. M. Schellinx (1995). Internodons as equivalence classes in genealogical networks: building-blocks for a rigorous species concept. Journal of Mathematical Biology 34: 110-122.

Mayr, E. (1942). Systematics and the Origin of Species. Dover, New York (reprinted 1964).

Mayr, E. (1957). Species Concepts and Definitions. In: Mayr, E. (Ed.). The Species Problem. American Association for the Advancement of Science, Washington, D.C. pp. 1-22.

Nelson, G. (1989). Cladistics and evolutionary models. Cladistics 5: 275-289.

Nelson, G. and N. I. Platnick (1981). Systematics and Biogeography. Columbia University Press, New York.

Nixon, K. C. and Q. D. Wheeler (1990). An amplification of the phylogenetic species concept. Cladistics 6: 211-223.

Ridley, M. (1989). The cladistic solution to the species problem. Biology and Philosophy 4: 1-16.

Rosen, D. E. (1979). Fishes from the uplands and intermontane basins of Guatemala: revisionary studies and comparative geography. Bulletin of the American Museum of Natural History 162: 267-376.

Simpson, G. G. (1961). Principles of Animal Taxonomy. Columbia University Press, New York (reprinted 1990).

Swofford, D. L. (1991). PAUP: Phylogenetic Analysis using Parsimony. Computer program distributed by the Illinois Natural History Survey, Champaign, Illinois.

Wiley, E. O. (1981). Phylogenetics: The Theory and Practice of Phylogenetic Systematics. John Wiley, New York.

Zandee, M. (1991). CAFCA, version 1.3c for Macintosh. User's manual distributed by the author. Institute of Theoretical Biology, Leiden.

D. J. Kornet, Philosophy of the Life Sciences Group, University of Leiden

James W. McAllister, Faculty of Philosophy, University of Leiden

6

The Wonderful Crucible of Life's Creation:
An Essay on Contingency versus Inevitability of
Phylogenetic Development

R. Hengeveld

ABSTRACT

In this paper I discuss the question of whether life processes are contingent or inevitable, particularly when viewed on a long, phylogenetic scale. In my opinion, this contrast does not exist. Rather, the perception of a dichotomy is the result of differences in how measurements are made or in the way data processing is carried out. Observations made in one way result in the conclusion that phylogenetic development is contingent and that process outcomes are, as a consequence, entirely unforeseeable. Clear trends could have shown up with different observations. Furthermore, differences in approach or in philosophical attitude could also result in life processes appearing to be either contingent or inevitable. Such diverse and complex processes can probably best be studied by adopting an integrated approach.

6.1 INTRODUCTION

In his book *Wonderful Life*, Gould (1989) states that life processes as seen on a broad, phylogenetic scale are contingent; replaying the tape of life would never again give the same result but always widely different ones. Conway Morris (1998) writing in his *Crucible of Creation* about the same geological period as Gould, the Early Cambrian, disagrees with Gould (see also his recent book *Life's Solution*, Conway Morris, 2003). In it, he states that life processes are, instead, constrained such that the outcome becomes inevitable; broadly, replaying the tape of life would give comparable results.

This paper discusses this contrast in opinion at three levels of understanding: 1) the technical level of measurement and data processing; 2) the methodological one on the choice of process studied; and 3) the philosophical one on the choice of the type of questions being asked. At the first two levels the contrast between Gould and Conway Morris seems more apparent than real. At the third, the philosophical one, they take the same position be it one that seems dated relative to the one currently prevailing in scientific discourse.

T.A.C. Reydon and L. Hemerik.,(eds.), Current Themes in Theoretical Biology,
129-157.
© 2005 *Springer. Printed in the Netherlands.*

As an ecologist as well as a biogeographer I have tried to unify these two disciplines (e.g. Hengeveld, 1989, 1990, 1993, 1994) often meeting with difficulties quite similar to the present one in palaeontology. Disagreements on the point of determinism *vs.* contingency often emerge from differences in measurement and data processing (Hengeveld, 2002). This experience prompts me to apply the findings from my own field also to that of palaeontology. Moreover many, if not all, of the processes at the various ecological and biogeographical scales of variation are also basic to those happening at the scale of phylogenetic development (e.g. Eldredge, 1989; Vrba, 1985). The controversy appears more general than the one between these two authors only; the debate between Gould and Conway Morris is, in fact, quite representative of ever-recurring debates not only within palaeontology but in population biology at large. I will illustrate my arguments by examples taken from a variety of fields of population biology.

6.2 TERMINOLOGY

The literature on the present subject is riddled with terms such as lawful, lawlike, contingency, stochasticity, determinism, inevitability, models, pluralism, holism, reductionism, etc. But terminology, of course, goes deeper than mere words: these often reflect basic concepts on the nature of the subject or of the methodology applied. As my use of terms may be idiosyncratic it seems necessary to spend some words on them. At this level already, basic agreements or disagreements show up as seen from my perspective.

When a physicist starts analysing a particular phenomenon, he isolates one or a few processes that may together have led to the phenomenon studied. Each of these processes is, as a next step, derived from a set of first principles. This set constitutes some parameters, such as F, m and a, which are thought to relate to each other in a certain way, expressed by the structure of the equation $F = ma$. Such a law or model can be formulated as a result of causal-analytical, often purely deductive, reasoning. The initial phenomenon can thus be reduced to the operation of one or a few such basic laws or models, thus exhibiting reductionistic reasoning.

When we look for biological laws we have to follow the same approach in asking for a basic equation that is mathematically derived from first principles. Of course, in biology the parameters as well as their mutual relationships involved must be biological ones to make the equation a biological law. At the population level, for example, Darwin's speciation model of selection through competition between species is, in principle, a biological law although its applicability depends on the occurrence of competition as its general driving process. Paterson's (1985) speciation model, although not formulated in mathematical terms, represents an alternative law, substituting competitive processes for those of mutual recognition between the two sexes. In these terms, Van den Bosch *et al.*'s (1992) mathematically derived invasion model is only in part a biological law, depending on biological parameters set within a more general framework of reaction-diffusion processes. Pure science is basically concerned with the search for and formulation of these kinds of (verbal) equations or models, previously and formally called laws.

Methodologically, the procedure is therefore rather easy and straightforward. Problems arise however, when we follow the opposite way of reasoning in trying "to put Humpty Dumpty together again". It is now the combined effect or the synthetic end product of the operation of one or more laws or rules we are interested in. Such effects or end products themselves can never be considered a law. Thus, a pebble, the solar system, or some organisms, species or phylogenetic processes do not themselves constitute laws; they result from their combined operation. Consequently, effects of different sets of laws operating together are studied in a different field of enquiry, in applied science, and are applied in various sorts of technology. Thus, from this viewpoint, individual organisms resulting from and operating according to a number of physical and biological laws are comparable to technological instruments, just like sewing machines, etc. (compare Rosenberg, 1985). They can be explained in terms of the constituent laws operating conjointly, resulting in pluralistic (Beatty, 1995) or supervenient (Carrier, 1995) explanations.

Such explanations are, therefore, descriptive rather than causal-analytical in terms of a basic model. They have an *ad hoc* character, as there are no general rules about which processes, each described as a law, are operating together or about what their weights are relative to each other, etc. (For example, there are many ways to build a bridge, both with respect to materials used (wood, rope, metal, concrete) as well as to the principles applied (Bailey bridges, arch bridges, suspension bridges, etc.).) When effects of interactions among the individual processes predominate over their individual effects, we consider the process or the end product as holistic. When the effect at some point in the process is given a central weight as if inevitably leading the process to this particular effect, the pertaining biological theories are considered finalistic or teleological, similar to historiography where they are known as Whig explanations. They are explanations by hindsight, usually taking the present as the final stage and are, in fact, unable to predict beyond the present, despite their reliance on the effect assumed. All these concepts and terms, alien to causal-analytic methodology, indicate the existence of two sharply contrasting research methodologies.

Conway Morris' idea on phylogenetic convergence contains this form of inevitability leading to the origin of man as the steward of nature (Conway Morris, 2003). As soon as life processes were set into motion, some 3.8 billion years ago, mankind would inevitably result at some stage, whatever external accidents the process would find on his way. Similarly to contingency, inevitability therefore concerns the outcome of a process, not necessarily a certain type of process (such as a deterministic process). In contrast, Gould (1989) estimates that these stochastic accidents, resulting in processes with contingent results, are happening continually by chance. They would have smaller or larger effects on the course of the process, but always overshadow the effects of any possible trend resulting from constraints internal or external to the process. Among such internal constraints one could think of genetic or morphogenetic constraints, or mechanical or energetic ones underlying allometric relationships. External ones could be biotic, such as processes bundled in the Red Queen hypothesis, or mechanical ones giving fast-swimming organisms their torpedo shape. Conway Morris (1998, 2003) estimates these constraints to overrun the stochastic influences of the environment.

When stochastic effects predominate over internal and external constraints in determining the end result, I consider the process to be contingent, contrary to inevitable processes in which stochastic effects do not predominate but in which the constraints do. Yet, neither Gould's conception of phylogenetic processes, nor Conway Morris' conception can be understood in terms of laws. The phylogenetic processes concerned follow descriptive rather than causal-analytical explanations not resulting directly from first principles.

Some examples clarify the link between terminology and methodology. The simplest physical system, the Solar System, already shows some problem resulting from the application of the only law it is based on: Newton's Law of Gravitation, derived causal-analytically. However, when applied to several planets circling with different periods and at different distances from the Sun their interactions result in non-linearity and this in turn in unpredictable, chaotic behaviour.

In biological systems the reductionistic tracing back the origin(s) or root(s) of some phenomena may become increasingly simple with the developing technology. Although synthetic understanding will often be far away, if at all feasible, just because life processes are defined by interactions between often great numbers of processes that take place at many levels.

Another problem arises when time (and, in fact, space as well) is involved as in phylogeny: the problem of understanding why one route has been taken at a branching point rather than another. This problem is not present when going down reductionistically to the root of some phenomenon, such as in that of tracing one or a few genes for eye colour. However, synthesising the processes happening from the molecular level upwards and this over ontogenetic time will suffer from both the number of interactions of these genes with other processes as well as from understanding their behaviour at switching points. What happened at switching points during phylogeny will usually remain hidden in the past. Conway Morris feels that within a family or within the human species as a whole traits will be kept and will eventually emerge. This may be true, but his generalisation beyond a species to phylogenetic processes among taxa is wrong, just because of the independent evolution of supra-specific taxa. This is the region where chance reigns. Rather than inevitable trends to develop (Conway Morris, 1998, 2003) we have to expect a highly unpredictable phylogenetic contingency *sensu* Gould (1989).

6.3 THE IMPACT OF THE SCALE OF OBSERVATION

Some examples taken from a wide variety of population-biological disciplines show how important it is to know how, under what conditions and on what scale the observations were made. All this determines whether or not the result of the process can be foreseen with a given degree of certainty. From this technical viewpoint of measurement and data processing there is, in principle, no distinction between stochastic and deterministic processes.

Rabid foxes

At the finest scale of statistical resolution Sayers *et al.* (1977) reconstructed the course of individual rabid foxes through fields and across bridges, whereas broader spatial patterns of rabies progressing across Central Europe resulted from coarser degrees of resolution. On the finest scale, therefore, the front of the invasion wave has to be described in stochastic terms, individual foxes being dependent on local patterns and temporal vagaries of the environment. Exactly where and when an individual fox will be found, and where it will go, is unpredictable. In contrast, on the coarser scale, the progression of its invasion should be described deterministically and the progress of rabies appears to be highly predictable (see Hengeveld, 1989, Chapter 9). It is therefore impossible to say that the invasion process as such is stochastic or deterministic and, hence, if the process outcome is contingent or inevitable.

Invading muskrats

On a broad, geographical scale, the rate of progression and way of invasion of the Muskrat, *Ondatra zibethicus L.*, into Europe can be described using only the deterministic diffusion parameter (Van den Bosch *et al.*, 1992). At such a broad scale the area can be considered homogeneous according to the model assumption. However, at the finer scale of a single country, Germany, within this geographical space the vagaries of local and temporal variation in humidity come into play, making the assumption of spatial uniformity inappropriate (Schröpfer and Engstfeld, 1983).

At this finer spatial scale on which the environment should be considered as ecologically non-uniform, the proportion of suitable biotope introduces chance mortality of predation, or of failing to reach the preferred habitat on time, etc. In fact, mortality can be so high that it stops the progression of the invasion wave altogether (Hengeveld and Van den Bosch, 1997). This mortality can also be due to a suite of other chance factors, such as the degree of clumping, the unfavourability of the biotopes to be crossed, missing the suitable ones, or physiological exhaustion, etc. Moreover, local conditions vary stochastically both within and between years (see Hengeveld and Hemerik, 2002).

On finer, regional spatio-temporal scales, therefore, the invasion rate varies stochastically whereas on a broader, geographical scale of variation it is deterministic. Moreover, the scale of measurement chosen requires either the inclusion or the exclusion of several parameters accounting for variation in the data from the same basic processing model (Hengeveld, 1999).

Measles in Iceland

The incidence of measles in Iceland has been unpredictable for most of the history of its human habitation, i.e. as long as its population size was small. When raging over the island the disease rapidly exhausted the pool of susceptibles, so that it soon died out (Cliff *et al.*, 1981). During that time measles was a contingent, epidemic disease, its incidence depending on its occasional arrival with infectious travellers. However, when the Icelandic population exceeded the critical minimum population size of ca. 200,000 (Black, 1966), the disease became endemic, going round from one susceptible part of the population to the next, as it also does in present megalopolises (Infantosi, 1986). The rate at which the population becomes susceptible again, enabling measles to be transmitted and thus to persist, depends on the number of new births relative to the number of individuals in the older age classes that are still immune. Between various parts of the population, it then depends on the deterministic contact rate whether a particular sub-population will be infected or not, which still depends on the chance arrival of infected individuals. Locally, the disease behaves stochastically as before but as seen over the whole population, at the scale of the whole of Iceland, it behaves deterministically (see also Anderson and May (1986) for simulation results).

Moreover, at the scale of individuals, infections can be described in terms of chance, the susceptibles having good or bad luck. Consequently, the infections in a small population vary independently of age, all age groups having roughly equal proportions of susceptibles. However, with greater family size or with higher densities during the endemic phase, the proportion of immune individuals in the higher age classes increases with the result that the age of susceptibles declines towards the lower age classes (see Cliff *et al.*, 1993). Eventually, the disease modifies from a general infectious disease into a child disease with a transmission rate depending on the contact rate among children during school attendance, etc. This effect can be described deterministically again.

Thus, depending on its spatio-temporal scale, on population size and density, as well as on the dispersal rate of the individuals and the uniformity of population dispersion the outcome of the process of a measles epidemic is foreseeable or not, that is, the process can be considered deterministic or stochastic.

Genetic drift

When, irrespective of their spacing, the number of individuals in a population drops below ca. 500 its genetic composition becomes dominated by stochastic sampling effects resulting from, for example, random meeting of individuals, a process called genetic drift (Wright, 1955). In these cases the impact of other processes, such as selection, are negligible relative to this stochastic effect.

However, for selection to have any effect it must be assumed that its intensity and direction is constant: an assumption that probably applies over short time intervals and within limited areas only. Over longer intervals and larger regions, though, both the intensity and the direction of selection will themselves vary stochastically. Thus, on a broader scale, the genetic composition of populations with sizes greater than 500 individuals does not remain predictable either (Lande, 1976). One might call the effect of this longer-term stochastic selection process ecological drift which is, to varying degrees, independent of population size (e.g. Grant and Grant, 2002). Yet, patterns of temporal variation on all scales become superimposed (Hengeveld, 1997; Mitchell, 1976) thus cancelling each other out. Taking long sampling intervals therefore hides the stochasticity of the process so that it seems that there is no evolutionary change or that phylogeny follows trends (Gingerich, 1983). In fact, constraints to following environmental stochasticity closely will be partly determined by the size of the population, by the longevity of the individuals, and by the overlapping of generations.

Therefore, selection varies with the scale of variation of the factor to which a certain trait responds. At the same time selection can also happen at a scale at which the factor concerned does not vary, but nevertheless exerts a significant mortality. Not varying, it easily passes by unnoticed. At the other extreme selection may occur without incurring any population dynamic effect, that is, when it is operating at a low intensity. For example after a life of selection experiments for other traits the flies of the *Drosophila* strain that Bakker used at Leiden University (pers. comm.) no longer escaped from their bottles although they had done so for decades before. Selective mortality must have been infinitesimally small despite its significant biological effect on a long term. Many macro-evolutionary trends, even at higher taxonomic levels, may be explained this way; micro-evolutionary studies being concerned with short-term processes having more significant impacts within a species or a local population.

Thus, depending on a combination of effects of population size and temporal and spatial scale, the development of the genetic composition of a population is contingent or not.

Within the context of the present debate, Conway Morris (1998, 2003) selects those longer time scales that show up predictable, inevitable trends. These scales, though, vary largely in extent covering the whole history of life, the phanerozoic only, or any period during which in hindsight Conway Morris recognizes convergent traits developing. This subjective procedure therefore is selective. Gould, in contrast, without being explicit about the temporal scale in which to look, selects an array of slightly finer scales, that is those on which effects of stochastic accidents are most prominent. The scale on which such accidents are found vary greatly in extent as well, again from those covering the entirety of the history of life to that of the day-to-day struggle for survival of individual organisms. In contrast to Conway Morris, though, Gould attempts to look forward, finding much reason for life forms to become extinct. Yet, his selection of finer scales similarly depends on a subjective selection criterion. This, of course, happens more often, but it does make the problem impossible to solve.

6.4 METHODOLOGY

A difference in emphasis

Technically, the risk of extinction of a population or a species is the complement of its survival probability. It now seems that part of the controversy between Gould and Conway Morris depends on the emphasis they lay either on extinction or on survival, respectively.

From early on, Gould (1967, 1977) has been very aware of constraints imposed by mechanical as well as by ontogenetic demands and processes. Yet Gould (1989) maintains that species extinction is largely contingent caused by the failure of a species' idiosyncratic combination of traits to match random environmental variation. Background extinction depends on minor chance events and mass extinction on those of considerable intensities (e.g. Raup, 1991). (See Willis and McElwain (2002) for an overview and discussion of percentages of extinct animal and plant taxa during the five major extinction events which gives an impression of the significance of extinction.) In contrast, survival or persistence depends on the chance of the proper combination of traits having come together during stochastic processes at molecular levels and those of the individual organism. Similarly, Conway

Morris (1998) states that morphospace is shaped both by the genetic make-up of the individuals of a species and by mechanical demands of the environment to which it adapts. These demands eventually allow the survivors certain properties. Depending on the similarity of environmental constraints, adaptations would converge. Thus, by convergence, fast-swimming fishes, aquatic reptiles and whales, for example, came to possess a similar torpedo-like shape of their bodies, adding to their survival (Conway Morris, 1998; see also Conway Morris, 2003, where he elaborates on this).

Central to Conway Morris' thinking is that similarity in environment-induced adaptive processes leads deterministically to the same outcome, character convergence. Thus, the origin of man and its stewardship over all other living beings would be inevitable (see Conway Morris and Gould, 1998). Gould for his part has always fought this anthropocentric conception, considering the origin of mankind a unique chance event, the trend leading to it in fact being contingent (see below). According to Gould each species is unique due to a specific combination of traits which justifies their selection as a unique evolutionary end product "towards which evolution converged". Human intelligence or consciousness, for example, are among an endless multitude of traits from which to choose; its arbitrary choice is based on a subjective, anthropocentric criterion. Gould (1999) thereby denounces the confounding of the religious and the scientific magisteria, apparent in Conway Morris' convergence to the supposed stewardship of humankind (see also Hengeveld, 2004). For Gould species, being unique, will respond idiosyncratically and momentarily to environmental variation even if the spatial constellation and temporal sequence of events would ever be similar.

Thus viewed, extinction is a matter of chance or ecological risk of living in the wrong place at the wrong time and survival and persistence a matter of matching inevitable, ever-present constraints biological properties impose. Together with an emphasis on extinction, therefore, goes that of the impact of biological constraints within a variable environment. With an emphasis on survival belongs an emphasis on the positive impact of possessing particular properties within a predictive environment. These aspects are, of course, logically connected to each other as are extinction and survival themselves. It then depends on the author whether a proneness to stochastic extinction (Gould) or deterministic survival (Conway Morris) is emphasized.

This distinction in approach coincides with the distinction that Lewontin (2000) made between backward and forward approaches in the study of forms of adaptation. Philosophically the backward approach, which has more emphasis on constraints forthcoming from gene interactions and pleiotropy, from complex characters and from ontogenetic developmental pathways agrees best with biological structuralism. And the forward approach, often adopting the random mutability of genes and hence of traits in order to match an altered niche space or other living conditions, agrees most with biological reductionism which Conway Morris (2003) loathes as materialistic or as ultra-Darwinian.

In fact, the significance of chance within a process makes the outcome unpredictable, thus justifying a historical approach. History is defined by chance events. Such events make the process disorderly and the outcome inevitably unique. Conversely, unique outcomes can only be evaluated by looking backward, that is by historical reconstruction, which is basic to Gould's thinking, expressed, for example by his criticism of the "adaptationist programme" (e.g Gould and Lewontin, 1979). However, they hamper prediction. Conway Morris, by emphasizing inevitable trends towards predictable end results, emphasizes orderliness and phylogenetic lawfulness, following the opposite forward-looking approach. Given his examples of the inevitable evolution of the torpedo shape among various fast-swimming vertebrates, of intelligence and consciousness, he emphasizes functionality over the significance of historical causes (Conway Morris, 2003). It seems, therefore, that they agree with respect to internal constraints but that they differ in their ecological outlook, Gould emphasizing the capriciousness of abiotic ecological causes and Conway Morris the steadiness of biotic requirements of the individuals.

In these debates in palaeontology, like those in ecology (Hengeveld and Walter, 1999; Walter and Hengeveld, 2000), each of the two approaches is shaped in the form of a coherent set of concepts, assumptions and techniques, such that the debaters appear to be talking different languages to a large extent. These sets of non-overlapping methodological approaches accord with Kuhn's (1962) concept of scientific paradigms which in the cases of both ecology and palaeontology coincide and intertwine rather than follow each other up in time. Yet, as looked at from this methodological viewpoint, the distinction between contingency and determinism is a matter of emphasising the one or the other aspect of the very same process, that of succeeding or failing to adapt to new conditions given a certain set of properties. Species can evolve only within a confined part of morphospace, thus having broadly predictable properties. However, they die out by chance when the intensity or nature of environmental variation changes according to chance, such changes predictably hitting all species that, by evolutionary chance, happen to occur in the same part of morphospace possessing or lacking certain traits (e.g. Stanley, 1987; Gould, 2002).

Testing

For a long time, it remained uncertain whether there is sufficient regularity in stochastic (and, hence, contingent) colonization and extinction processes allowing them to be described in deterministic terms and, if so, whether such a regularity has a biological background. This is a question of statistical testing, which can be done in a qualitative way by examining graphical representations of process outcomes, or in a quantitative way by performing the statistical test.

Graphical representation of stochastic process outcomes

Settlement distances

Exactly where in Europe, relative to their parent's nest, the invading young birds of the Collared Dove *Streptopelia decaocto* will settle for rearing their own young cannot be predicted with any precision. Yet, it is clear that the majority of birds nests in the vicinity of where they grew up. Only occasionally do birds form bridgeheads far ahead of the invasion front by moving appreciable distances before settling.

At first, it seemed that two biologically distinct dispersal processes, short- and long-distance dispersal, were involved (Hengeveld, 1989). Short-distance dispersal would account for a steady rate of progression of a closed invasion front, whereas long-distance dispersal would lead to a rapid invasion through the saturation of biotopes between the isolated bridgeheads (e.g. Hengeveld, 1989). However, at closer analysis all distances together appeared to form a continuous frequency distribution - a contact distribution - of settling in space (Hengeveld, 1993). At present, it appears to be a generalized distribution consisting of two superimposed chance distributions. An exponential distribution represents distance decay around the parent's nest due to exhaustion of the dispersing birds, etc., and a second distribution represents their Brownian way of movement (Hengeveld and Hemerik, 2002). Although the generalized distance distribution is regular, showing hardly any scatter, the exact location of an individual bird settling cannot be predicted. When the long-distance jumps prevail in the contact distribution, the invasion front is scattered and its progression irregular, whereas the front is closed and progresses steadily when the short-distance movements dominate (Mollison, 1977). This gives the invasion process a qualitatively different appearance, although the invasion can result from the same dispersal process but with different tail lengths of the contact distribution characterizing the dispersal movements.

Waiting times extinctions

Similarly, larger (mass) and smaller (background) extinction events, due to various geological, climatic, or environmental processes as well as to astronomical ones, are distinguished from each other by periods of different duration (see e.g. Stanley, 1987). Mass extinction events and background ones were felt to have different causes, the first sometimes being explained by (periodical) extraterrestrial or global causes, and the second mainly by random terrestrial ones at finer scales. As to mass extinction events neither a periodicity (e.g. Elliott, 1986; Raup, 1986; Raup and Jablonski, 1986; Raup *et al.*, 1973) could be proven to exist (e.g. Connor, 1986) nor could their astronomical cause be found. After having worked on the possibility of a periodicity (Raup, 1986), Raup (1991) eventually found that the various extinction events form one single, J-shaped frequency distribution of intensities relative to their temporal separation, showing hardly any scatter. As rare long-distance jumps, the occasional mass extinction events form the tail end of this distribution curve, whereas the background extinction events are represented by the bulk of it. Apparently, a single chance process causes them, the combined outcome of which Raup called a kill curve.

Therefore, from this geological chance process of extinction events, we cannot predict exactly when an extinction event of a particular intensity will happen although the pattern of all events together accurately predicts the relative frequencies at which they occur. Thus, the impact of a bolide of a certain size and speed can be calculated with deterministic precision, although the appearance of each individual geological event is unpredictable.

Summary

The graphical analysis of these two, spatial and temporal, processes shows that the relative frequencies of their individually contingent events, their chances of occurrence, are, by contrast, predictable almost without any uncertainty. Still, as individual events with relatively great intensities, those few events in the tail end of the distribution have a distinct and very unpredictable impact on the course and rate of both the spatial and the evolutionary process. Thus, the shape of the distribution of the individual temporal and spatial events can be described in deterministic terms, whereas the course of the process and its impact remain contingent.

The two stances as the statistical null hypothesis and the statistical alternative hypothesis

Methodology

The graphic representations discussed above allow a qualitative examination of the shape of a single frequency distribution. Their interpretation, though, in terms of one or more known chance processes concerning individual organisms or events is based on a quantitative comparison between the observed frequency distributions and theoretically expected ones exclusively based on chance. A possible difference between the observed distribution and the one based on chance can be interpreted as biologically forced. Such quantified comparisons suggesting a particular interpretation are called statistical tests. The present controversy on how to interpret the palaeontological record requires, in principle, a quantitative comparison of two interpretations of the same data, either in terms of some biological process or in those of a non-biological, statistical process. The biological process is represented by the alternative hypothesis, and the statistical one by the null hypothesis. As such they are not independent theories but methodologically connected testing alternatives (Hengeveld, 2002).

Thus, when the two frequency distributions of the previous section fit known chance distributions exactly, there is no reason to explain them in any specific biological, or in some geological or astronomical way, respectively, as in the distance distribution or in the kill curve mentioned. Instead, they should be explained in terms of chance only. Similarly, Gould's statistical interpretation of the geological record represents the null hypothesis whereas the one by Conway Morris in terms of biological convergence gives the alternative hypothesis of explanation. Therefore, methodologically it is not possible to argue in favour of either the one or the other "theory". Rather, one has to collect data at the appropriate scale of variation in order to carry out the proper statistical test to make a justified choice between these two alternatives. From the viewpoint of testing theory, the apparent controversy can be unreal; the two opinions are methodologically connected alternative explanations one of which has to be discarded only after testing and this each time for the scale of observation concerned. The debate, as discussed so far, concerns the interpretation and applicability of either hypothesis even before any test has been done.

Some tests

Gould (1996) discusses some tests in his book *Full House* which was, in fact, specifically meant as the statistical counterpart of his earlier book *Wonderful Life* (1989) which was concerned with the interpretation of possible phylogenetic trends. This later, 1996 book, together with some studies it discusses (e.g. McShea, 1993, 1994), seem to have been missed by Conway Morris (1998, 2003) as he fails to refer to this literature. Gould (1996) distinguished three aspects of some supposed phylogenetic trends, the first of which concerns their generality and hence representativity. A trend towards consciousness or intelligence as observed in mankind is unique as seen from the viewpoint of the evolutionary development of all life forms including that of the bacteria. As such, it cannot be considered a representative example of trends in the direction of specifically these two traits.

A second aspect concerns the physical (im)possibility of developing in another, alternative way than in the one observed. For example, broadening the range of variation, given a beginning of the evolutionary process at some minimum value, automatically results in a shift of both the average as well as the extreme values away from this minimum starting value. A trend would occur if the minimum value shifts along with the average and the maximum value. Such a process, though, is not observed; the minima of the taxa analysed remain the same. Only the variance among the taxa increases thus generating a pseudo-trend in the average and maximum values. Similarly, the initial prokaryotic or multicellular taxa obviously score minimally at the point of consciousness or intelligence, so that the mean values and variance of their descendants can only remain the same or increase, whilst their still existing forms stay at these minimum values.

The third aspect concerns the sequence of individual taxa followed through evolutionary time. Thus, the question is do the subsequently developing taxa, evolving from particular ancestral ones, all develop in the same direction, or does this direction vary randomly, a taxon with large individuals, for example, being followed unpredictably by one either with yet larger, or with smaller individuals? In fact, the frequency distributions of increases or decreases in size found so far are symmetric around a stationary value. The data sets analysed at present therefore do not substantiate Conway Morris' (1998) hypothesis of inevitable phylogenetic trends.

Particularly this third aspect concerns a test of the existence of a direction in evolutionary development whereas the former two concern possibilities of statistical bias. The outcome of the test argues in favour of Gould's (1989) evolutionary contingency, whereas the two causes of bias argue in favour of biased observation and measurement underlying the possibility that evolutionary development is deterministic. These forms of bias can lead to the erroneous acceptance of the existence of phylogenetic trends. Yet another cause of bias, unknown from this analysis, concerns the choice of time interval of measurement, which could be too long, thus suggesting a trend where in fact there is none (Gingerich, 1983).

6.5 MEASUREMENT, DATA PROCESSING, RESEARCH AIMS, AND MODELLING METHODOLOGY

The first tier of understanding, that of measurement and data processing of the first chapter interacts with the second tier, that of the methodological approach chosen. As to extinction processes, they can cover long periods of time and can be global as one aspect of macro-evolutionary development. But often they are also rapid and local happening at ecological scales of space and time. Then, Gould's methodological stance concerning extinction leads him to emphasise short-term, stochastic processes similar to those of speciation (e.g. Gould *et al.*, 1977). Conway Morris, on his part, emphasises survival concentrating on long-term, deterministic processes of environmentally constrained adaptation (compare Van Valen, 1973 and Vermeij, 1987). We need to realize the distinction between these types of processes and their implications as to assumptions to be made. In fact, investigating these processes opens the way to integration of the two sides of the controversy.

Two types of models

Deterministic models assume that there is no significant additional variation entering at any one point during the process concerned that can influence its course and result. The initial conditions fully determine its course and outcome. Deterministic models also apply when, for example, great numbers of independently operating individuals are involved, smaller numbers leading to small chance deviations, as in genetic drift. This differs in stochastic models: added variation is assumed to play an essential role in the process. The difference between these two types of models shows up most clearly in a particular type of deterministic models, models of deterministic chaos. These are non-linear deterministic models in which even a slight variation in initial conditions greatly inflates through many interactions happening during the process, such that its outcome can only be described in statistical terms, despite the fact that the process itself is fully deterministic throughout its course (e.g. May, 1976).

Consider, for example, some billiard balls on a table. As it is practically impossible to hit a ball twice in exactly the same spot, in the same direction and with the same force, even the slightest variation in starting it off results in a different course towards a second ball. Hence, in its impact, giving this second ball another quite different impact and direction on a third ball, etc. if this second one hits it at all. Thus, no game can ever be exactly the same; the effect of even the smallest difference in the initial hit is rapidly inflated through the subsequent interactions of the balls in an uncontrollable way. Still, each aspect of the process of rolling and impacting is entirely deterministic; no variation is added along the way despite its increase. Variation would be added when, for example, the surface of the table would be bumpy and when the pattern of bumps and hollows would continually be altered at random. Apart from such "environmental" sources of variation the balls themselves can also be thought to change continually in size, weight, elasticity, etc.

Under field conditions, these latter random changes happen all the time: the individuals differ among each other, change their behaviour with weather conditions, etc. Moreover, this applies not only with respect to initial conditions as in the models of chaotic behaviour but variation is added throughout the process and this differently from place to place, individuals changing with age or with local feeding conditions, for example. This changeability applies not only to the individuals, but also to the components of their environment, some areas suddenly becoming too wet, for example. All this variation, continually being added over time, is such that usually no distinction can be made between initial conditions with a slight variation and that of the end result with a large variation; the sources of variation during the process are continuous without a beginning or end. Therefore, this variation is accounted for in the model structure representing the process. Moreover, this increasing variation cannot be described by an increase in variance relative to a stationary mean value; instead this value also varies and the process as a whole evolves. In these cases the process is stochastic; both the input of the process as well as its structure happen to vary throughout its course. Its outcome is completely determined by chance. This is clearly the sort of process Gould had in mind from early on (e.g. Gould *et al.*, 1977).

However, when many species are taken together and looked at on a global scale and subjected to some stringent and constant selection factors over long stretches of time, deterministic trends like that in character convergence may occasionally show up. Apparently, this is the type of process Conway Morris is talking about, not bothering about the exact processes happening at finer scales (Conway Morris, 2003).

The difference between these two opinions, therefore, is that Gould assumes a few stringent and generally important external factors to be present but that the evolutionary process itself is stochastic and non-directional. Conway Morris, in contrast, assumes some internal factor to operate, driving adaptations deterministically towards improved adaptation with respect to such factors. Next he interprets this internal driving force in religious terms (Conway Morris, 2003).

Research aims and interpretation

The search for phylogenetic trends can be motivated by pre-scientific views, occasionally resulting in biased observation as indicated above. Thus, Gould (1996) mentions trends supposedly existing in body size (Cope's Law), in body shape (the torpedo shape of fast swimming vertebrates Conway Morris (1998) mentioned), in complexity, intelligence, or towards consciousness. (Interestingly, Conway Morris attacked Gould on his supposed Marxist leanings whilst siding with Friedrich Engels on the occurrence of inevitable, recurrent trends toward human intelligence to which Gould objected.) Conway Morris (in Conway Morris and Gould, 1998) also mentioned the given or revealed stewardship of mankind over all other creatures on earth as one of those traits. These would, therefore, constitute a continuously required precondition for adaptations to develop. In a way, some non-existing trait precedes the steps to be taken during the evolutionary process like a Platonic idea or, rather, ideal. Gould (1996) for his part fights these research aims underlying the interpretation of supposed trends. In fact, by doing so and by accepting the current ecological paradigm underlying evolution theory he has still not freed himself from the same methodological mistake. After all competition resulting in optimising adaptation, or in the competitive replacement of ill-adapted individuals by fitter ones hinges on the very same idea, inevitably leading to the gradual improvement of species and the structure of the communities they constitute (see for a similar image Eldredge, 1989). Ecological population dynamics in fact represents one of the last strongholds of static Platonism within biology emphasising stationary mean values (equilibrium states), optimization processes within and between species, or even the prevalence of models over observational data (Hengeveld and Walter, 1999; Walter and Hengeveld, 2000). Typically, Hardin's (1960) competitive exclusion principle was formulated within an ecological context and interpreted in evolutionary terms (e.g. Mayr, 1963). It is one of the ecological underpinnings of the development of phylogenetic trends (e.g. Brown and Maurer, 1986). The process as such aims at particular results and competition as its main driving force would have a quantitative effect on its course. Yet, Gould (1996) recognizes that more recently evolved species have not at all replaced earlier ones or those evolving in parallel but the latter categories keep dominating the more recently evolved biota numerically.

Simpson (1953) recognized progression in phylogenetic development but defined it explicitly as an independent accumulation of sequential, newly-formed traits. As the reverse of Dollo's Law, such new traits determine a new course in evolutionary development which lasts as long as the trait exists. Similarly, competition or any evolutionary replacement of traits need not have any numerical effect when this is small and long-lasting. Evolutionary change, therefore, can be completely aimless not operating towards some qualitative or quantitative optimum and one of its driving forces, competition, need not have any demographic or population-dynamic impact.

The fact that trends would depend on deterministic processes includes a potential weakness for the existence of clear trends. When ecological responses under evolutionary change are non-linear, their results may become chaotic, thus deviating from being aimed at optimum values. The evolutionary pathway can therefore, assuming such a model structure, be capricious with stochastic end results. From this viewpoint steady, long-term trends operating in a predetermined direction are therefore difficult to conceive. However, despite similar stochastically varying end results such a deterministic model structure may be inapplicable to be replaced by a stochastic one. As several ecological variables operate not only stochastically but also partly independently on different scales, the results may become fractal (Mandelbrot, 1977).

This great variation in the impact of ecological variables, though, does not necessarily result in an extreme capriciousness of evolutionary development. Below, I shall give some reasons for viewing living systems themselves considered to be rigid, being stabilized in many ways by a multitude of mechanisms. Here, I argue that stabilization at the population level results particularly from the large number of variables operating on many traits simultaneously, as known for example from preventive medicine. For the same reason, using a seat belt is beneficial for individual drivers but has no effect whatever on general mortality figures. Thus, improving conditions may help individual cases but it does not in the least affect values population parameters take (Rose, 1992). Such effects being one or a few out of many are swamped by those of many other factors. Similarly, enhancing or reducing effects of particular ecological causes of mortality on individual organisms need not have any selective value within a population. Precisely the number of ecological interactions prevents selection from operating, and therefore evolutionary change, from being capricious. The dampened outcome should not fool us, although by mistake making us interpret the operating stochastic processes to be deterministic.

6.6 PHILOSOPHICAL BACKGROUND

Gould (1989) and Conway Morris (1998) may seem to disagree technically and methodologically but they agree in a philosophical way. Yet, by doing so they may actually be sharing a weak point in their reasoning.

Van Peursen (1970) distinguished three broad periods in philosophical reasoning during the history of mankind. First came the animistic period, which was followed by the ontological one which, in turn, was succeeded by the functional period. (This may bear some resemblance to Comte's tripartition of human thinking into the theological, metaphysical and scientific eras.) According to Van Peursen, our time, roughly beginning with the rise of science in the 16th Century, experiences the transition from the ontological way of thinking to the functional way. Galileo, for example, by measuring the speed of falling bodies, opened the way to earthly and astronomical mechanics. This had not been feasible following the previous, Aristotelian, ontological concept of falling as the resumption of some natural place; not knowing from independent information what this natural place is, you cannot do anything but accept or reject his concept. Aristotle defined what falling is, whereas Galileo estimated how it works. The present, functional way of thinking is most apparent in the physical complementarity principle, which holds that no choice can be made between treating atomic and subatomic phenomena either as corpuscles or as waves. One chooses the most convenient one for the problem at hand. The same way of thinking underlies modelling in general such as commonly applied in population biology. In ontology, questions are asked about the nature of some thing or event (what it is) whereas the functional way, using operational concepts, models and definitions, concentrates on their operation (how it works and how to measure it). The functional approach, making statements testable, is methodologically preferable; definitions play only *ad hoc* roles during model construction.

The advantage of a philosophically functional approach is that we can test and improve the applicability of their underlying assumptions. This is not possible with ontological definitions. For example, it does not matter what a species is, but it is interesting how species originated and maintain their existence. Not the definitions or concepts of species are of biological interest but the very biological processes are, differentiating sets of individuals in space and time. Similarly, we should not try and define the (ontological) nature of the phylogenetic process, i.e. whether it is contingent or deterministic. This distracts our attention from the process itself towards that in concepts and definitions, these having an *ad hoc* meaning and relevance only within the context of an explanatory model on the operation of the process mechanisms. In the context of philosophically functional approaches in which testability takes an important position, it is relevant to realize that Gould's contingency in fact represents the null hypothesis and Conway Morris' determinism the alternative hypothesis (see above).

6.7 INTEGRATION: LIFE AS A CONTINGENCY-REDUCING PROCESS

Life is conservative

For most terrestrial organisms life is a hazard; small insects, for example, can drown in a raindrop or lose their food source when blown from the leaf on which they are feeding. At that level life is contingent, effectively never giving exactly the same result twice. Technically, dying because of ecological hazards like these should be described in stochastic terms concerning the risks individual organisms run within their lifetime.

In fact, stochastic processes are found all the way from interactions between atoms and molecules within and around cells to those between individuals or species. This means that at all those levels there is, in mathematical terms, an inherent tendency towards chaotic inflation, if they contain any deterministic components on the pertinent scale of variation at all. Since life processes typically depend on (variance inflating) interactions, their results are inherently contingent (compare Monod, 1971). This tendency towards biological chaos in all those often highly intricate interactions, therefore, should be controlled by all means in order to keep the inflating effects within bounds of variation for the biological systems to operate. Not the slightest variation can be tolerated, and the smallest deviations have to be countered instantly. Biological variation is but the result of entropic decay being counteracted through a multitude of biological mechanisms operating at all levels. Its presumed biological significance is relevant only in idealistic models of speciation and phylogeny but not in those viewing these processes as incidental and hence contingent due to the occasional failing of some counteracting mechanisms.

Biological adaptation effectively reduces environmental risks and those due to entropic decay. During a shower of rain, individuals can wait for the sun to return which happens in behavioural homeostasis when, for example, lizards move in and out of the sun, thereby maintaining a constant body temperature. And over the year, they are adapted to hibernating or aestivating, etc. Thus, by avoiding intervening, randomly or regularly occurring unfavourable conditions they experience favourable ones all the time and thus operate deterministically to some extent. By producing many offspring, by increasing the life span of individuals, or by developing a great phenological flexibility, groups of individuals can also operate more deterministically. The same applies to spatial adaptation when individuals track their preferred habitat conditions (Hengeveld and Hemerik, 2002). At all scales their individual chance of survival, because of some biological adaptation, is reduced and may even become predictable to some degree. Biological systems are as much as possible rigid systems; when mechanisms counteracting various sources of decay fail, speciation and adaptive evolution follow as side-effects.

Biochemical mechanisms of organizational conservation

The genetic system arose both as a standard reference (genes as reaction norms) for keeping control over the continually operating, highly interactive metabolic processes, and as an information-storing system for keeping control over their transmission during cell replication. From initial interactions between triplets of nucleic acids and amino acids, fulfilling this reference function, first the more stable RNA arose and from this through dropping oxygen the even more stable DNA (e.g. Berezovsky and Trifonov, in prep.; Miles and Davies, 2000). Thus strands were formed in RNA enhancing stability and operating as mutual reference points. This was kept in the DNA molecule (Maritan et al., 2000). Furthermore, the proteins that at present form the bulk of the ribosome, buttress the ancient ribosomal RNA as its active part in assembling proteins from amino acids since the oldest times, stabilizing them mechanically (see Moore and Steitz, 2002; and Hengeveld and Fedonkin, 2004). Moreover, an elaborate system of repair enzymes guards against the occurrence of molecular lesions in one or both strands of the DNA molecule which, as a macromolecule, is mechanically buttressed by histones. Finally, introns greatly increase exon shuffling in the genome resulting either in duplications or in new combinations in the modular assembly of proteins (protein domains), in principle keeping the basic molecular structure of these domains intact (e.g. Miklos and Campbell, 1994). Thus modules can duplicate or change positions relative to each other. Thereby obtaining different functions (e.g. Miles and Davies, 2000) without structural changes happening that might disrupt the fine-tuned, highly intricate network of biochemical and metabolic functions. Shuffling of sections of the DNA strings can be an important mechanism in speciation (King, 1993). Since many things can still go wrong, a number of checkpoints have to be passed before cell division thus reducing the chance of replication of mistakes.

All these adaptations concern improvements enhancing static biochemical structures in the cell. Structures can also be stabilized dynamically by giving biochemical compounds a particular turnover rate depending on their function. Thus, proteins are produced and broken down at certain rates by yet other proteins, etc. which restrict malfunctioning and enhance metabolic homeostasis.

Still later, with the development of multicellularity in eukaryotic evolution, the modular system of inheritance was further elaborated by a hierarchically set up system of temporally controlled gene expression; this control being differently executed in different parts of the organism. Thus, at different times and places in the development of the organism the same gene and thus one and the same protein could execute different functions, up into the hundreds or thousands. This temporal and spatial organization of gene expression once more allowed phenotypic variation without altering the modules as building blocks (e.g. Carroll, 1997, 2000; Raff, 1996). This modular construction is repeated at the level of organs within multicellular organisms (e.g. Carroll, 2001).

The significance of such a modular mechanism of duplication, reshuffling and modification is that the relatively small number of basic building blocks remained the same, right from the origin of nucleic acids and amino acids, and from the few triplets that these formed (Trifonov, 1999). Similarly, the consequent ribosomes, proteins and basic metabolic processes form a highly conservative system of fine-tuned reactions having hardly changed over the billions of years since their origin (Hengeveld and Fedonkin, 2004). The principle of this set-up is that the construction mechanisms remain the same so that the new components minimally disturb the remainder of the metabolic or organismic system, if at all, contrary to what a succession of mutations may do. This set up has two consequences: it results in 1) great phylogenetic variability with 2) a minimum of structural change. Modularity opened the way of unsurpassed, contingent variation in a basically conservative, if not rigid, at certain levels deterministically operating interactive system. At these levels of organization both sides of the present controversy are tightly integrated.

These are only a few of the many mechanisms known at the cellular level that control the contingency in the highly interactive (and, hence, non-linear) systems that constitute life (e.g. Kolodner *et al.*, 2002). Particularly the number and complexity of these fine-tuned, interacting mechanisms that constitute the whole system, requires an extremely high degree of stability.

Consequently life is conservative at the utmost. The initial systems of metabolism and inheritance still being basically the same as when they originated ca. 3.8 billion years ago (see e.g. Eck and Dayhoff, 1966; Trifonov, 1999). Because of the elaborate and intricate biological control mechanisms inserted in this system in all parts and at all levels of organization against contingent deviations, the outcome of the metabolic and reproductive processes became deterministic.

Speciation as a consequence of organizational conservatism

Despite the minute chance of mistakes still slipping through this tightly integrated, conservative system, its conservative nature was tightened even more by the addition of the process of crossing over during meiosis that developed in the eukaryotes. This resulted in species as natural units of stabilization.

The process concerned implies a continuous insertion of DNA strands from outside the cell, that is, from other independently varying individuals. As the alleles thus originate from two different cells or individuals, each with an independent history, their combination considerably reduces the chance of coincident mistakes occurring in two alleles. Thus, bringing their genomes together results once more in a highly efficient way of checking for any deviations and thus of keeping this source of variation within bounds (see Bernstein and Bernstein, 1997). Moreover, basic to this new way of genomic matching, at the level of the cell and later on at that of the multicellular organism, a system of chemical, behavioural and ecological matching (signalling and recognition) of cells or organisms developed. On top of all previous stabilizing processes, this recognition system resulted in relatively uniform clusters of organisms with more or less the same properties and behaviour, known as species (Paterson, 1985). As such, species inevitably accord with other mechanisms and processes reducing chance variation. Buss (1987) showed that, on top of this, within the multicellular eukaryotes the distinction and spatial separation between germ cells and somatic ones shifted gradually towards the earliest developmental stages of the organism, thus reducing the impact of somatic variation on replication stability. (To my knowledge, whether or not this represents a real trend or a pseudo trend has not been tested so far.)

According to Paterson's (1985) recognition concept, species are, in fact, expected to remain the same under their normal environmental conditions defining a period of evolutionary stasis. The mechanism here consists of three independent processes of stabilizing selection that operate in concert for effective reproduction to take place. These are: 1) a rather precise habitat choice; and 2) timing of the life cycles of the sexes, together enabling the mating partners to meet; and 3) an accurate chemical, morphological and behavioural matching (all matching types falling under the general heading of recognition) of the mating partners and their cells and physiologies. Individuals in which any of these requirements fall short are unlikely to reproduce. Together, they are thus characterized by their specific ecological requirements and biochemical and behavioural adaptations maintained by stabilizing selection operating at the level of the individual organism.

The effectiveness of keeping the variability at the species level at an absolute minimum shows by the feasibility of taxonomy, in contrast to the rapid change shown in mtDNA, which does not take part in the sexual recognition mechanism and results, even at fine scales, in contingent, non-anastomosal and hierarchical variation (e.g. Avise, 2000). However, as soon as genomic interchange stops because of spatially reduced gene flow (allopatry), the individuals of the two or more spatially independent clusters can deviate freely under the local conditions resulting in speciation. Speciation rates could be high given the fact that evolutionary rates appear to be high when measured at the proper time scale (Grant and Grant, 2002; Gingerich, 1983). Thus, periods of stasis are expected to alternate with those of rapid change during periods of spatial fragmentation (Paterson, 1985; see also Vrba, 1985, who elaborated this idea in her turnover-pulse hypothesis), together constituting Eldredge and Gould's (1972) punctuated evolution. Therefore, the stasis periods do not concern the real problem, but those of the breakdown of the stabilization mechanism do. We should therefore analyse the mechanisms of punctuation to find out how these mechanisms occasionally fail to operate during speciation; their failure leads to contingent variation and this to evolutionary radiation.

Thus, variation is not enhanced by biologically functional mechanisms such as mutation or recombination for evolution to take place as, for example, Mayr (1963) proposed. Nor is speciation intended to enhance species diversity through niche differentiation at the ecological level and niche diversity, in turn, through the erection of ecological barriers (Dobzhansky, 1937, 1951, 1970; Hutchinson 1959). Rather, speciation or evolutionary development at large is incidental, being due to small failures of an elaborate control system still squeaking through a specifically conservative system. Living systems are, in fact, distinguished from the non-living part of the world by control mechanisms operating at all levels; they form the essence of life.

Integrating contingency with determinism, and its consequences

According to this model, species as relatively rigid units, adapting to changing conditions in space rather than adapting genetically, track the variation in their environment found at various spatio-temporal scales individualistically rather than as community members (e.g. Hengeveld, 1997;

Jackson and Overpeck, 2000). Moreover as habitat tracking operates through stochastic movements of individual organisms congregating statistically in temporally favourable sites or regions, the resulting "population" is a dynamic entity. It has a certain turnover rate of individuals rather than a spatially static entity with an ecological integrity characterized by demographic attributes like density, birth rate, etc. (Hengeveld and Hemerik, 2002). Their dynamism makes one wonder if an alternative model, that of ecosystem collapse, which is supposed to cause rapid speciation through mass extinction (e.g. Eldredge, 1989, 1991) does occur as an ecologically internal process. And if so, whether supra-individual entities such as communities or ecosystems can exist at all (Hengeveld, 1990; Walter and Paterson 1995).

According to processes internal or inherent to the system, the evolutionary development of life would be gradual and deterministic, rather than capricious and contingent. Alternatively processes external to such communities and operating directly on individual organisms should be held responsible for both evolutionary punctuation and extinction. From this latter, ecologically individualistic perspective the origin of new species is basically contingent and hence incidental rather than law-like and following rules (Hengeveld and Walter, 1999). However, contingent development may be concentrated during periods of punctuation, whereas more conservative and hence more deterministic response processes dominate during those of stasis.

Summary

Treating the juxtaposition of contingency and determinism as unreal and accepting the conservative nature of life thus leads to a very different methodological approach to both evolutionary (Paterson, 1985), as well as to underlying ecological processes (Hengeveld and Walter, 1999; Walter and Hengeveld 2000; Walter and Paterson 1995). Each type of process being based on the ecologically individualistic behaviour of organisms.

Overall our methodological problem is not whether evolutionary processes are either contingent or inevitable. The problem is, in fact, much more interesting than this. It concerns the origin and functioning as well as the occasional failure of all sorts of correction mechanisms existing in living systems and keeping the inevitable contingency in all its interactive processes within bounds. Their failure results in the species either evolving or dying out. We have to integrate these two perspectives into a scientifically more exciting and productive approach.

6.8 CONCLUSION

Can the development of life be considered contingent or deterministic? What does an answer to this question actually solve, what can it add to our understanding, and can it further our research? Both Gould and Conway Morris have an open eye for the processes happening, both within the organism as well as in its environment. But their search seems misdirected seeking ontological generalities. Rather than formulating comprehensive, typifying concepts like the contingency or determinism of phylogenetic development, they should concentrate on the process parameters. These parameters have, in turn, to be represented in specific model structures in order to weigh their relative effects. And the model results should be tested against new observations made according to the functional thinking in present-day science.

What I have tried to show is that two independent, alternative theories of phylogenetic development, one based on contingency and the other on trends of inevitability happening over geological time do not exist. Neither from the viewpoint of scale-dependent variation nor from a methodological one involving statistical testing can their existence be maintained. They should be integrated following a methodologically functional approach, which allows us a better view of phylogenetic phenomena at all levels of organization. Their continued distinction prevents us from solving the most basic processes that have happened during the development of life whereas their integration clarifies our views.

ACKNOWLEDGEMENTS

Here, I would like to thank Hugh Paterson for all the attention he gave to the manuscript of this paper, as well as for the many improvements he suggested.

REFERENCES

Anderson, R. M. and R. M. May (1986). The invasion, persistence and spread of infectious diseases within animal and plant communities. Philosophical Transactions of the Royal Society B314: 533-570.

Avise, J. C. (2000). Phylogeography. Harvard University Press, Cambridge, Massachusetts.

Beatty, J. (1995). The evolutionary contingency thesis. In: Wolters, G. and J. G. Lennox (Eds). Concepts, Theories, and Rationality in the Biological Sciences. Universitätsverlag Konstanz, Konstanz. p. 44-81.

Berezovsky, I. N. and E. N. Trifonov (in prep). Evolutionary aspects of protein structure and folding.

Bernstein, C. and H. Bernstein (1997). Aging, Sex and DNA Repair. Academic Press, New York.

Black, F. L. (1966). Measles endemicity in insular populations: critical community size and its evolutionary implication. Journal of Theoretical Biology 11: 207-211.

Brown, J. H. and B. A. Maurer (1986). Body size, ecological dominance, and Cope's Rule. Nature 324: 248-250.

Buss, L. W. (1987). The Evolution of Individuality. Princeton University Press, Princeton.

Carrier, M. (1995). Evolutionary change and lawlikeness. Beatty on Biological Generalizations. In: Wolters, G. and J. G. Lennox (Eds). Concepts, Theories, and Rationality in the Biological Sciences. Universitätsverlag Konstanz, Konstanz. p. 83-97.

Carroll, R. L. (1997). Patterns and Processes of Vertebrate Evolution. Cambridge University Press, Cambridge.

Carroll, R. L. (2000). Towards a new evolutionary synthesis. Trends in Ecology and Evolution 15: 27-32.

Carroll, S. B. (2001). Chance and necessity: the evolution of morphological complexity and diversity. Nature 409: 1102-1109.

Cliff, A., P. Haggett, J. K. Ord and G. R. Versey (1981). Spatial Diffusion. Cambridge University Press, Cambridge.

Cliff, A., P. Haggett and M. Smallman-Raynor (1993). Measles. Blackwell, Oxford.

Connor, E.F. (1986). Time series analysis of the fossil record. In: Raup, D. M. and D. Jablonski (Eds). Patterns and Processes in the History of Life. Springer, Heidelberg. pp. 119-148.

Conway Morris, S. (1998). The Crucible of Creation. Oxford University Press, Oxford.

Conway Morris, S. (2003). Life's Solution. Inevitable humans in a lonely world. Cambridge University Press, Cambridge.

Conway Morris, S. and S. J. Gould (1998). Showdown on the Burgess Shale. Natural History 107: 48-55.

Dobzhansky, Th. (1937). Genetics and the Origin of Species. Columbia University Press, New York.

Dobzhansky, Th. (1951). Genetics and the Origin of Species. Columbia University Press, New York, 3rd edition.

Dobzhansky, Th. (1970). Genetics and the Origin of Species. 4th edition. Columbia University Press, New York.

Eck, R. V. and M. O. Dayhoff (1966). Evolution of the structure of ferredoxin based on living relics of primitive amino acid sequences. Science 152: 363-366.

Eldredge, N. (1989). Macroevolutionary Dynamics. McGraw-Hill, New York.

Eldredge, N. (1991). The Miner's Canary. Prentice Hall, New York.

Eldredge, N. and S. J. Gould (1972). Punctuated equilibria: an alternative to phyletic gradualism. In: Schopf, T. J. M. (Ed.). Models in Palaeobiology. Freeman and Cooper, San Francisco. pp. 82-115.

Elliott, D. K. (Ed.) (1986). Dynamics of Extinction. Wiley, New York.

Gingerich, P. D. (1983). Rates of evolution: effects of time and temporal scaling. Science 222: 159-161.

Gould, S. J. (1967). Evolutionary patterns in pelycosauran reptiles: a factor-analytic study. Evolution 21: 385-401.

Gould, S. J. (1977). Ontogeny and Phylogeny. Harvard University Press, Cambridge, Massachusetts.

Gould, S. J. (1989). Wonderful Life. Hutchinson Radius, London.

Gould, S. J. (1996). Full House, The spread of Excellence from Plato to Darwin. Harmony Books, New York.

Gould, S. J. (1999). Rocks of Ages: Science and Religion on the Fullness of Life. Ballantine, New York.

Gould, S. J. (2002). The Structure of Evolutionary Theory. Harvard University Press, Cambridge, Massachusetts.

Gould, S. J. and R. C. Lewontin (1979). The spandrels of San Marco and the Panglossian paradigm: a critique of the adaptationist programme. Proceedings of the Royal Society of London B 205: 581-589.

Gould, S. J., D. M. Raup, J. J. Sepkoski, T. J. M. Schopf and D. S. Simberloff (1977). The shape of evolution: a comparison of real and random clades. Paleobiology 3: 23-40.

Grant P. R. and B. R. Grant (2002). Unpredictable evolution in a 30-year study of Darwin's finches. Science 296: 707-711.

Hardin, G. (1960). The competitive exclusion principle. Science 131: 1292-1297.

Hengeveld, R. (1989). Dynamics of Biological Invasions. Chapman and Hall, London.

Hengeveld, R. (1990). Dynamic Biogeography. Cambridge University Press, Cambridge.

Hengeveld, R. (1993). What to do about the North American invasion by the Collared Dove? American Field Ornithologist 64: 477-489.

Hengeveld, R. (1994). Biogeographical ecology. Journal of Biogeography 21: 341-351.

Hengeveld, R. (1997). Impact of biogeography on a population-biological paradigm shift. Journal of Biogeography 24: 541-547.

Hengeveld, R. (1999). Modelling de impact of biological invasions. In: Sandlund, O. T., P. J. Schei and A. Viken (Eds). Invasive Species and Biodiversity Management. Kluwer, Dordrecht. pp. 127-138.

Hengeveld, R. (2002). Methodology going astray in population biology. Acta Biotheoretica 50: 77-93.

Hengeveld, R. (2004). Book Review: Conway Morris' Inevitable Solution. Acta Biotheoretica 52: in press.

Hengeveld, R. and M. A. Fedonkin (2004). Causes and consequences of eukaryotization through mutualistic endosymbiosis and compartmentalization. Acta Biotheoretica 52: 105-154.

Hengeveld, R. and L. Hemerik (2002). Biogeography and dispersal. In: Bullock, J., R. E. Kenward and R. S. Hails (Eds). Dispersal Ecology. Blackwell Publishing, Oxford. pp. 303-324.

Hengeveld, R. and F. Van den Bosch (1997). Invading into an ecologically non-uniform area. In: Huntley, B., W. Cramer, A. V. Morgan, H. C. Prentice and J. R. M. Allen (Eds). Past and Future Rapid Environmental Changes. Springer, Berlin. pp. 217-227.

Hengeveld, R. and G. H. Walter (1999). The two co-existing ecological paradigms. Acta Biotheoretica 47: 141-170.

Hutchinson, G. E. (1959). Homage to Santa Rosalia, or why are there so many species? American Naturalist 93: 145-159.

Infantosi, A. F. C. (1986). Interpretation of case studies in two communicable diseases using pattern analysis techniques. PhD Thesis, University of London.

Jackson, S. T. and J. T. Overpeck (2000). Responses of plant populations and communities to environmental changes of the late Quaternary. Paleobiology 26 (Supplement): 194-220.

King, M. (1993). Species Evolution. Cambridge University Press, Cambridge.

Kolodner, R. D., C. D. Putnam and K. Myung (2002). Maintenance of genome stability in *Saccharomyces cerevisiae*. Science 297: 552-557.

Kuhn, T. (1962). The Structure of Scientific Revolutions, University of Chicago Press, Chicago.

Lande, R. (1976). Natural selection and random drift in phenotypic evolution. Evolution 30: 314-334.

Lewontin, R. (2000). The Triple Helix. Harvard University Press, Cambridge, Massachusetts.

Mandelbrot, B. B. (1977). The Fractal Geometry of Nature. Freeman, New York.

Maritan, A., C. Micheletti, A. Trovato, and R. B. Banavar (2000). Optimal shapes of compact strings. Nature 406: 287-290.

May, R. M. (1976). Simple mathematical models with very complicated dynamics. Nature 261: 459-467.

Mayr, E. (1963). Animal Species and Evolution. Harvard University Press, Cambridge Massachusetts.

McShea, D. W. (1993). Evolutionary change in the morphological complexity in the mammalian vertebral column. Evolution 47: 730-740.

McShea, D. W. (1994). Mechanisms of large-scale evolutionary trends. Evolution 48: 1747-1763.

Miklos, G. L. G. and K. S. W. Campbell (1994). From protein domains to extinct phyla: Reverse-engineering approaches to the evolution of biological complexities. In: Bengtson, S. (Ed.). Early Life on Earth. Columbia University Press, New York. pp. 501-516.

Miles, E. W. and D. R. Davies (2000). On the ancestry of barrels. Science 289: 1490.

Mitchell, J. M. (1976). An overview of climatic variability and its causal mechanisms. Quaternary Research 6: 481-493.

Mollison, D. (1977). Spatial contact models for ecological and epidemic spread. Journal of the Royal Statistical Society B 39: 283-326.

Monod, J. (1971). Chance and Necessity. Knopf, New York.

Moore, P. B. and T. A. Steitz (2002). The involvement of RNA in ribosome function. Nature 418: 229-235.

Paterson, H. E. H. (1985). The recognition concept of species. In: Vrba, E. (Ed.). Species and Speciation. Transvaal Museum, Pretoria. pp. 21-29.

Raff, R. A. (1996). The Shape of Life. University of Chicago Press, Chicago.

Raup, D. M. (1986). The Nemesis Affair. Norton, New York.

Raup, D. M. (1991). Extinction: Bad Genes or Bad Luck? Norton, New York.

Raup, D. M. and D. Jablonski (Eds) (1986). Patterns and Processes in the History of Life. Springer, Heidelberg.

Raup, D. M., S. J. Gould, T. J. M. Schopf and D. S. Simberloff (1973). Stochastic models of phylogeny and the evolution of diversity. Journal of Geology 81: 525-542.

Rose, G. (1992). The Strategy of Preventive Medicine. Oxford University Press, Oxford.

Rosenberg, A. (1985). The Structure of Biological Science. Cambridge University Press, Cambridge.

Sayers, B. McA., B. G. Mansourian, T. Phan Tan and K. Bögel (1977). A pattern analysis study of a wildlife rabies epizootic. Medical Informatics 2: 11-34.

Schröpfer, R. and C. Engstfeld (1983). Die Ausbreitung des Bisams (*Ondatra zibethicus* Linne,1977, Rodentia, Arvicolidae) in der Bundesrepublik Deutschland. Zeitschrift für angewandte Zoologie 70: 13-37.

Simpson, G. G. (1953). The Major Features of Evolution. Columbia University Press, New York.

Stanley, S. M. (1987). Extinction. Freeman, New York.

Trifonov, E. N. (1999). Elucidating sequence codes: three codes for evolution. Annals of the New York Academy of Science 870: 330-338.

Van den Bosch, F., R. Hengeveld and J. A. J. Metz (1992). Analysing the velocity of animal range expansion. Journal of Biogeography 19: 135-150.

Van Peursen, C. A. (1970). De Strategie van de Cultuur. Elsevier, Amsterdam.

Van Valen, L. (1973). A new evolutionary law. Evolution Theory 1: 1-30.

Vermeij, G. J. (1987). Evolution and Escalation. Princeton University Press, Princeton.

Vrba, E. S. (1985). Environment and evolution: alternative causes of the temporal distribution of evolutionary events. South African Journal of Science 81: 229-236.

Walter, G. H. and R. Hengeveld (2000). The structure of the two ecological paradigms. Acta Biotheoretica 48: 15-46.

Walter, G. H. and H. E. H. Paterson (1995). Levels of understanding in ecology: interspecific competition and community ecology. Australian Journal of Ecology 20: 463-466.

Willis, K. J. and J. C. McElwain (2002). The Evolution of Plants. Oxford University Press, Oxford.

Wright, S. (1955). Classification of the factors of evolution. Cold Spring Harbor Symposia of Quantitative Biology 20: 16-24D.

R. Hengeveld
Department of Ecotoxicology, Vrije Universiteit Amsterdam

7

The Symbiontic Nature of Metabolic Evolution

S. A. L. M. Kooijman and R. Hengeveld

ABSTRACT

We discuss evolutionary aspects of metabolism, right from the beginning of life to the present day at various levels of organization, thereby including quantitative aspects on the basis of the Dynamic Energy Budget (DEB) theory. We propose a scheme for the evolution of the central metabolism with archaeal as well as eubacterial roots. After an extended initial phase of prokaryotic diversification, cycles of exchange of metabolites between partners in a symbiosis, integration of partners into new individuals and new specializations led to forms of symbiosis of various intensity ranging from loosely living together in species aggregates to several forms of endosymbiosis. While the prokaryotic metabolism evolved into a considerable chemical diversity, the eukaryotic metabolic design remained qualitatively the same but shows a large organizational diversity. Homeostasis of biomass evolved, introducing stoichiometric constraints on production and excretion of products that can be re-utilized; carbohydrates and inorganic nitrogen being the most important ones. This stimulates the formation of symbioses, since most are based on syntrophy, which is probably the basis of the huge biodiversity. A remarkable property of DEB theory for metabolic organization is that organisms of two species that exchange products, and thereby follow the DEB rules, can together follow a symbiogenic route such that the symbiosis behaves as a new organism that itself follows the DEB rules. This property of the reserve dynamics in the DEB theory also explains a possible evolutionary route to homeostasis. The reserve dynamics in DEB theory also plays a key role in linking the kinetics of metabolic pathways to needs of metabolites at the cellular level. Moreover, reserve kinetics, in combination with other DEB elements, explains how metabolic performance depends on body size and why such relationships work out differently within and between species. Apart from the key role of reserves, the dynamic interaction between surface areas and volumes is a basic feature of the DEB theory at all levels of organization (molecules, individuals, ecosystems). The explicit mass and energy balances of the DEB theory facilitates ecosystem modelling as it depends on nutrient exchange. The theoretical interest in this topic concerns the huge range in space-time scales that is involved in understanding the significance of the actions of life within the context of metabolic organization.

T.A.C. Reydon and L. Hemerik.,(eds.), Current Themes in Theoretical Biology,
159-202.

7.1 INTRODUCTION

Underlying the metabolic organization in individuals is a long evolutionary history of acquisition and loss of new metabolic pathways, as well as a recombination of existing pathways. The boundaries of individuals are frequently crossed in symbioses that span the full range from loosely coupled populations, to a fully integrated individual that is hard to recognize as a consortium of individuals of different species. The metabolic requirements of life can be energetic ones, or they can concern particular nutrients, or both. A proper understanding of metabolic organization cannot be achieved without exploring its historic roots.

The metabolism of individuals has adapted over time to overcome the consequences of changing living conditions. The question here is how this might have happened in interaction with the environment. One possibility is through changing the system itself by mutation and selection. This is a very slow process, but essential for building up a basic diversity in metabolic performance between different species. This explains the slow start of evolution. Much faster is the exchange of plasmids that evolved among prokaryotes (Doolittle, 1999), which is further accelerated by the process of symbiogenesis, typical for eukaryotes. The latter also duplicate DNA and reshuffle parts of their genome, giving adaptive change even more acceleration. Mutation still continues, of course, but the reshuffling of metabolic modules occurs at rates several orders of magnitude higher. The response to changes in the environment is further accelerated by the development of food webs, and therefore of predation, which enhances selection. Owing to their advanced locomotory and sensory systems, animals play an important role in food webs, and so in the acceleration of evolutionary change.

Basic to these processes is the question of how pliable complex metabolic systems are, i.e. how much they can be dropped, added, or altered without harm? How much of the initial structures are kept right from the beginning or from stages developed soon afterwards? Or, should symbiogenesis be understood in terms, not of changeability of the systems but in principle from those of their rigidity?

Aim

The aim of the present paper is to integrate existing ideas on quantitative aspects of symbiotic interactions based on syntrophic relationships (Kooijman et al., 2004) with ideas on the chemical evolution of metabolism (Hengeveld and Fedonkin, 2004). The topics that we discuss are widely scattered in the specialized literature. By bringing them together into one framework, we hope to stimulate an important field of research that crosses the traditional boundaries between various specializations. We believe that barriers to communication between molecular biology, physiology, microbiology, population biology, ecosystem ecology, and earth systems science hamper the development of a quantitative theory for metabolism at the various levels of organization.

Our view is that interactions among species can frequently be understood from their metabolic requirements, and that quantitative aspects of the metabolism of individuals can be understood from interactions between larger biochemical modules, in ways that are not too different from those between individuals. This also holds for systems of metabolically interacting species. Yet, basic differences exist between the various levels of organization. At the level of the individual, metabolic performance is studied as a dynamic system, *given* the concentrations of substrates, nutrients and/or food. At the ecosystem level such concentrations are not given but are part of the dynamic system that evolves interactively, which naturally leads to the study of nutrient cycles at this level.

When life first emerged, its quantitative impact on the environment cannot have been substantial. It need not have taken long, though, before considerable amounts of biomass built up to such levels that could have affected geochemical cycling. Precambrian cyanobacteria (stromatolites) in coastal areas testify to an increasing impact of life on its environmental conditions. Geochemical cycling cannot be studied without considering climate (temperature and water), substantially affecting (metabolic) rates, which makes it such that metabolism at larger spatial and temporal scales cannot be studied without involving climate and biogeochemical recycling in a holistic way (Kooijman, 2004).

We first present a brief introduction to the central metabolism of eukaryotes and then discuss its evolutionary history, starting with the first cells, the invention of phototrophy, diversification and interaction. So far, our discussion concerns prokaryotic evolution at the sub-organismic level that resulted in a substantial *chemical* diversity. After this, we consider the emergence of eukaryotes and their *organizational* diversity in the form of multicellularity, and the various direct and indirect syntrophic interactions at the supra-organismic level. Finally, we discuss quantitative aspects of metabolic organization.

Context of DEB theory

Every now and then, we will refer to the Dynamic Energy Budget (DEB) theory for quantitative aspects of the metabolic organization at the level of the individual (Kooijman, 2000, 2001; Nisbet *et al.*, 2000). These references not only serve to point to opportunities of understanding particular aspects of evolution quantitatively, but also to demonstrate that the DEB theory has significance for understanding evolutionary processes. A property that sets this theory apart from its (presently available) alternatives is the decomposition of biomass into reserves and structure and the special type of kinetics of the reserve, which quantify the metabolic memory of the system.

Perhaps contrary to what the term suggests, reserves are *not* characterized by "compounds set apart for later use"; their constituent compounds can have quite active metabolic functions. Each particular compound can belong to both reserve and structure. Most ribosomal RNA, for example, belong to the reserve, which implies that the rRNA content of the body increases with the growth rate (Elser, 2004), since abundant reserve comes with a large use of reserve (Vrede *et al.*, 2004). The dynamics of rRNA as part of the reserve is parsimonious given the role of rRNA in the elongation of peptides; if growth is low, there is little need for peptide elongation so less need for rRNA. The co-variation of reserve density and growth rate only holds for single-reserve systems and for the (most) limiting reserve in multiple-reserve systems. Non-limiting reserves typically show the opposite pattern of being more abundant for low growth rates.

7.2 THE CENTRAL METABOLIC PATHWAY

The idea that eukaryotes developed out of prokaryote assemblages, with or without a hypothetical "Urkaryote", is now widely accepted. Their evolutionary history implies that the metabolism of eukaryotic cells arose from several interacting prokaryotic modules. We can only hope to understand eukaryotic organization from that of prokaryotic ancestors, plus an appreciation of the interaction between the modules of the evolving eukaryotic cell. Therefore, let us focus on the organization of the central metabolic pathway first.

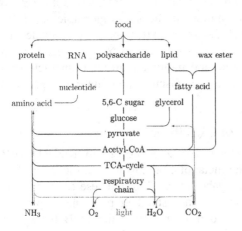

Figure 7.1. The very much simplified design of the central metabolism of eukaryotes and many prokaryotes, in which the Pentose Phosphate cycle, the glycolysis, the TriCarboxylic Acid cycle and the respiratory chain have a central position in the conversion of polymers. Heterotrophs use food (organic compounds) as a source for energy and building blocks; photo-autotrophs use light and nutrients for these purposes.

Four modules of central metabolism

The central metabolic pathway of many prokaryotes and almost all eukaryotes (Figure 7.1) consists of four main modules:

- The *Pentose Phosphate (PP) Cycle* comprises a series of extra-mitochondrial transformations by which glucose-6-phosphate is oxidized with the formation of carbon dioxide, reduced NADP and ribulose 5-phosphate. Some of this latter compound is subsequently transformed to sugar phosphates with 3 to 7 or 8 carbon atoms, whereby glucose-6-phosphate is regenerated. Some ribulose 5-phosphate is also used in the synthesis of nucleotides and amino acids. Higher plants can use the same enzymes also in reverse, thus running the reductive pentose phosphate cycle. The PP cycle is primarily used to interconvert sugars as a source of precursor metabolites and to produce reductive power. Theoretical combinatorial optimization analysis indicated that the number of steps in the PP cycle is evolutionarily minimized (Meléndez-Hevia and Isidoro, 1985; Meléndez-Hevia, 1990), which maximizes the flux capacity (Heinrich and Schuster, 1996; Waddell *et al.*, 1997).

- The *Glycolytic Pathway* (aerobically) converts glucose-6-phosphate to pyruvate or (anaerobically) to lactate, ethanol or glycerol, with the formation of 2 ATP. The transformations occur extra-mitochondrially in the free cytoplasm. However, in kinetoplastids they are localized in an organelle, the glycosome, which is probably homologous to the peroxisome of other organisms (Bakker, 1998; Cavalier-Smith, 2002b). The flux through this pathway is under control by phospho fructokinase and by hormones. Heinrich and Schuster (1996) studied some design aspects of the glycolytic pathway. Most pyruvate is converted to acetyl and bound to coenzyme A.

- The *TriCarboxylic Acid (TCA) Cycle,* also known as the citric acid or the Krebs cycle, oxidises (without the use of dioxygen) the acetyl group of acetyl coenzyme A to two carbon dioxide molecules, under the reduction of 4 molecules NAD(P) to NAD(P)H. In eukaryotes that contain them, these transformations occur within their mitochondria. Some plants and micro-organisms have a variant of the TCA cycle, the glyoxylate cycle, which converts pyruvate to glyoxylate and to malate (hence a carbohydrate) with another pyruvate. Since pyruvate can also be obtained from fatty acids, this route is used for converting fatty acids originating from lipids into carbohydrates. Some plants possess the enzymes of the glyoxylate cycle in specialized organelles, the glyoxysomes.

- The *Respiratory Chain* oxidizes the reduced coenzyme NAD(P)H, and succinate with dioxygen, which leads to ATP formation through oxidative phosphorylation. Similarly to the TCA cycle it occurs inside mitochondria. Amitochondriate eukaryotes process pyruvate through pyruvate-ferredoxin oxidoreductase rather than through the pyruvate dehydrogenase complex. If the species can live anaerobically, the respiratory chain can use fumarate, nitrate, or nitrite as electron acceptors in the absence of dioxygen (Tielens *et al.*, 2002).

In combination with nutrients (phosphates, sulphates, ammonia, iron oxides, etc), the first three pathways of the central metabolic pathway provide almost all the essential cellular building blocks, including proteins, lipids, and RNA. The universality of this central metabolic pathway is partly superficial or, if you like, the result of convergent evolution because the enzymes running it can differ substantially. This diversity in enzymes partly results from the modular make-up of the enzymes themselves. Some variation occurs in the intermediary metabolites as well.

The central role of carbohydrates

Obviously, glucose plays a pivotal role in the central metabolism. However, its accumulation as a monomer for providing a metabolism with a permanent source of substrate would give all sorts of problems, such as osmotic ones. This also applies to metabolic products. To solve these problems, cells typically store the supplies in polymeric form (polyglucose (i.e. glycogen), starch, polyhydroxyalkanoate, polyphosphate, sulphur, proteins, RNA), which are osmotically neutral. Their storage involves so-called inclusion bodies, the inherent solid/liquid interface of which controlling their utilization dynamics.

Quantitative aspects of metabolism

The understanding of the quantitative aspects of the central metabolism calls for kinetic modelling that is based on the availability of the interface between essential polymers and the cytosol, rather than of their amounts or concentrations. The concept of concentration hardly applies to polymers as a basis for kinetics. It is also very problematic for low concentrations of monomers given the complex spatial structure of a cell in which membrane-linked transformations dominate. Transporter proteins cause further deviations from the law of mass action on which classic enzyme kinetics is based. This is why DEB theory uses an alternative for classic enzyme kinetics as it is based on fluxes rather than concentrations. This alternative, synthesizing unit kinetics, is used to quantify simultaneous limitations and adaptations in assimilation, maintenance and growth (Kooijman, 1998, 2000; Kooijman *et al.*, 2004; Brandt, 2002; Kuijper *et al.*, 2003). It can deal with the dynamic interactions between surface areas and volume at all levels of organization. These interactions are a basic feature of the DEB theory

Almost all metabolites have a dual function as building blocks or as an energy source. It seems that reserves are required for modelling the regulation of these functions, where some of the enzyme molecules are part of the reserve, whereas the growth rate depends on the amount of reserve (Kooijman and Segel, 2003). Pyruvate that is sent to mitochondria in eukaryotes, for instance, is partly used to generate ATP and reducing power, and partly for the synthesis of the intermediary metabolites of the TCA cycle, e.g. succinate and fumarate. The nine different enzymes of the TCA cycle are spatially organized in a super-macromolecule and the interaction between these enzymes controls the fate of intermediary metabolites. The problem is that the ratio of the cell's requirements for building blocks *versus* energy depends on the growth rate. So, the need for products and intermediary metabolites depends on the growth rate. If the growth rate varies, the amounts of enzymes vary in a very special way. As shown by the application of a model for pathway kinetics that is based on synthesizing units, this can have the effect that the varying metabolic needs at the cellular level are exactly matched. Without reserves, so without the possibility of varying enzyme concentrations, it will be difficult, if not impossible, to deal with these varying needs in a theoretically satisfactory way.

7.3 HOW DID METABOLIC SYSTEMS EVOLVE?

Since the central metabolic pathway involves the operation of a large number of enzymes, its evolution must have taken many steps. Dioxygen was rare, if not absent, during the time life emerged on earth which classifies the respiratory chain as an advanced feature. We doubt that glucose could have been that central during the remote evolutionary origins of life, since its synthesis and degradation typically involves dioxygen. Early life forms must probably be sought among the anaerobic chemolithoautotrophic bacteria (Wächtershäuser, 1988). Like all phototrophic eukaryotes, most of these bacteria fix inorganic carbon in the form of carbon dioxide through the Calvin cycle. At present, this cycle is part of the phototropic machinery, a rather advanced feature in metabolic evolution which is not found in any archaea (Schönheit and Schafer, 1995). It has glucose as its main product, which suggests that the central position of glucose and, therefore, of carbohydrates, evolved only after oxygenic phototrophy evolved. Like the Calvin cycle, eukaryotic and eubacterial glycolysis (the Embden-Meyerhof pathway) is not found in archaea either; hyperthermophilic archaea possess the Embden-Meyerhof pathway in modified form (Schönheit and Schafer, 1995; Selig *et al.*, 1997), and generally do not use the same enzymes (Martin and Russell, 2003). This places the pyruvate processing TCA cycle at the origin of the central metabolism. However, if we leave out the glycolysis as a pyruvate-generating device, what process was generating pyruvate?

Early cells

Interestingly, the eubacteria *Hydrogenobacter thermophilus* and *Aquifex* use the TCA cycle in reverse, binding and transforming CO_2 into building blocks (lipids, cf. Lengeler *et al.*, 1999), including pyruvate. Both species are *Knallgas* bacteria, extracting energy from the oxidation of dihydrogen. The green sulphur bacterium *Chlorobium,* as well as the archaea *Sulfolobus* and *Thermoproteus* (Madigan *et al.*, 2000) also run the TCA cycle in reverse for generating building blocks. Hartman (1975), Wächtershäuser (1990) and Morowitz *et al.* (2000) hypothesized the reverse TCA cycle to be one of the first biochemical pathways.

The interest in hydrogen bacteria relates to the most likely energy source for the first cells on earth. *Hydrogenobacter* optimally thrives at 70-75°C in Japanese hot springs. It is an aerobic bacterium, using ammonia and nitrate, but not nitrite and possesses organelles (mesosomes). Several enzymes of the PP cycle and the glycolytic pathway are present although their activities are low (Staley *et al.*, 1989). The togobacterium *Aquifex* is even more interesting since its metabolism might still resemble that of an early cell. Although it is also aerobic, it tolerates only very low dioxygen concentrations, which may have been present when life emerged (Holland, 1994; Kasting, 2001; Anbar and Knoll, 2002). Growing optimally at 85°C in marine thermal vents, it utilizes H_2, S^0 or $S_2O_3^-$ as electron donors and O_2 or NO_3^- as electron acceptors. With a genome size of only 1.55 Mbp, its genome amounts to only one third of that of *E. coli,* which is really small for a non-parasitic prokaryote. The archaeon *Nanoarchaeum equitans*, which lives symbiotically with the H_2-producing and sulphur-reducing archaeon *Ignicoccus,* has a genome size of 0.5 Mbp (Huber *et al.*, 2002), one of the smallest known genomes for a non-parasitic bacterium. The phototrophic cyanobacterium *Prochlorococcus* has 1.7 Mbp (Fuhrman, 2003). These small genome sizes illustrate that autotrophy is metabolically not more complex than heterotrophy (see Discussion section).

The TCA cycle seems to be remarkably efficient, which explains its evolutionary stability. Moreover, it is reversible, which directly relates to its efficiency and the inherent small steps in chemical potential between subsequent metabolites. Yet, with its nine transformations, the TCA cycle is already rather complex and must have been preceded by simpler CO_2-binding pathways (Orgel, 1998, 2000) such as the (linear) acetyl-CoA pathway of homoacetogens: $2 CO_2 + 4 H_2 + CoASH \quad \rightarrow \quad CH_3COSCoA + 3 H_2O$ (Hugenholtz and Ljungdahl, 1990; Ljungdahl, 1994). Apart from H_2, electron donors for acetogenesis include a variety of organic and C_1-compounds. Coenzyme A, which plays an important role in the TCA cycle, is a ribonucleotide and the main substrate for the synthesis of lipids, a remembrance of the early RNA world (Stryer, 1988). Several eubacteria and archaebacteria employ the acetyl-CoA pathway; they include autotrophic homoacetogenic and sulphate-reducing bacteria, methanogens, *Closterium*, *Acetobacterium*, and others. The RNA-world is generally thought to predate the protein/DNA-world. RNA originally catalyzed all cellular transformations; protein evolved later to support RNA in this role. Many protein enzymes still have RNA-based cofactors (e.g. ribosomes and spliceozomes), while RNA still has catalytic functions. DNA evolved as a chemically more stable archive for RNA, probably in direct connection with the evolution of proteins. The step from the RNA to the protein/DNA world came with a need for the regulation of transcription.

The hyperthermophilic methanogens, such as *Methanococcus*, *Methanobacterium* or *Methanopyrus*, have also been proposed as contemporary models for early cells (Lindahl and Chang, 2001); they have the acetyl-CoA pathway, which they run in both the oxidative and the reductive direction (Simpson and Whitman, 1993). Like *Aquifex*, they are thermophilic and taxonomically close to the archaea/eubacteria fork (eukaryotes have some properties of both roots), have a small genome (*Methanococcus jannaschii* has 1.66 Mbp, coding for only 1700 genes), and they utilize H_2 as electron donor.

Intermezzo: Before the first cells

A possible exergonic process generating energy in the initial stages of life involves the formation of makinawite crusts at the interface of mildly oxydizing, iron-rich acidulous ocean water above basaltic floors from which alkaline seepages arose (e.g. Russell *et al.*, 1994). These crusts consist of FeS layers allowing free electron flow from the reducing environment beneath, generated by the activation of hydrothermal hydrogen. Thus, energy was constantly supplied which, moreover, could easily be tapped at the steep gradient formed by the crust. FeS can spontaneously form cell-like structures on a solid surface (Russell and Hall, 1997, 2002; Boyce *et al.*, 1983; Cairns-Smith *et al.*, 1992), and has a high affinity for the ATP ingredients organophosphates and formaldehyde (Rickard *et al.*, 2001), which can form ribulose (see Bengtson, 1994: 81). The released energy could stimulate the formation of larger molecules at each inner surface, such as phosphorus or nitrogen compounds. The chemically labile energy-rich inorganic pyrophosphate compounds could have served as energy-transferring molecules (Baltscheffsky, 1996; Baltscheffsky *et al.*, 1999), whereas the nitrogen-containing molecules on the inner surface of the crust could have developed into nucleic acids or, later, into larger peptides. Of these, the peptides, in turn, could have combined with iron and sulphur complexes in the crust, thus initiating the formation of ferredoxins, or they could have nested themselves within the crust, thus forming the second step in the formation of membranes (Russell and Hall, 2002).

The membranes of membrane-bound vesicles are at the basis of transformations typical for life (Segré *et al.*, 2001). Membranes need membranes (plus genes) for propagation; genes only are not enough (Cavalier-Smith, 2000). Strong arguments in favour of the hypothesis "cells before metabolism" include the abiotic abundance of amphiphilic compounds (even on arriving meteorites), the self-organization of these compounds into membranes and vesicles, and their catalytic properties (Deamer and Pashley, 1989). This argument only works if amphiphilic compounds tend to accumulate in very specific micro-environments; otherwise they will be too dilute. The modifications of substrates that are taken up from the environment to compounds that function in metabolism were initially probably small and gradually became substantial. Compartmentalization is essential for the accumulation of metabolites and for any significant metabolism. Norris and Raine (1998) suggest that the RNA world succeeded the lipid world, which is unlikely because the archaebacterial lipids consist of isoprenoid ethers, while eubacterial lipids consist of fatty acids (acyl esters) with completely different enzymes involved in their turnover (Kates, 1979; Kandler, 1998; Wächtershäuser, 1988). Lipids were probably synthesized first from pyruvate, the end product of the acetyl-CoA pathway and the reverse TCA cycle, before the extensive use of carbohydrates.

Koga *et al.* (1998) hypothesized that the eubacterial taxa made the transition from non-cellular ancestors to cellular forms independently from the archaebacteria (see also Martin and Russell, 2003). This seems unlikely, however, because they are similar in the organization of their genes (e.g. in operons) and genomes, and in their transcription and translation machinery (Olsen and Woese, 1996; Cavalier-Smith, 1998). Eubacteria do have a unique DNA replicase and replication initiator proteins however. These properties apply especially to cells, rather than to pre-cellularly existing forms, and are complex enough to make it very unlikely that they evolved twice. Woese (2002) hypothesized that lateral gene transfer could have been intense in proto-cells with a simple organization; diversification through Darwinian mutation and selection could only occur after a given stage in complexity had been reached, that is when lateral gene transfer could have been much less intense. The eubacteria, archaebacteria and eukaryotes would have crossed this stage independently. Since all eukaryotes once seem to have possessed mitochondria (Roger, 1999; Gupta, 1998; Keeling, 1998; Embley and Hirt, 1998), this origin is unlikely for them. Cavalier-Smith (2002a) argued that archaebacteria and eukaryotes evolved in parallel from eubacteria since about 850 Ma ago, and that eukaryotes have many properties in common with actinomycetes. However the differences in, for example, lipid metabolism and many other properties between eubacteria and archaebacteria are difficult to explain in this way. Moreover, carbon isotope differences between carbonates and organic matter of 2.8-2.2 Ga ago are attributed to archaean methanotrophs

(Knoll, 2003). Although so far the topic remains speculative, a separate existence of eubacteria and archaebacteria before the initiation of the lipid metabolism and before the origin of eukaryotes through symbiogenesis with mitochondria seems to be the least-problematic sequence explaining metabolic properties among these three taxa.

A hypothetical energy-generating scheme involving the consumption of dihydrogen and sulphur is based on the overall exergonic reaction $FeS + S \rightarrow FeS_2$ (Taylor et al., 1979; Wächtershäuser, 1988; Madigan et al., 2000).

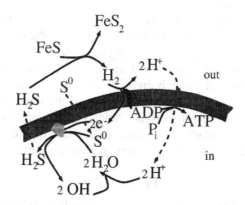

Figure 7.2. A possible early ATP generating transformation, based on pyrite formation, that requires a membrane and three types of enzyme: proto-hydrogenase, proto-ATP-ase and S^0-reductase; modified from (Madigan et al., 2000). Sulphur has to be imported in exchange for H_2S.

The scheme of Figure 7.2 may have applied to the initial cellular life forms because of the availability of the substrates in the deep ocean (van Dover, 2000), and few enzymes are required. Keefe et al. (1995) however, argue that the oxidation of FeS gives insufficient energy to fix carbon dioxide through the inverse TCA cycle. Yet, this fixation may have occurred along other pathways using accumulated ATP. Schoonen et al. (1999) demonstrated that the energy of this reaction diminishes sharply at higher temperatures. Contrary to pyrite, greigite ($Fe_5Ni_6S_8$) has structural moieties that are similar to the active centres of certain metallo-enzymes, as well as to electron transfer agents (see, for example, Russell and Hall, 2002), and catalizes the transformation $2 CO_2 + CH_3SH + 8 [H] \rightarrow CH_3COSCH_3 + 3 H_2O$.

Concerning homeostatic membranes, transformations of substrates and products, occurring in the enclosed vesicle and catalyzed by membrane-bound enzymes, depend on the size of the vesicle, that is on the amount of enzyme proportional to the amount of membrane and therefore to the surface area of the cell. The transformation rate involves the ratio of surface area to volume, which constitutes a measure of length. The change in this ratio naturally leads to the cell cycle, that is a cyclic pattern in the metabolism of the cell, and represents one of the cornerstones of the DEB theory. This theory implies that the turnover rate of reserve density, that is the ratio of the amounts of reserve and structure, is inversely proportional to a length measure in isomorphs, i.e. organisms that do not change in shape when they grow. The crucial parameter in reserve turnover, the energy conductance with the dimension of length per unit of time, testifies to the basic role of surface area-volume interactions in metabolic rate control.

A natural implication of the reversal of the TCA cycle is that the direction of glycolysis was initially reversed as well, and served to synthesize building blocks for e.g. carbohydrates. Comparing the carbohydrate metabolism among various bacterial taxa, Romano and Conway (1996) concluded that originally glycolysis must indeed have been reversed. Thus, the reversed glycolytic pathway probably developed as an extension of the reversed TCA cycle, and they both reversed to their present standard direction upon linking to the Calvin cycle, which produces glucose in a phototrophic process. So, what could have been the evolutionary history of photoptrophy?

Phototrophy

Phototrophy developed early in evolution; some workers even think that it has been present right at the origin of life (Woese, 1979; Cavalier-Smith, 1987b; Hartman, 1998; Blankenship and Hartman, 1992, 1998). In an anoxic atmosphere, and therefore without ozone, UV damage must have been an important problem for the early phototrophs though and protection and repair mechanisms against UV damage must have evolved in parallel with phototrophy (Dillon and Castenholz, 1999). The green non-sulphur bacterium *Chloroflexus* probably resembles the earliest phototrophs and is unique in lacking the Calvin cycle, as well as the reverse TCA cycle. In the hydroxypropionate pathway, it reduces two CO_2 to glyoxylate, using many enzymes also found in the thermophilic non-phototrophic archaeon *Acidianus*. Its photoreaction centre is similar to that of purple bacteria. The reductive dicarboxylic acid cycle of *Chloroflexus* is thought to have evolved into the reductive tricarboxylic acid cycle as found in *Chlorobium*, and further into the reductive pentose phosphate cycle, which is, in fact, the Calvin cycle (Hartman, 1998).

Like sulphur and iron-oxidizing chemolithotrophs, aerobic nitrifying bacteria use the Calvin cycle for fixing CO_2. The substrate of the first transformation of the monophosphate pathway for oxidizing C_1-compounds, such as methane, is very similar to the C_1-acceptor of the Calvin cycle, which suggests a common evolutionary root of these pathways (Madigan *et al.*, 2000). The first enzyme in the Calvin cycle, RubisCO is present in most chemolithotrophs and phototrophs and even in some hyperthermophilic archaea. It is the only enzyme of the Calvin cycle of which (some of) the code is found on the genome of chloroplasts. The enzymes that are involved in the Calvin cycle show a substantial diversity among organisms and each has its own rather complex evolutionary history (Martin and Schnarrenberger, 1997). This complicates the finding of its evolutionary roots (see Figure 7.3).

The thermophilic bacterium *Chlorobium tepidum* has a reverse TCA cycle and a RubisCO-like gene. In combination with the observations mentioned above, this suggests that the present central glucose-based metabolism evolved when the Calvin cycle became functional in CO_2 binding, and the glycolysis and the TCA cycle reversed to their present standard direction, operating as a glucose and pyruvate processing devices, respectively (see Figure 7.3).

Most phototrophs use the Calvin cycle for fixing CO_2 in their cytosol in combination with a pigment system in their membrane for capturing photons. Archaea use a low-efficient retinal-protein and are unable to sustain true autotrophic growth; five of the 11 eubacterial phyla have phototrophy. Bacterio-chlorophyll in green sulphur bacteria is located in chlorosomes, organelles bound by a non-unit membrane, attached to the cytoplasmic membrane.

Green non-sulphur and purple bacteria utilize photosystem (PS) II; green sulphur and Gram-positive bacteria utilize PS I, whereas cyanobacteria (including the prochlorophytes) utilize both PS I and II (Zubay, 2000). The cyanobacterium *Oscillatoria limnetica* can utilize their PS I and II in conjunction, thus being able to split water and to produce dioxygen. In the presence of H_2S as an electron donor, it uses only PS I, an ability pointing to the anoxic origin of photosynthesis. This anoxic origin appears to be ancient (Xiong *et al.*, 2000). Oxygenic photosynthesis is a complex process that requires the co-ordinated translocation of four electrons. It evolved more than 2.7 Ga ago (Bjerrum and Canfield, 2002). Based on the observation that bicarbonate serves as an efficient alternative for water as an electron donor, Dismukes *et al.* (2001) suggested the following evolutionary sequence for oxygenic photosynthesis, starting from green non-sulphur bacterial phytosynthesis that uses organic substrates as electron donor:

Electron Donation	Pigment	Reaction Centre	Photo-synthesis
Oxalate $\rightarrow$ Oxalate$^+$	BChl-a		Anoxygenic
$Mn_2(HCO_3)_4 \rightarrow Mn_2(HCO_3)_4^+$	BChl-a		Anoxygenic
$2\,HCO_3^- \rightarrow O_2 + 2\,CO_2 + 2\,H^+$	BChl-g	$Mn_4O_x(HCO_3)_y$	Oxygenic
$2\,H_2O \rightarrow O_2 + 4\,H^+$	Chl-a	$CaMn_4O_x(HCO_3)_yY_z$	Oxygenic

The phototrophic machinery eventually allowed the evolution of the respiratory chain (the oxidative phosphorylation chain), which uses dioxygen that is formed as a waste product of photosynthesis, as well as the same enzymes in reversed order. If the respiratory chain initially used sulphate, for example, rather than dioxygen as electron acceptor, it could well have evolved simultaneously with the phototrophic system.

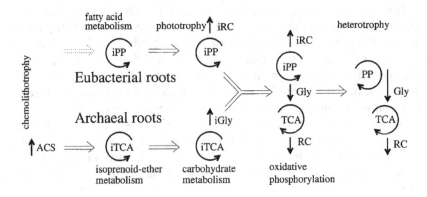

Figure 7.3. Evolution of the central metabolism among prokaryotes that formed the basis of eukaryotic organization of the central metabolism. ACS = acetyl-CoA Synthase pathway, iPP = inverse Pentose Phosphate cycle (= Calvin cycle), PP = Pentose Phosphate cycle, iTCA = inverse TriCarboxylic Acid cycle, TCA = TriCarboxylic Acid cycle (= Krebs cycle), iGly = inverse Glycolysis, Gly = Glycolysis, iRC = inverse Respiratory Chain, RC = Respiratory Chain. The arrows indicate the directions of synthesis to visualize where they reversed. All four main components of eukaryote's heterotrophic central metabolism originally ran in the reverse direction to store energy and to synthesize metabolites.

Figure 7.3 summarizes the broad pattern of the possible evolution of the central metabolism as it took place in prokaryotes and that formed the basis for the eukaryotes. It implies considerable conjugational exchange between the archaea and eubacteria, but given the long evolutionary history, such exchanges might have been very rare. The exchange must have been predated by a symbiontic coexistence of archaea and eubacteria to tune their very different metabolic systems. The production of dioxygen during phototrophy, which predates the oxidative phosphorylation, changed the earth (e.g. Dismukes *et al.*, 2001; Lane, 2002).

The availability of a large amount of energy and reducing power effectively removed energy limitations; primary production in terrestrial environments is mainly water-limited, that in aquatic environments nutrient-limited. This does not imply however, that the energetic aspects of metabolism could not be quantified usefully; energy conservation also applies in situations where the energy supply is not rate-limiting.

Nutrients may have run short of supplies because of oxidation by dioxygen; this would have slowed down the rate of evolution (Anbar and Knoll, 2002). First, sulphur precipitated out, followed by iron and towards the end of the Precambrian by phosphate and, since the Cambrium revolution, by calcium as well. Also, under aerobic conditions, nitrogen fixation became difficult, which makes biologically required nitrogen unavailable, despite its continued great abundance of dinitrogen in the environment (see Bengtson, 1994: 41).

Since the Calvin cycle produces fructose 6-phosphate, those autotrophic prokaryotes possessing this cycle are likely to have a glucose-based metabolism. Indeed, the presence of glucose usually suppresses all autotrophic activity. Several obligate chemolithotrophic prokaryotes, such as sulphur-oxidizers, nitrifiers, cyanobacteria and prochlorophytes contain this cycle in specialized organelles, the carboxysomes, which are tightly packed with RubisCO. Facultative autotrophs, like purple anoxyphototrophs, use the Calvin cycle for fixing CO_2, although they lack the carboxysomes.

Diversification and interactions

The prokaryotes as a group evolved a wide variety of abilities for the processing of substrates, whilst remaining rather specialized as species (e.g. Amend and Shock, 2001). The nitrogen cycle in Figure 7.4 illustrates this variety, as well as the fact that the products of one group are the substrate of another.

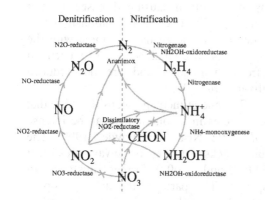

Figure 7.4. Conversions of inorganic nitrogen species by prokaryotes. The compound CHON stands for biomass. Modified from Schalk (2000).

Some of the conversions of inorganic nitrogen species can only be done by a few taxa. The recently discovered anaerobic oxidation of ammonia is only known from the planctobacterium *Brocadia anammoxidans* (Schalk, 2000) (nonetheless, it might be responsible for the removal of one-half to one-third of the global nitrogen in the deep oceans (Dalsgaard *et al.*, 2003)); the aerobic oxidation of ammonia to nitrite is only known from *Nitrosomonas*, the oxidation of nitrite into nitrate is only known from *Nitrobacter*; and the fixation of dinitrogen can only be done by a few taxa, such as some cyanobacteria, *Azotobacter*, *Azospirillum*, *Azorhizobium*, *Klebsiella*, *Rhizobium,* and some other ones (Sprent, 1987).

If the composition of structural mass, i.e. a combination of proteins, lipids, carbohydrates, etc., does not change too much, we have stoichiometric constraints on growth. These constraints are revealed when the nutrient concentrations in the environment change relative to each other. The DEB theory holds that growth happens at the expense of reserves rather than at that of nutrients in the environment. Also, nutrient uptake is a function of the nutrient concentration in the environment and the amount of structural mass only and is not a function of the amount of reserve. The consequence is that (some of) the utilized reserves that are not immediately used for maintenance or growth must be excreted in one form or another, which links homeostasis to excretion (see, for example, Smith and Underwood, 2000). The excretion of polysaccharides (carbohydrates) and other organic products by nutrient-limited photosynthesizers (such as cyanobacteria), stimulated heterotrophs to decompose these compounds through the anaerobically operating glycolytic pathway. Thus, other organisms came to use these excreted species-specific compounds as resources, and a huge biodiversity resulted.

Apart from the use of each other's products, prokaryotes, such as the proteobacteria *Bdellovibrio* and *Daptobacter,* invented predation on other prokaryotes. When the eukaryotes emerged, many more prokaryote species turned to predation, with transitions to parasitism causing diseases in their eukaryotic hosts. Predators typically have a fully functional metabolism, while parasites use building blocks from the host, reducing their genome with the codes for synthesizing these building blocks. The smallest genomes occur in viruses which probably evolved from their hosts and are not reduced organisms (Hendrix *et al.,* 1999; Sullivan *et al.,* 2003).

Prokaryotic mats on intertidal mud flats and at methane seeps illustrate that the exchange of metabolites between species in a community can be intense (van den Berg, 1998; Michaelis *et al.,* 2002; Nisbet and Fowler, 1999). The occurrence of multi-species microbial flocks, such as in sewage treatment plants (Brandt and Kooijman, 2000; Brandt, 2002) further illustrates an exchange of metabolites among species. The partners in such syntrophic relationships sometimes live epibiotically, possibly to facilitate exchange. Internalization further enhances such exchange (Kooijman *et al.*, 2003). The gradual transition of substitutable substrate to become complementary is basic to the formation of obligate syntrophic relationships. The mathematical framework for such a smooth transition is discussed in Brandt *et al.* (2003) and in Kooijman *et al.* (2003).

7.4 THE EMERGENCE OF THE EUKARYOTES

Symbiontic origins of mitochondria

Eukaryotes may have emerged from the internalization of a fermenting, facultative anaerobic H_2- and CO_2-producing eubacterium into an autotrophic, obligatory anaerobic H_2- and CO_2-consuming methanogenic archaebacterium (Martin and Mueller, 1998), the host possibly returning organic metabolites (see Figure 7.5). Once the H_2-production and consumption had been cut out of the metabolism, aerobic environments became available, where the respiratory chain of the symbiont kept the dioxygen concentration in the hosts' cytoplasm at very low levels. The internalization of (pro)mitochondria might be a response to counter the toxic effects of dioxygen. This hypothesis for the origin of eukaryotes explains why the DNA replication and repair proteins of eukaryotes resemble that of archaea, and not that of eubacteria. Notice that the eukaryotization, as schematized in Figure 7.5, just represents a recombination and compartmentation of existing modules of the central metabolism (cf. Figure 7.3).

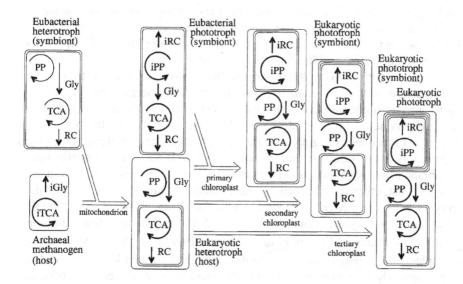

Figure 7.5. Scheme of symbiogenesis events; the first two primary inclusions of prokaryotes (to become mitochondria and chloroplasts respectively) were followed by secondary and tertiary inclusions of eukaryotes. Each inclusion comes with a transfer of metabolic functions to the host. The loss of endosymbionts is not illustrated. See Figure 7.3 for the meaning of the codes for the modules of the central metabolism and for the ancestors of the mitochondria and chloroplasts.

It can be shown that such forms of syntrophy can easily lead to homeostatic assemblages, where the relative abundance of the partners become independent of variations of the primary resources in the environment (Kooijman *et al.*, 2003). Moreover, it can also be shown that this merging of initially independently living populations, each following the rules of the DEB theory, can be such that the integrated assemblage again follows these rules (Kooijman *et al.*, 2003). This remarkable property poses stringent constraints on reserve dynamics which the DEB model appears to satisfy. Given the common occurrence of symbiogenesis in evolutionary history, this property is required for any model that is not species-specific. Most (if any) alternative models will not have this property which makes them species-specific. Syntrophic associations between methanogens and hydrogenosomes are still abundant; ciliates can have methanogens as endosymbionts and interact in the exchange (Fenchel and Finlay, 1995).

Much discussion exists about which metabolites may have been exchanged between the pro-mitochondrial symbionts and their hosts; some workers believe that both were aerobic heterotrophs, although they do not give clues about the nature of the compounds being exchanged (e.g. Kurland and Andersson, 2000). Part of the problem is that mitochondria and hosts exchanged quite a few genes, and the genome of mitochondria reduced considerably, down to 1% of its original bacterial genome (Fenchel, 2002). The mitochondrial DNA in kinetoplasts, however, is amplified and can form a network of catenated circular molecules (Lee *et al.*, 2000).

Cavalier-Smith (1987a, 2002b) argued that eukaryotes descend from some actinobacterium that engulfed a phototrophic posibacterium (an α-proteobacterium) as mitochondrion, which later lost phototrophy, and used it as a slave to produce ATP. The ability to phagotise is central to his reasoning. Actomyosin mediates phagocytosis and actinobacteria have proteins somewhat related to myosin, although they do not phagotise. If he is right that the outer membrane of mitochondria is derived from the original posibacterium, and not from the host, there is little need for the existence of phagocytosis prior to the entry of a posibacterium to become a mitochondrion. At least one example exists of prokaryotic endosymbiosis (β-proteobacteria that harbour γ-proteobacteria, von Dohlen *et al.*, 2001) in absence of phagocytosis. More examples exist of penetration through the membrane without killing the victim instantaneously (e.g. Guerrero, 1991). His present view, shared by others, is that it happened only once and the logical implication is just in a single individual. If phagocytosis would have been well established prior to the entry of a mitochondrion, it is hard to understand why it did not occur more frequently. It seems more likely that eukaryotic membrane transport (with applications in phagocytosis), the cytoskeleton (with applications in cilia) and the Endoplasmatic Reticulum (ER, including the nuclear envelope) became operational somewhere between the entries of mitochondria and chloroplasts, which do have host-derived envelopes. The origin of eukaryotes is possibly some 1.5 Ga (Knoll, 2003) or 2.0 Ga (Raven and Yin, 1998) or 2.7 Ga (Brocks *et al.*, 1999) ago. The rhodophytes were among the first eukaryotes having chloroplasts; their fossil record goes back to 1.2 Ga (Knoll, 2003) ago.

We agree with Cavalier-Smith on the need to understand the evolution of phagocytosis which is still enigmatic. A weak element in his reasoning is that phagocytotic entry was prior to enslavement to produce ATP for the host. The development of exchange systems for metabolites doubtlessly took many generations, while the endosymbiosis must have been operational right from the moment of penetration into the host cell for (more or less) co-ordinated cell growth and duplication. We cannot see how this is possible without a prior (epibiontic) existence of a syntrophic relationship between host and symbiont (Kooijman *et al.*, 2004). Moreover, the relationship between mitochondria and their host is much more complex than the delivery of ATP in exchange for pyruvate, ADP and P from the host. Kooijman and Segel (2003) argue that the delivery of intermediary metabolites by the mitochondria is at least as essential for the host.

No eukaryotes are known with plastids but are lacking mitochondria, which suggests that possessing mitochondria was compulsory for cyanobacteria to move in into the symbiotic relationship. Genes that moved from mitochondria to the genome of their host reveal that some eukaryotes (also) lost their mitochondria. As mentioned before, recent studies suggest that all eukaryotes once possessed mitochondria (Simpson and Roger, 2002; Stechmann and Cavalier-Smith, 2002), despite the many taxa that presently lack mitochondria.

The amitochondriate pelobiont *Pelomyxa palustris* has intracellular methanogenic bacteria that may have comparable functions. Other members of the α-group of purple bacteria (from which the mitochondria arose; Andersson *et al.*, 1998) such as *Agrobacterium* and *Rhizobium*, can also live inside cells, and usually function in dinitrogen fixation; *Rickettias* became parasites, using their hosts' building blocks and reducing their own genome to viral proportions.

Symbiontic origins of chloroplasts

The process of internalization of a cyanobacterium of uncertain phylogenetic origin probably occurred only once in eukaryotic history (Delwiche, 1999; McFadden, 2001; Cavalier-Smith, 2002a), where the plastids of glaucophytes retained most of their genome and properties, whereas that of rhodophytes and chlorophytes became progressively reduced by transfer of thousands of genes to the nucleus (Martin *et al.*, 2002) and by gene loss. Secondary endosymbioses of red algae occurred in cryptophytes, haptophytes, heterokonts, dinoflagellates and apicomplexans and those of green algae occurred in euglenoids and chlorarachniophytes. The presence of plastids in the parasitic Kinetoplastids and of cyanobacterial genes in the heterotrophic percolozoans (= Heterolobosea) suggests that secondary endosymbiosis did not take place in the euglenoids, but much earlier in the common ancestor of all excavates, where chloroplasts became lost in the percolozoans (Andersson and Roger, 2002). Alveolates (including dinoflagellates and ciliates) have a more dynamic association with plastids. Even weaker associations evolved between phototrophic dinoflagellates and chlorophytes on the one hand and heterotrophs on the other, such as fungi (lichens), foraminiferans, radiolarians, and animals (sponges, coelenterates, molluscs, platyhelmintes).

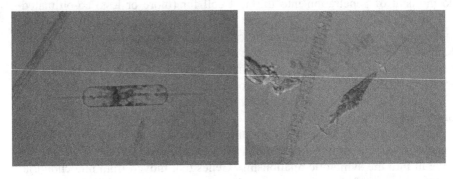

Figure 7.6. Chloroplasts of the marine diatom *Ditylum brightwellii* disperse at low light levels, and aggregate at high ones. They move in a co-ordinated way.

The intra-cellular dynamics of mitochondria and plastids is still poorly known (Osteryoung and Nunnari, 2003). Growth and division are usually only linked to the cell cycle. Mitochondria move actively through the cell and can easily fuse with each other (Kooijman *et al.,* 2003), in yeasts and chlorophytes even forming networks. Their numbers can range from a single one to many depending on species and conditions. In some algae, the single mitochondrion can cyclically divide into many small ones and fuse to a single one again. Moreover, the host cell can kill mitochondria and lysosomes can decompose the remains. Likewise chloroplasts can move through the cell sometimes in a co-ordinated way (see Figure 7.6). They can reversibly lose their chlorophyll and fulfil non-photosynthetic tasks, which are permanent in the kinetoplasts (e.g. the endoparasite *Tripanosoma*) and in heterotrophic plants (*Triurdaceae,* some *Orchidaceae, Burmanniaceae,* prothallium-stage of Lycopods and Ophioglossids), in parasitic plants (*Orobanchaceae, Rafflesiaceae, Balanophoraceae,* some *Convolvulaceae*), and in predatory plants (some *Lentibulariaceae*), for instance. (This list of exclusively heterotrophic plants suggests that heterotrophy might be more important among plants than is generally recognized.) Eukaryotes also had to master the control of transmission of mitochondrial and chloroplast genomes. Most use a system in meiosis where these genomes come from a single parent; stochastic models for mitotic genome segregation seem to be most effective (Birky, 2001).

Features unique to eukaryotes

Eukaryotes have many properties not known from prokaryotes, which challenges the view that they are "simply" prokaryotic chimaeras. An example is the production of clathrin, a protein which plays a key role in the invagination of membranes such as during endocytosis. We are just beginning to understand the complex processes involved in membrane deformation (Bigay *et al.*, 2003). No prokaryote seems to be able to form vesicles, while membrane transport (including phagocytosis and pinocytosis, vesicle mediated transport) is basic in eukaryotes (de Duve, 1984; Gruenberg, 2001), and essential for endosymbiotic relationships. Today, only a single endosymbiotic relationship among prokaryotes is known (von Dohlen *et al.*, 2001), but the endosymbionts are probably not surrounded by a membrane of the hosts. Eukaryotes also have ATP-fuelled cytoplasmatic mobility driven by myosin and dynein.

Another example of a property unknown in prokaryotes is the vacuole (Leigh and Sanders, 1997), which is used for storing nutrients in ionic form and carbohydrates; sucrose, a precursor of many other soluble carbohydrates, typically occurs in vacuoles. This organelle probably evolved to solve osmotic problems that came with storing substrates. The storage of water in vacuoles allowed plants to invade the terrestrial environment; almost all other organisms depend on plants in this environment. The DEB theory predicts that the storage capacity of energy and building-blocks scales with volumetric length to the power of four; since eukaryotic cells are generally larger than prokaryotic ones, storage becomes more important to them. Diatoms typically have extremely large vacuoles, which occupy more than 95% of the cell volume, allowing for a very large surface area (the outer membrane, where the carriers for nutrient uptake are located), relative to their structural mass that requires maintenance. In some species, the large chloroplast wraps around the vacuole like a blanket. Since, according to the DEB theory, reserve does not require maintenance, the large ratio of surface area to structural volume explains why diatoms are ecologically so successful, and also why they are the first group of phytoplankton to appear each spring. Archaebacteria and posibacteria do have gas vacuoles but their function is totally different from that of eukaryotic vacuoles.

The Golgi apparatus, a special set of flat, staked vesicles, called dictyosomes, develops after cell division from the endoplasmatic reticulum. They appear and disappear repeatedly in the amitochondriate metamonad *Giardia*. The nuclear envelope can disappear in part of the cell cycle in some eukaryotic taxa and it is also formed by the endoplasmatic reticulum. The amitochondriate parabasalid *Trichomonas* does not have a nuclear envelope while the planctobacterium *Gemmata oscuriglobus* has one. The possession of a nucleus itself is therefore not a basic requisite distinguishing between prokaryotes and eukaryotes. The situation is quite a bit more complex than molecular biology textbooks suggest; e.g. the macronuclei (sometimes more than one) in ciliates are involved in metabolism, while the micronuclei deal with sexual recombination.

Although some prokaryotic cells, such as the planctobacteria, are packed with membranes, eukaryotic cells are generally more compartmentalized, both morphologically and functionally. Compounds can be essential in one compartment, and toxic in another (Martin and Schnarrenberger, 1997). Eukaryotic cilia differ in structure from the prokaryotic flagella, and are therefore called undulipodia to underline the difference (Margulis, 1970). The microtubular cytoskeleton of eukaryotes is possibly derived from protein constricting the prokaryotic cell membrane during fission, as both use the protein tubulin (van den Ent *et al.*, 2001).

Another feature particular to eukaryotes concerns the organization of their genome into chromosomes (Chela-Flores, 1998), with a spindle machinery for genome allocation to daughter cells and telomerase guide RNA. Chromosomes are linked to the evolution of reproduction, which includes cell-to-cell recognition, sexuality and mating systems. Moreover, many eukaryotes have haploid as well as diploid life stages and two or more (fungi, rhodophytes) sexes (Kirkpatrick, 1993). Although reproduction may seem to have little relevance to metabolism at the level of the individual, metabolic rates at the population level depend on the amount of biomass and, hence, on rates of propagation. Eukaryotes also have a unique DNA topoisomerase I, which is not related to type II topoisomerase of the archaea (Forterre *et al.*, 1996) which further questions their origins.

Despite all their properties, the eukaryotic genome size can be small; the genome size of the acidophilic rhodophyte *Cyanidoschyzon* is 8 Mbp, only double the genome size of *E. coli* (Chela-Flores, 1998); the chlorophyte *Ostreococcus tauri* has a genome of only 10 Mbp, and the yeast *Saccaromyces cerivisiae* of 12 Mbp (Derelle *et al.*, 2002).

This list of metabolic differences between prokaryotes and eukaryotes largely concerns biochemical and morphological ones. The following sections will focus on their organizational differences: multicellularity and syntrophy.

7.5 MULTICELLULARITY AND BODY SIZE

Multicellularity evolved many times in evolutionary history, even among the prokaryotes but particularly among the eukaryotes. It allows a specialization of cells to particular functions, and the exchange of products is inherently linked to specialization. Think, for instance, of filamental chains of cells in cyanobacteria where heterocysts specialize in N_2 fixation. To this end, specialization requires adaptations for the exclusion of dioxygen and the production of nitrogenase. The existence of dinitrogen-fixation unicellular cyanobacteria shows that all metabolic functions can be combined within a single cell, which is remarkable as its photosynthesis produces dioxygen, inhibiting dinitrogen fixation. A temporal separation of the processes solves the problem, but restricts dinitrogen fixation during darkness; specialization can be more efficient under certain conditions. The mixobacterium *Chondromyces* and the proteobacteria *Stigmatella* and *Mixococcus* have life cycles that remind us of those of cellular slime moulds, involving a multicellular stage, whereas acetinobacteria, such as *Streptomyces* resemble fungal mycelia (e.g. Dworkin, 1985).

Pathogens, such as viruses can kill individual cells without killing the whole organism, which is an important feature of multicellularity, and is basic to the evolution of defence systems.

Cell differentiation is minor in poriferans, reversible in coelenterates and plants, and irreversible in vertebrates. The number of cells of one organism very much depends on the species, and can be up to 10^{17} in whales (Rizzotti, 2000), which requires advanced communication. Many larger organisms, including opisthokonts (fungi plus animals), tracheophytes, rhodophytes and phaeophytes, evolved elaborate transport systems to facilitate exchange of metabolites among the cells and with the environment. Animals evolved advanced locomotory abilities, which require accurate co-ordination by a nervous system. This latter system not only took tasks in information exchange and processing but also in metabolic regulation. Animals also evolved an immune system which supplements chemical defences to fight pathogens.

Differentiation and cellular communication

Multicellularity has many implications. Cells can be organized into tissues and organs, which gives metabolic differentiation once more an extra dimension. It comes with a need for regulation of the processes of growth and apoptosis of cells in tissues (Rothenberg and Jan, 2003), in which communication between cells plays an important role. Animals (from cnidarians to chordates) use gap junctions between cells of the same tissue, where a family of proteins called connexins form tissue-specific communication channels. They appear early in embryonic development (in the eight-cell-stage in mammals) and are used for nutrient exchange, cell regulation, conduction of electrical impulses, development and differentiation. Together with the nervous and endocrine systems, gap junctions serve to synchronize and integrate activities. When cell-to-cell communication systems fail, tumours can develop; only a small fraction of tumours result from DNA damage (van Leeuwen and Zonneveld, 2001). Plants use plasmodesmata to interconnect cells, which are tubular extensions of the plasma membrane of 40-50 nm in diameter, that traverse the cell wall and interconnect the cytoplasm of adjacent cells into a symplast. Higher fungi form threads of multi-nucleated syncytia, known as mycelia; sometimes septa are present in the hyphae, but they have large pores. Otherwise the cells of fungi only communicate via the extracellular matrix (Moore, 1998). Rhodophytes have elaborate pit connections between the cells (Dixon, 1973), which have a diameter in the range 0.2-40 μm, filled with a plug that projects in the cytoplasm on either side. Ascomycetes and Basidiomycetes have similar pit connections, but lack the plug structure and the cytoplasm is directly connected, unlike the situation in rhodophytes.

The cell-individual-population continuum

The boundaries between cells, individuals, colonies, societies and populations are not sharp at all. Fungal mycelia can cover up to 15 hectares as in the basiodiomycete *Armillaria bulbosa*, but they can also fragment easily. Cellular slime moulds (dictyostelids) have a single-celled free-living amoeboid stage, as well as a multicellular one; the cell boundaries dissolve in the multicellular stage of acellular slime moulds (eumycetozoa), which can now creep as a multi-nucleated plasmodium over the soil surface. The mycetozoans are not the only amoebas with multi-nuclear stages; *Mastigamoeba* (a pelobiont) is another example (Bernard *et al.,* 2000). Many other taxa also evolved multi-nucleated cells, plasmodia or stages, e.g. ciliates, Xenophyophores, Actinophryids, Biomyxa, Loukozoans, Diplomonads, Gymnosphaerida, Haplosporids, Microsporidia, Nephridiophagids, Nucleariidae, Plasmodiophorids, Pseudospora, Xanthophyta (e.g. Vaucheria), most classes of Chlorophyta (Chlorophyceae, Ulvophyceae, Charophyceae (in mature cells) and all Cladophoryceae, Bryopsidophyceae and Dasycladophyceae)) (Patterson, 1999; van den Hoek *et al.,* 1995); the Paramyxea have cells inside cells. Certain plants, such as grasses and sedges, can form runners that give off many sprouts and cover substantial surface areas; sometimes, these runners remain functional in transporting and storing resources such as tubers, whereas in other cases they soon disintegrate. A similar situation can be found in, for example, corals and bryozoans where the tiny polyps can exchange resources through stolons. Behavioural differentiation between individuals, such as between those in syphonophorans, invites one to consider the whole colony an integrated individual, whereas the differentiation in colonial insects and mammals is still that loose that it is recognized as a group of co-ordinated individuals.

These examples illustrate the vague boundaries of multicellularity and even those of individuality. A sharpening of definitions or concepts may reduce the number of transition cases to some extent, but this cannot hide the fact that we are dealing here with a continuum of metabolic integration in the twilight-zone between individuals and populations. This illustrates that organisms, and especially eukaryotes, need each other metabolically.

The implications of body size and shape

Although some individual cells can become quite large, with inherent consequences for physiological design and metabolic performance (Hope and Walker, 1975; Raven and Brownlee, 2001), multicellularity can also lead to really large body sizes. According to the DEB theory, the ultimate body size is determined by the ratio of the assimilation flux, coupled to its surface area, and the maintenance requirements, which are coupled to body volume. Surface area is proportional to volume$^{2/3}$ in isomorphs (organisms that do not change in shape during growth), which explains why maximum body size has an upper boundary, even in the presence of abundant food; an insight that goes back to A. R. Wallace in 1865 and results in a von Bertalanffy growth curve at constant food density. This does not hold for e.g. V1-morphs, where surface area is proportional to volume^{1}; they can really grow to large sizes, as shown by some fungal mycelia, which can cover some 15 hectares (Smith et al., 1992). Their growth curve is typically exponential at constant food density. Growing crusts, such as lichens on a rocky surface, can be conceived as a dynamic mixture between a V1-morph (the outer annulus) and a V0-morph (the centre), where the surface area is proportional to volume^{0} (i.e. constant). The implication being that the diameter of a crust grows linearly in time at constant food density. These seemingly different growth patterns demonstrate their common feature: the significance of surface area-volume relationships at the individual level; we already discussed these relationships at the molecular level.

The DEB theory implies that the inter-specific storage capacity for reserves of organisms is proportional to the (maximum) structure's volumetric length to the power of four (Kooijman, 1986, 2000). Thus, the physiological condition of large organisms, therefore, follows environmental changes only slowly; maximum starvation times increase with body length. Body size has a bearing on the transport of material (feeding, respiration, excretion); moreover, body size and capacity for metabolic memory are interdependent. Many physiological properties are linked to transport rates and storage capacity. A classic topic of this concerns the respiration rate, also known as the metabolic rate, which is less than proportional to the weight of an organism. According to the DEB theory, this scaling is because reserves do not require maintenance costs and body weight has contributions of structural mass and of reserve; the latter contributions are relatively more important for large-bodied species. Freshly-produced eggs or seeds beautifully illustrate that reserves do not require maintenance; they consist entirely of reserve and hardly respire. Structure is growing at the expense of reserve, which explains why total embryonic mass decreases, while its respiration rate increases. Once the feeding process is initiated at birth, we see the reversed pattern: respiration rate is increasing with body mass.

The basic difference between reserve and structure in the context of the DEB theory is in the turnover of their compounds. All compounds in the reserve have the same turnover rate, which is inversely proportional to body length in isomorphs, so the turnover rate decreases during growth. Compounds in the structure can have compound-specific turnover rates, independent of the size of the organism, due to the somatic maintenance efforts. Since large-bodied species have relatively more reserves, more of the compounds in their body have the same (low) turnover time.

Physical arguments indicate that energy costs for movement (walking, swimming, flying) scale with surface area, while energy investment scales with volume (mass). Mean travelling distance, therefore, scales with length. This implies that the diameter of the home range scales with length. The fact that maximum starvation times scale with length and feeding rate with surface area all match beautifully. In an ecosystem with many organisms of widely different body sizes, the smaller organisms live at other space-time scales than the larger ones, which has profound consequences for ecosystem dynamics as well as the stability of species diversity. Small organisms can live in locally homogeneous environments whereas large ones cannot. As a result, for instance, we have to quantify resource availability for bacteria and micro plankters in terms of concentrations (amounts per volume), but for large herbivores (such as cows) in terms of amounts per surface area. This has profound consequences for ecosystem modelling.

DEB theory makes a fundamental distinction between intra- and inter-species scaling relationships. When a young (small) organism is compared with a large (fully grown) con-specific, its specific respiration rate will be higher, just as expected for inter-species comparisons, but for a totally different reason: Overhead costs of growth make the difference. Fully grown mice and elephants, to the contrary, both do not grow.

Temperature

Temperature directly affects metabolic rates, expressed quantitatively by the Arrhenius relationship. In a variety of species, large body sizes have also led to the control of body temperature for accelerating metabolic functions, which is impossible for unicellular organisms. Some *Arum* species can elevate the temperature of their flowers metabolically, which helps volatile smells to escape and thus to attract insects and to ripen the fruit. Also, in preparation for flying, many insects warm up their body to allow them to generate enough energy (Heinrich, 1993); they generate heat metabolically, as well as by their movements. Similarly, tuna fish can increase their body temperature up to ten degrees above that of seawater; whereas birds and mammals are well known for regulating their body temperature, often within narrowly defined boundaries. Behavioural mechanisms for controlling body temperature are known in many animals and some plants (such as in the mountain avens *Dryas*). The regulation of body temperature is just another example of achievements made possible by multicellularity.

From supply to demand systems

By feeding on organisms, which body composition varies to a limited extent only, animals achieved a relatively high level of homeostasis; the dynamics of their metabolism is well described by the DEB theory using a single reserve (consisting of a generalized compound, which is a mixture of many compounds). A small group of animals, mainly birds and mammals, pushed the condition of a (relatively) constant internal environment one step further by using extensively neuronal and hormonal regulation systems that allowed them to become more independent of their environment. These systems are probably intimately linked to their high degree of endothermy. Organisms can be ranked on a scale from supply to demand systems, where supply systems are characterized by "eating what is available" (constrained by the food processing capacity, of course), and demand systems by "eating according to the needs" (where the needs are more or less "pre-programmed"). Organisms excluding animals and sea anemonies are examples at the supply end of the spectrum and birds and mammals at the demand end; other animals taking an intermediate position. Supply systems typically have a much larger ratio between the maximum feeding rate and the minimum one to stay alive; they can easily cease growth in response to food shortage without adverse effects on their health. Demand systems typically suffer from serious health problems if food intake no longer covers growth needs. The maximum feeding rate of demand systems can be regulated to temporary needs; up-regulation occurs prior to migration, and during egg synthesis or pregnancy and lactation. (Down-regulation, such as during hibernation and even deeper forms of turpor, follows more complex patterns in the supply-demand spectrum.) Demand systems also tune their diet more finely to their needs than supply systems. While the adult chicken does well on seed, for instance, the young and fast-growing chicken favours protein-rich insects for food.

The reserve dynamics of the DEB theory is special in its capacity to reduce the number of reserves from many (the native state of supply systems) down to one (the evolutionary advanced state of demand systems) in a smooth way, using incremental changes of parameter values. The mechanism is the same as for the integration of a syntrophic symbiosis to a single system as discussed earlier. The stepwise integration of a metabolic system is of a much wider significance than for symbiogenesis only.

Respiration and feeding

The processes of aging develop gradually in multicellular organisms, rather than binary for individual cells (Kooijman, 2000). Free radicals, which play an important role in aging, might be used to accelerate changes in the genome across generations (Kooijman, 2000). Since free radical generation is linked to respiration, and that to food intake, the processes of aging and tumour induction in organisms with irreversible cell differentiation have intimate links with energetics, and so with body size (van Leeuwen and Zonneveld, 2001; van Leeuwen et al., 2002). High food intake levels generally reduce life spans, although the patterns are complex in detail. An important function of sleeping in animals seems to be in the repair of neuronal damage by free radicals (Siegel, 2003). The time allocated to sleeping among animal species of different body size does follow the pattern of metabolic rate per body mass, which means that small-bodied species take more sleeping time. This observation links sleeping to energetics and aging.

The DEB theory uses conservation of time to quantify feeding rate as a function of food density in the environment. In its simplest formulation the maximum feeding rate follows from the time required to process food items; this not only includes the mechanical handling but also digestion and further metabolic transformation to reserve(s). The switching between searching and handling causes food intake to depend hyperbolically on the rate of encounters with food items under a wide range of "details" for the various sub-processes. From a more abstract point of view the feeding process has a lot in common with the mechanism of enzyme kinetics. The conservation of time argument can be used to account for behavioural modification of this feeding process, where time allocation to social interaction (such as territorial defence or sexual behaviour) and sleeping reduce the time allocated to searching for food. Since time allocated to sleeping relates to food intake, as discussed, rather complex patterns in feeding rates can emerge. The details of the feeding process are further complicated by diurnal cycles and the synchronization of these cycles between predator prey species.

7.6 SYNTROPHY

Product formation and excretion are basic to metabolism. The dynamic difference in the context of the DEB theory between product formation and the excretion of unusable reserves is that product formation is linked with fixed weighting coefficients to the basic fluxes of assimilation, maintenance, and growth, whereas the excretion of unusable reserves depends on the amounts of reserves, relative to structure, which can vary much more dynamically over time. The reason for reserves becoming unusable so that they must be excreted, is to be found in homeostatic relationships; excretion must occur when the product has a fixed composition and requires substrates in fixed proportions, whereas the proportions in arriving substrates actually vary. The DEB theory does not distinguish between waste products (e.g. faeces production which is associated with assimilation only and urine production which also has contributions from growth and maintenance) and other products (such as penicillin production by some fungi and secondary metabolite production by plants), since waste products also can have vital functions for the organism, while the functions of some products are not always clear.

In syntrophic relationships one organism lives off the products and/or excretions of another one. However, excretion can also be toxic to some other organism, such as the nitrogen-containing domoic acid excreted by the diatom *Pseudonitzschia*, which is neurotoxic to most fish. Such an excretion particularly occurs when the silicon reserve of *Pseudonitzschias* becomes depleted, causing the diatom to get rid of its nitrogen reserve, as quantified by the DEB theory. Another type of toxic interactions occurs in bacteria, which first transform readily degradable organic compounds into acetate, resulting in a lowering of the pH which, in turn, has a negative effect on competing species. When only acetates are left they use these as a substrate.

In the next two sections, we first discuss syntrophic interactions between autotrophs and heterotrophs, which both evolved from specialization of mixotrophs, after which we concentrate on aspects of food and nutrient recycling.

Direct symbiotic syntrophy

The demand of nutrients and energy in the form of carbohydrates has led to many syntrophic relationships between carbohydrate-supplying photo-autotrophs and nutrient-supplying heterotrophs. Pure photoautotrophs are probably rare, if they exist at all; either they have mixotrophic capabilities or they form associations with heterotrophs. Being able to move independently and over considerable distances, jellyfish, for example, are able to commute between anaerobic conditions at lower water strata for nitrogen intake and higher ones for photosynthesis by their dinozoan endosymbionts supplying them with energy stored in carbohydrates. Dinozoans are engaged in similar relationships with hydropolyps (corals) and molluscs; extensive reefs testify of the evolutionary success of this association.

A close relationship between chlorophytes (or cyanobacteria) and fungi (mainly ascomycetes) evolved relatively recently, i.e. only ca. 450 million years ago, in the form of lichens and *Geosiphon* (Schüßler, 2002). The fungal partner specialized in decomposing organic matter, which releases nutrients for the algae in exchange for carbohydrates not unlike the situation in corals. Similarly, mycorrhizas exchange nutrients against carbohydrates with plants which arose in the same geological period. The endomycorrhizas (presently recognized as a new fungal phylum, the glomeromycetes) evolved right from the beginning of the land plants; the ectomycorrhizas (ascomycetes and basidiomycetes) evolved only during the Cretaceous. These symbioses seemed to have been essential for the invasion of the terrestrial environment (Selosse and Le Tacon, 1998). Some plants can also fix dinitrogen with the help of bacteria, encapsulated in specialized tissues. A single receptor seems to be involved in endosymbiontic associations between plants on the one hand and bacteria and fungi on the other (Stracke *et al.*, 2002), but the recognition process is probably quite complex (Parniske and Downie, 2003) and not yet fully understood. Associations between the dinitrogen-fixation cyanobacterium *Nostoc* and the fern *Azolla* have been known for some time, but the association with the bryophyte *Pleurozium schreberi* has only recently been discovered (DeLuca *et al.*, 2002); this extremely abundant moss covers most soil in boreal forests and in the taiga. The cyanobacteria are localized in extra-cellular pockets in these examples, but in some diatoms they live intracellularly. See Rai *et al.* (2000) for a review of symbioses between cyanobacteria and plants.

Heterotrophs not only have syntrophic relationships with photoautotrophs, but also with chemolithoautotrophs. A nice example concerns the gutless tubificid oligochaete *Olavius algarvensis*, with its sulphate-reducing and sulphide-oxidizing endosymbiontic bacteria (Dubilier *et al.*, 2001). These symbionts exchange reduced and oxidized sulphur; the fermentation products of the anaerobic metabolism of the host provide the energy for the sulphate reducers, whereas the organic compounds produced by the sulphide oxidizers fuel the (heterotrophic) metabolism of the host. Taxonomic relationships among hosts can match that among symbionts (van Dover, 2000), which suggest considerable co-evolution in syntrophic relationships.

When tree leaves fall on the forest floor, fungi release nutrients locked in them by decomposition; the soil fauna accelerates this degradation considerably (van Wensum, 1992). Without this activity by fungi and the soil fauna, trees soon deplete the soil from nutrients, as most leaves last for only one year, even in evergreen species. As mentioned trees, and plants in general, also need mycorrhizas to release nutrients from their organic matrix. Moreover, most of them also need insects, birds or bats and other animals to be pollinated (e.g. Proctor and Yeo, 1973; Barth, 1991), and yet other animals for seed dispersal. Thus, berries, for example of *Caprifoliaceae, Solanaceae* and *Rosaceae*, are "meant" to be eaten (Snow and Snow, 1988); some seeds have edible appendices (e.g. *Viola*) to promote dispersal, but others have no edible parts in addition to the seed, such as *Adoxa* and *Veronica*, and germinate better after being eaten by snails or birds and ants, respectively. Still other seeds stick to animals (e.g. *Boraginaceae, Arctium*) for dispersal. Fungi, such as the stinkhorn *Phallus* and the truffle *Tuber,* also interact with animals for their dispersal. By shading and evaporation, trees substantially affect their microclimate and thereby allow other organisms to live there as well. This too can be seen as an aspect of metabolism.

As mentioned, non-photosynthesizing plastids are still functional in plants; such plants can still have arbuscular mycorrhizas as are found in the orchid *Arachnitis uniflora* (Hibbett, 2002). Although the plant cannot transport photosynthetically produced carbohydrate to their fungal partner *Glomus*, it is obviously quite well possible that other metabolites are involved in the exchange. The complex role of plastids shows that the plant is not necessarily parasitizing the fungus.

Like plants, animals need other organisms (e.g. for food). The processing of food requires symbiosis too. We briefly discuss some aspects.

Many animals feed on cellulose-containing phototrophs but no animal can itself digest cellulose. Most animals have associations with prokaryotes, amoebas and flagellates to digest plant-derived compounds (Smith and Douglas, 1987). These micro-organisms transform cellulose to lipids in the anaerobic intestines of their host animal; the lipids are transported to the aerobic environment of the tissues of the animal for further processing. Attine ants even culture fungi to extract cellulases (Martin, 1987). Many symbioses are still poorly understood, such as the *Trichomycetes*, which live in the guts of a wide variety of arthropods in all habitats (Misra and Lichtwardt, 2000); the role of smut fungi (*Ustilaginales*) in their symbioses with plants also seems more complex than just a parasitic relationship (Vánky, 1987).

Faeces, especially that of herbivores, represent nutritious food for other organisms. This is because proteins often limit food uptake, implying that other compounds must be excreted; protein supplements to the grass diet of cows can greatly reduce the amount of grass they need. Organisms specialized on the use of faeces as a resource are known as coprophages. Examples are the bryophyte *Splachnum*, which lives off faeces of herbivores (*S. luteum* actually lives off that of the moose *Alces alces*); the fly *Sarcophaga* which lives off cattle dung; the fungus *Coprinus* which lives off mammalian faeces, similar to beetles of the dung beetle family *Scarabaeidae*.

Dead animals are processed by a variety of other animals; burrowing beetles of the family *Silphidae* specialize in this activity, for instance. Almost all animal taxa engage in carrion feeding, since the chemical make up of organisms does not differ that much; because of their great capacity of moving around, animals are often the first to arrive at the feast. Many examples illustrate that it is just a small step from feeding off dead corpses to that of living off live ones. Predation, a specialization of most animals, has many consequences and some can actually be "beneficial" for the prey: nutrient recycling, selection of healthy individuals, reduction of competition by weak individuals, reduction of transmission of diseases and enhancing the co-existence of prey species are all implications of predation (Kooi and Kooijman, 2000; Kooi et al., 2004; Kooijman et al., 2004).

Intricate relationships between organisms evolved, especially in prey-predator interactions, such as those between insects and plants (e.g. Schoonhoven et al., 1998). A low predation pressure on symbiotic partners enhances their stable co-existence (Kooi et al., 2004), whereas co-existence becomes unstable at a high pressure and easily leads to the extinction of both prey and predator. This points to a co-evolution of parameter values quantifying the dynamics in prey-predator systems. The time scale of the effects on fitness is essential; short-term positive effects can go together with long-term negative effects of behavioural traits on fitness. Time scales and indirect side effects that operate through changes in food availability are important aspects that are usually not included in the literature on evolutionary aspects of life history strategies.

Indirect symbiontic syntrophy

In this section, we only give some examples of the many indirect trophic relationships that exist between species.

Phytoplankters bind nutrients in the photic zone of the oceans, sink below it, die and are degraded by bacteria. Subsequently, a temporary increase in wind speed brings some of the released nutrients back to the photic zone by mixing and enables photosynthesis to continue. The sinking of organic matter is accelerated by grazing zooplankters. The result of this process is that, over time, phytoplankters build up a nutrient gradient in the water column, that CO_2 from the atmosphere becomes buried below the photic zone, and that organic resources are generated for the biota living in the dark waters below this zone and on the ocean floor. Mixing by wind makes phytoplankters commute between the surface, where they can build up and store carbohydrates by photosynthesis, and the bottom of the mixing zone, where they store nutrients. Reserves are essential here for growth, because no single stratum in the water column is favourable for growth; their reserve capacity must be large enough to cover a commuting cycle, which depends on wind speed. Although nutrient availability controls primary production ultimately, wind is doing so proximately.

The rain of dead or dying phytoplankters fuels the dark ocean communities, not unlike the rain of plant leaves fuelling soil communities, but then on a vastly larger spatial scale. Little is known about the deep ocean food web;

recent studies indicate that cnidarians (jellyfish) form a major component (Dennis, 2003).

When part of this organic rain reaches the anoxic ocean floor, the organic matter is decomposed by fermenting bacteria (many species can do this); the produced hydrogen serves as substrate for methanogens (i.e. archaeans), which convert carbon dioxide into methane. This methane can accumulate in huge deposits of methane hydrates, which serve as substrate for symbioses between bacteria and a variety of animals, such as the ice worm *Hesiocoeca*, a polychaete. The total amount of carbon in methane hydrates in ocean sediments is more than twice the amount to be found in all known fossil fuels on Earth. If the temperature rises in the deep oceans, the hydrates become unstable and result in a sudden massive methane injection into the atmosphere. This happened e.g. 55 Ma years ago (e.g. Zachos *et al.*, 2003), the Paleocene-Eocene Thermal Maximum (PETM) event, which induced massive extinctions globally (Kroon *et al.*, 2001). Methanogens are involved in similar synthrophic relationships with chemolithotrophic bacteria in the deep underground (>1 km), that release hydrogen in the transformation $FeO + H_2O \rightarrow H_2 + FeO_2$ (Madigan *et al.*, 2000). We are just beginning to understand the significance of these communities on ocean floors and deep underground.

The colonization of the terrestrial environment by plants may in fact have allowed reefs of brachiopods, bryozoans and molluscs (all filter feeders) to flourish in the Silurian and the Devonian; the reefs in these periods were exceptionally rich (Wood, 1999). With the help of their bacterial symbionts, the plants stimulated the conversion from rock to soil, which released nutrients that found their way to the coastal waters, stimulated algal growth, and, hence, the growth of zooplankton, which the reef animals, in turn, filtered out of the water column. The reefs degraded gradually during the time Pangea was formed towards the end of the Permian, which reduced the length of the coastline considerably and thereby the nutrient flux from the continents to the ocean. Moreover, large continents come with long rivers, and more opportunities for water to evaporate rather than to drain down to the sea; large continents typically have salt deposits. When Pangea broke up, new coastlines appeared. Moreover, this coincided with a warming of the globe, which brought more rain, more erosion, and high sea levels, which caused covering of large parts of continents by shallow seas. This combination of factors caused planktontic communities to flourish again in the Cretaceous, and completely new taxa evolved, such as the coccolithophorans and the diatoms. This hypothesis directly links the activities of terrestrial plants to the coastal reef formation through nutrient availability. Although plants reduce erosion on a time scale of thousands of years, they promote erosion on a multi-million years time scale in combination with extreme but very rare physical forces that remove both vegetation and soil (Kooijman, 2004).

The geological record of the Walvis Ridge suggests that the mechanism of physical-chemical forces that remove the vegetation, followed by erosion and nutrient enrichment of coastal waters in association with recolonization of the rocky environment by plants might also have been operative in e.g. the 0.1 Ma recovery period following the PETM event (Kroon, personal communication).

A direct quantitative relationship exists between the fossil carbohydrates (methane hydrates, coal, oil, gas, all of biotic origin) and dioxygen in the atmosphere. Although dioxygen, a by-product of oxygenic photosynthesis, was doubtlessly very toxic for most organisms when it first occurred freely in the atmosphere; today most life is dependent upon it, both directly, as well as indirectly, such as the ozon shield against UV radiation. So phototrophs generate dioxygen that is used by heterotrophs; again a form of syntrophy.

A discussion of the interactions between biota and climate is beyond the scope of this paper, see e.g. Kooijman (2004) for further discussion. The example illustrates the dynamic interaction between surface areas (where erosion takes place) and volumes (in which nutrients are diluted) at the ecosystem level. We have already discussed the importance of these interactions at the individual and molecular levels.

7.7 DISCUSSION

The red thread through our presentation is that the evolutionary invention of homeostasis comes with stoichiometric constraints on production and with the excretion of metabolic products, which promote syntrophy and the formation of symbioses that are based on syntrophy. Organisms became increasingly connected metabolically; loose forms of symbiosis can evolve into tight forms and even into a full integration, processes that happened frequently and repeatedly throughout the evolutionary history of prokaryotes and especially of eukaryotes. Syntrophy is the basis of biodiversity and supplements Darwin's notion of survival of the fittest, which is based on competitive exclusion (Ryan, 2003).

We do realize that our account of metabolic evolution is sketchy at best, and controversial in places. Contrary to our present evaluation, for instance, fermentation is still widely seen as the origin of metabolism (Alberts *et al.*, 2002; Heinrich and Schuster, 1996; Fenchel, 2002). The main motivation is probably that few steps seem to be required to convert organic compounds into "biomass". As genome size already suggests, the extracellular extraction of energy from organic matter is not necessarily simpler than from inorganic compounds, though; energy supply by chemolithoautotrophy still allows the uptake of organic building blocks. Each heterotrophic bacterial species can handle only a very limited number of organic substrates. Moreover, contemporary fermenting prokaryotes have a glucose-based metabolism, which must have been an advanced feature. The ionic strength of cytoplasm equals that of seawater, which suggests that life arose in the sea. It seems unlikely that pre-biotic organic compounds could accumulate in concentrations that allowed the emergence of life in ocean water without separation and containment. Forterre and Philippe (1999) argue that the eukaryotes are at the root of life, from which prokaryotes developed by simplification. Although cladistic analysis of the properties of archaea, eubacteria and eukaryotes still allow multiple interpretations for the roots of life, given the metabolic uniformity of eukaryotes and the advanced nature of their heterotrophic metabolism, we find it hard to accept that life would originate with eukaryotes in a geochemical context.

A proper qualitative understanding of metabolism at the molecular level involves ecological, evolutionary and geochemical aspects. While theory on competition and predation dominates population ecology, our aim has been to reveal the increasing importance in evolution of metabolic interdependence of the various forms of life based on an exchange of nutrients and metabolites. Organisms did not only become increasingly dependent on each other, but the interaction with geochemical cycling of macro nutrients, and with the climate system also became stronger (Kooijman, 2004).

A proper quantitative understanding of metabolic organization involves a holistic setting. It is essential, though, to delineate proper modules that represent entities with similar time scales, and to nest modules for keeping the models relatively simple, and thereby useful for developing a better understanding of life at both the organismic as well as the supra-organismic level. The DEB theory is useful in linking the various levels of organization, where surface area-to-volume interactions are operative at all levels, and reserves are basic in the understanding of metabolic organization.

ACKNOWLEDGEMENTS

We would like to thank Bill Martin, Mike Russell, Ad Stouthamer, Peter Westbroek, Tiago Domingos, Lothar Kuijper, Tineke Troost, John Speakman and Cor Zonneveld for helpful discussions and comments.

REFERENCES

Alberts, B., A. Johnson, J. Leweis, M. Raff, K. Robert and P. Walter (2002). Molecular Biology of the Cell. Garland Science, New York.

Amend, J. P. and E. L. Shock (2001). Energetics of overall metabolic reactions of thermophilic and hyperthermophilic archaea and bacteria. FEMS Microbiological Reviews 25: 175-243.

Anbar, A. D. and A. H. Knoll (2002). Proterozoic ocean chemistry and evolution: A bioinorganic bridge? Science 297: 1137-1142.

Andersson, J. O. and A. J. Roger (2002). A cyanobacterial gene in nonphotosybthetic protests – an early chloroplast acquisition in eukaryotes? Current Biology 12: 115-119.

Andersson, S. G. E., A. Zomorodipour, J. O. Andersson, T. Sicheritz-Pontén, U. C. M. Alsmark, R. M. Podowski, A. K. Näslund, A.-S. Eriksson, H. H. Winkler and C. G. Kurland (1998). The genome sequence of *Rickettsia prowazekii* and the origin of mitochondria. Nature 396: 133-143.

Bakker, B. (1998). Control and Regulation of Glycolysis in Trypanosoma brucei. PhD Thesis, Vrije Universiteit, Amsterdam.

Baltscheffsky, H. (1996). Energy conversion leading to the origin and early evolution of life: did inorganic pyrophosphate precede adenosine triphosphate? In: Baltscheffsky, H. (Ed.). Origin and Evolution of Biological Energy Conversion. VCH Publishers, Cambridge. pp. 1-9.

Baltscheffsky, M., A. Schultz and H. Baltscheffsky (1999). H+-P pases: a tightly membrane-bound family. FEBS Letters 457: 527-533.

Barth, F. G. (1991). Insects and Flowers. Princeton University Press, Princeton.

Bengtson, S. (1994). Early Life on Earth. Columbia University Press, New York.

Bernard, C, A. G. B. Simpson and D. J. Patterson (2000). Some free-living flagellates (Protista). from anoxic habitatats. Ophelia 52: 113-142.

Bigay, J., P. Guonon, S. Robineau and B. Antonny (2003). Lipid packing sensed by ArfGAP1 couples COPI coat disassembly to membrane bilayer urvature. Nature 426: 563-566.

Birky, C. W. (2001). The inheritance of genes in mitochondria and chloroplasts: Laws, mechanisms, and models. Annual Review of Genetics 35: 125-148.

Bjerrum, C. J. and D. E. Canfield (2002). Ocean productivity before about 1.9 Gyr ago limited by phosphorus adsorption onto iron oxides. Nature 417: 159-162.

Blankenship, R. E. and H. Hartman (1992). Origin and early evolution of photosynthesis. Photosynthesis Research 33: 91-111.

Blankenship, R. E. and H. Hartman (1998). The origin and evolution of oxygenic photosynthesis. Trends in Biochemical Sciences 23: 94-97.

Boyce, A. J., M. L. Coleman and M. J. Russell (1983). Formation of fossil hydrothermal chimneys and mounds from Silvermines, Ireland. Nature 306: 545-550.

Brandt, B. W. (2002). Realistic Characterizations of Biodegradation. PhD Thesis. Vrije Universiteit, Amsterdam.

Brandt, B. W. and S. A. L. M. Kooijman (2000). Two parameters account for the flocculated growth of microbes in biodegradation assays. Biotechnology and Bioengineering 70: 677-684.

Brandt, B. W., I. M. M. van Leeuwen and S. A. L. M. Kooijman (2003). A general model for multiple substrate biodegradation. Application to co-metabolism of non structurally analogous compounds. Water Research 37: 4843-4854.

Brocks, J. J., G. A. Logan, G. A. Logan, R. Buick and R. E. Summons (1999). Archean molecular fossils and the early rise of eukaryotes. Science 285: 1033-1036.

Cairns-Smith, A. G., A. J. Hall and M. J. Russell (1992). Mineral theories of the origin of life and an iron sulphide example. Origins Life and Evolution of the Biosphere 22: 161-180.

Cavalier-Smith, T. A. (1987a). The simulataneous origin of mitochondria, chloroplasts and microbodies. Annals of the New York Academy of Sciences 503: 55-71.

Cavalier-Smith, T. A. (1987b). The origin of cells, a symbiosis between genes, catalysts and membranes. Cold Spring Harbor Symposia on Quantitative Biology 52: 805-824.

Cavalier-Smith, T. A. (1998). A revised six-kingdom system of life. Biological Reviews 73: 203-266.

Cavalier-Smith, T. A. (2000). Membrane heredity and early chloroplast evolution. Trends in Plant Sciences 5: 174-182.

Cavalier-Smith, T. A. (2002a). Chloroplast evolution: Secondary symbiogenesis and multiple losses. Current Biology 12: R62-64.

Cavalier-Smith, T. A. (2002b). The phagotrophic origin of eukaryotes and phylogenetic classification of Protozoa. International Journal of Systematic and Evolutionary Microbiology 52: 297-354.

Chela-Flores, J. (1998). First step in eukaryogenesis: Physical phenomena in the origin and evolution of chromosome structure. Origins of Life and Evolution of the Biosphere 28: 215-225.

Dalsgaard, T., D. E. Canfield, J. Petersen, B. Thamdrup and J. Acuna-Gonzalez (2003). N-2 production by the anammox reaction in the anoxic water column of Golfo Dulce, Costa Rica. Nature 422: 606-608.

de Duve, C. (1984). A Guided Tour of the Living Cell. Scientific American Library, New York.

Deamer, D. W. and R. M. Pashley (1989). Amphiphilic components of the Murchison carbonaceous chondrite; surface properties and membrane formation. Origins of Life and Evolution of the Biosphere 19: 21-38.

DeLuca, T. H., O. Zackrisson, M.-C. Nilsson and A. Sellstedt (2002). Quantifying nitrogen-fixation in feather moss carpets of boreal forests. Nature 419: 917-920.

Delwiche, C. F. (1999). Tracing the thread of plastid diversity through the tapestry of life. American Naturalist 154: S164-S177.

Dennis, C. (2003). Close encounters of the jelly kind. Nature 426: 12-14.

Derelle, E., C. Ferraz, P. Lagoda, S. Eychenié, R. Cooke, F. Regad, X. Sabau, C. Courties, M. Delseny, J. Demaille, A. Picard and H. Moreau (2002). DNA libraries for sequencing the genome of Ostreococcus tauri (Chlorophyta, Prasinophyceae): the smallest free-living eukaryotic cell. Journal of Phycology 38: 1150-1156.

Dillon, J. G. and R. W. Castenholz (1999). Scytonemin, a cyanobacterial sheath pigment, protects against UVC Radiation: implications for early photosynthetic life. Journal of Phycology 35: 673-681.

Dismukes, G. C., V. V. Klimov, S. V. Baranov, Yu. N. Kozlov, J. DasGupta and A. Tyryshkin (2001). The origin of atmospheric oxygen on earth: the innovation of oxygenic photosynthesis. Proceedings of the National Academy of Sciences of the USA 98: 2170-2175.

Dixon, P. S. (1973). Biology of the Rhodophyta. Oliver and Boyd, Edinburgh.

Doolittle, W. F. (1999). Phylogenetic classification and the universal tree. Science 284: 2124-2128.

Dubilier, N., C. Mulders, T. Felderman, D. de Beer, A. Pernthaler, M. Klein, M. Wagner, C. Erséus, F. Thiermann, J. Krieger, O. Giere and R. Amann (2001). Endosymbiotic sulphate-reducing and sulphide-oxidizing bacteria in an oligochaete worm. Nature 411: 298-302.

Dworkin, M. (1985). Developmental Biology of the Bacteria. Benjamin-Cummings Publishing Company, California.

Elser, J. J. (2004). Biological stoichiometry: a theoretical framework connecting ecosystem ecology, evolution, and biochemistry for application in astrobiology. International Journal of Astrobiology: to appear.

Embley, T. M. and R. P. Hirt (1998). Early branching eukaryotes? Current Opinion in Genetics and Development 8: 624-629.

Fenchel, T. (2002). Origin and Early Evolution of Life. Oxford University Press, Oxford.

Fenchel, T. and B. L. Finlay (1995). Ecology and Evolution in Anoxic Worlds. Oxford University Press, Oxford.

Forterre, P. and H. Philippe (1999). Where is the root of the universal tree of life? BioEssays 21: 871 - 879.

Forterre, P., A. Bergerat, P. Lopez-Garvia (1996). The unique DNA topology and DNA topoisomerase of hyperthermophilic archaea. FEMS Microbiology Reviews 18: 237-248.

Fuhrman, J. (2003). Genome sequences from the sea. Nature 424: 1001-1002.

Gruenberg, J. (2001). The endocytic pathway: a mosaic of domains. Nature Reviews 2: 721-730.

Guerrero, R. (1991). Predation as prerequisite to organelle origin: Daptobacter as example. In: Margulis, L. and R. Fester (Eds). Symbiosis as a Source of Evolutionary Innovation. MIT Press, Cambridge, Mass.

Gupta, R. S. (1998). Protein phylogenies and signature sequences: a reappraisal of evolutionary relationships among Archaebacteria, Eubacteria, and Eukaryotes. Microbiology and Molecular Biology Reviews 62: 1435-1491.

Hartman, H. (1975). Speculations on the origin and evolution of metabolism. Journal of Molecular Evolution 4: 359-370.

Hartman, H. (1998). Photosythesis and the origin of life. Origins of Life and the Evolution of the Biosphere 28: 515-521.

Heinrich, B. (1993). The Hot-Blooded Insects. Harvard University Press, Cambridge, Massachusetts.

Heinrich, R. and S. Schuster (1996). The Regulation of Cellular Systems. Chapman and Hall, New York.

Hendrix, R. W., M. C. Smith, R. N. Burns, M. E. Ford and G. F. Hatfull (1999). Evolutionary relationships among diverse bacteriophages: All the world's phage. Proceedings of the National Academy of Sciences of the USA 96: 2192-2197.

Hengeveld, R. and M. A. Fedonkin (2004). Causes and consequences of eukaryotization through mutualistic endosymbiosis. Acta Biotheoretica 52: 105-154.

Hibbet, D. S. (2002). When good relationships go bad. Nature 419: 345-346.

Holland, H. D. (1994). Early proterozoic atmospheric change. In: Bengtson, S. (Ed.). Early life on earth. Columbia University Press, New York: pp 237-244.

Hope, A. B. and N. A. Walker (1975). The Physiology of Giant Algal Cells. Cambridge University Press, Cambridge.

Huber, H., M. J. Hohn, R. Rachel, T. Fuchs, V. C. Wimmer and K. O. Stetter (2002). A new phylum of Archaea represented by nanosized hyperthermophilic symbiont. Nature 417: 63-67.

Hugenholtz, J. and L. G. Ljungdahl (1990). Metabolism and energy generation in homoacetogenic Clostridia. FEMS Microbiology Reviews 87: 383-389.

Kandler, O. (1998). The early diversification of life and the origin of the three domains: a proposal. In: Wiegel, J. and M. W. W. Adams (Eds). Thermophiles: The keys to molecular evolution and the origin of life. Taylor and Francis, Washington. pp 19-31.

Kasting, J. F. (2001). Earth's early atmosphere. Science 259: 920-925.

Kates, M. (1979). The phytanyl ether-linked polar lipids and isopreniod neutral lipids of extremely halophilic bacteria. Lipids 15: 301-342.

Keefe, A. D., S. L. Miller and G. Bada (1995). Investigation of the prebiotic synthesis of amino acids and RNA bases from CO_2 using FeS/H_2S as a reducing agent. Proceedings of the National Academy of Sciences of the USA 92: 11904-11906.

Keeling, P. J. (1998). A kingdom's progress: Archaezoa and the origin of eukaryotes. BioEssays 20: 87-95.

Kirkpatrick, M. (Ed.). (1993). The evolution of Haploid-Diploid life cycles. Lectures on Mathematics in the Life Sciences 25. American Mathematical Society, Providence, Rhode Island.

Knoll, A. H. (2003). Life on a Young Planet; The First Three Billion Years of Evolution on Earth. Princeton University Press, Princeton.

Koga, Y., T. Kyuragi, M. Nishhihara and N. Sone (1998). Did archaeal and bacterial cells arise independently from noncellular precursors? A hypothesis stating that the advent of membrane phospholipids with enantiomeric glycerophosphate backbones caused the separation of the two lines of decent. Journal of Molecular Evolution 46: 54-63.

Kooi, B. W. and S. A. L. M. Kooijman (2000). Invading species can stabilize simple trophic systems. Ecological Modelling 133: 57-72.

Kooi, B. W., L. D. J. Kuijper and S. A. L. M. Kooijman (2004). Consequences of symbiosis on food web dynamics in an open system. Journal of Mathematical Biology: to appear.

Kooijman, S. A. L. M. (1986). Energy budgets can explain body size relations. Journal of Theoretical Biology 121: 269-282.

Kooijman, S. A. L. M. (1998). The synthesizing unit as model for the stoichiometric fusion and branching of metabolic fluxes. Biophysical Chemistry 73: 179-188.

Kooijman, S. A. L. M. (2000). Dynamic Energy and Mass Budgets in Biological Systems. Cambridge University Press, Cambridge.

Kooijman, S. A. L. M. (2001). Quantitative aspects of metabolic organization; a discussion of concepts. Philosophical Transactions of the Royal Society - Series B 356: 331-349.

Kooijman, S. A. L. M. (2004). On the coevolution of life and its environment. In: Miller, J., P. J. Boston, S. H. Schneider and E. Crist (Eds). Scientists on Gaia: 2000. MIT Press, Cambridge, Massachusetts, Chapter 30, to appear.

Kooijman, S. A. L. M. and L. Segel (2003). How growth affects the fate of metabolites. (to appear).

Kooijman, S. A. L. M., T. R. Andersen and B. W. Kooi (2004). Dynamic energy budget representations of stoichiometric constraints on population dynamics. Ecology 85: 1230-1243.

Kooijman, S. A. L. M., P. Auger, J. C. Poggiale and B. W. Kooi (2003). Quantitative steps in symbiogenesis and the evolution of homeostasis. Biological Reviews 78: 435-463.

Kroon, D., R. D. Norris and A. Klaus (2001). Western North Atlantic Palaeogene and Cretaceous Palaeoceanography, Geological Society Special Publication 183: 1-319.

Kuijper, L. D. J., T. R. Anderson and S. A. L. M. Kooijman (2003). C and N gross efficiencies of copepod egg production studies using a Dynamic Energy Budget model. Journal of Plankton Research 26: 213-226.

Kurland, C. G. and S. G. E. Andersson (2000). Origin and evolution of the mitochondrial proteome. Microbiology and Molecular Biology Reviews 64: 786-820.

Lane, N. (2002). Oxygen, the Molecule that made the World. Oxford University Press, Oxford.

Lee, J. J., G. F. Leedale and P. Bradbury (2000). An Illustrated Guide to the Protozoa. Society of Protozoologists, Lawrence, Kansas.

Leigh, R. A. and D. Sanders (1997). The Plant Vacuole. Academic Press, San Diego.

Lengeler, J. W., G. Drews and H. G. Schlegel (1999). Biology of the Prokaryotes. Thieme Verlag, Stuttgart.

Lindahl, P. A. and B. Chang (2001). The evolution of acetyl-CoA synthase. Origins of Life and Evolution of the Biosphere 31: 403-434.

Ljungdahl, L. G. (1994). The acetyl-CoA pathway and the chemiosmotic generation of ATP during acetogenesis. In: Drake, H. L. (Ed.). Acetogenesis. Chapman and Hall, New York. pp 63-87.

Madigan, M. T., J. M. Martinko and J. Parker (2000). Brock Biology of Micro-organisms. Prentice Hall International, New Jersey.

Margulis, L. (1970). Origins of Eukaryotic Cells. Freeman, San Francisco.

Martin, M. M. (1987). Invertebrate-Microbial Interactions; Ingested Fungal Enzymes in Arthropod Biology. Comstock Publishers & Associates, Ithaca.

Martin, W. and M. Muller (1998). The hydrogen hypothesis for the first eukaryote. Nature 392: 37-41.

Martin, W. and M. Russell (2003). On the origins of cells: a hypothesis for the evolutionary transitions from abiotic geochemistry to chemoautotrophic prokaryotes, and from prokaryotes to nucleated cells. Philosophical Transactions of the Royal Society - Series B 358: 59-85.

Martin, W. and C. Schnarrenberger (1997). The evolution of the Calvin cycle from prokaryotic to eukaryotic chromosomes: a case study of functional redundancy in ancient pathways through endosymbiosis. Current Genetics 32: 1-18.

Martin, W., T. Rujan, E. Richly, A. Hansen, S. Cornelsen, T. Lins, D. Leister, B. Stoebe, M. Hasegawa and D. Penny (2002). Evolutionary analysis of Arabidopsis, cyanobacterial, and chloroplast genomes reveals plastid phylogeny and thousands of cyanobacterial genes in the nucleus. Proceedings of the National Academy of Sciences of the USA 99: 12246-12251.

McFadden, G. I. (2001). Primary and secondary endosymbiosis and the origin of plastids. Journal of Phycology 37: 951-959.

Meléndez-Hevia, E. (1990). The game of the pentose phosphate cycle: a mathematical approach to study the optimization in design of metabolic pathways during evolution. Biomedica Biochemica Acta 49: 903-916.

Meléndez-Hevia, E. and A. Isidoro (1985). The game of the pentose phosphate cycle. Journal of Theoretical Biology 117: 251-263.

Michaelis, W., R. Seifert, K. Nauhaus, T. Treude, V. Thiel, M. Blumenberg, K. Knittel, A. Gieseke, K. Peterknecht, T. Pape, A. Boetius, R. Amann, B. B. Jørgensen, F. Widdel, J. Peckmann, N. V. Pimenov and M. B. Gulin (2002). Microbial reefs in the Black Sea fuelled by anaerobic oxidation of methane. Science 297: 1013-1015.

Misra, J. K. and R. W. Lichtwardt (2000). Illustrated Genera of Trichomycetes; Fungal Symbionts of Insects and Other Arthropods. Science Publishers, Enfield, NH.

Moore, D. (1998). Fungal Morphogenesis. Cambridge University Press, Cambridge.

Morowitz, H. J., J. D. Kostelnik, J. Yang and G. D. Cody (2000). The origin of intermediary metabolism. Proceedings of the National Academy of Sciences of the USA 97: 7704-7708.

Nisbet, E. G. and C. M. R. Fowler (1999). Archaean metabolic evolution of microbial mate. Proceedings of the Royal Society of London - B - Biological Sciences 266: 2375-2382.

Nisbet, R. M., E. B. Muller, K. Lika and S. A. L. M. Kooijman (2000). From molecules to ecosystems through Dynamic Energy Budget models. Journal of Animal Ecology 69: 913-926.

Norris, V. and D. J. Raine (1998). A fission-fusion origin for life. Origins of Life and Evolution of the Biosphere 28: 523-537.

Olsen, G. J. and C. R. Woese (1996). Lessons from an Archaeal genome: what are we learning from Methanococcus jannaschii? Trends in Genetics 12: 377-379.

Orgel, L. E. (1998). The origin of life – a review of facts and speculations. Trends in Biochemical Sciences 23: 491-495.

Orgel, L. E. (2000). Self-organizing biochemical cycles. Proceedings of the National Academy of Sciences of the USA 97: 12503-12507.

Osteryoung, K. W. and J. Nunnari (2003). The division of endosymbiotic organelles. Science 302: 1698-1704.

Parniske, M. and J. A. Downie (2003). Lock, keys and symbioses. Nature 425: 569-570.

Patterson, D. J. (1999). The diversity of eukaryotes. American Naturalist 154: S96-S124.

Proctor, M. and P. Yeo (1973). The Pollination of Flowers. Collins, London.

Rai, A. N., E. Soderback and B. Bergman (2000). Cyanobacterium-plant symbioses. New Phytologist 147: 449-481.

Raven, J. A. and C. Brownlee (2001). Understanding membrane function. Journal of Phycology 37: 960-967.

Raven, J. A. and Z. H. Yin (1998). The past, present and future of nitrogenous compounds in the atmosphere, and their interactions with plants. New Phytologist 139: 205-219.

Rickard, D., I. B. Butler and A. Olroyd (2001). A novel iron sulphide switch and its implications for earth and planetary science. Earth and Planetary Science Letters 189: 85-91.

Rizzotti, M. (2000). Early Evolution. Birkhauser Verlag, Basel.

Roger, A. J. (1999). Reconstructing early events in eukaryotic evolution. American Naturalist 154: S146-S163.

Romano, A. H. and T. Conway (1996). Evolution of carbohydrate metabolic pathways. Reseach in Microbiology 147: 448-455.

Rothenberg, M. E. and Y.-N. Jan (2003). The hyppo hypothesis. Nature 425: 469-470.

Russell, M. J. and A. J. Hall (1997). The emergence of life from iron monosulphide bubbles at a submarine hydrothermal redox and pH front. Journal of the Geological Society of London 154: 377-402.

Russell, M. J. and A. J. Hall (2002). From geochemistry to biochemistry; chemiosmotic coupling and transition element clusters in the onset of life and photosynthesis. The Geochemical News 133/October: 6 - 12.

Russell, M. J., R. M. Daniel, A. J. Hall and J. A. Sherringham (1994). A hydro-thermally precipitated catalytic iron sulfide membrane as a first step toward life. Journal of Molecular Evolution 39: 231-243.

Ryan, F. (2003). Darwin's Blind Spot. Texere, New York.

Schalk, J. (2000). A Study of the Metabolic Pathway of Anaerobic Ammonium Oxidation. PhD Thesis, University of Delft.

Schönheit, P. and T. Schafer (1995). Metabolism of hyperthermophiles. World Journal of Microbiology and Biotechnology 11: 26-57.

Schoonen, M. A. A., Y. Xu and J. Bebie (1999). Energetics and kinetics of the prebiotic synthesis of simple organic and amino acids with the FeS-H2S/FeS2 redox couple as a reductant. Origins of Life and Evolution of the Biosphere 29: 5-32.

Schoonhoven, L. M., T. Jermy and J. J. A. van Loon (1998). Insect-Plant Biology; From Physiology to Ecology. Chapman and Hall, London.

Schüßler, A. (2002). Molecular phylogeny, taxonomy, and evolution of *Geosiphon pyriformis* and arbuscular mycorrhizal fungi. Plant and Soil 244: 75-83.

Segré, D., D. Ben-Eli, D. W. Deamer and D. Lancet (2001). The lipid world. Origins of Life and Evolution of the Biosphere 31: 119-145.

Selig, M., K. B. Xavier, H. Santos and P. Schönheit (1997). Comparative analysis of Embden-Meyerhof and Entner-Doudoroff glycolytic pathways in hyper-thermophilic archaea and the bacterium Thermotoga. Archives of Microbiology 167: 217-232.

Selosse, M.-A. and F. Le Tacon (1998). The land flora: a phototroph-fungus partnership? Trends in Ecology and Evolution 13: 15-20.

Siegel, J. M. (2003). Why we sleep. Scientific American, Nov. 2003: 72-77.

Simpson, A. G. B. and A. J. Roger (2002). Eukaryotic evolution: getting to the root of the problem. Current Biology 12: R691-693.

Simpson, P. G. and W. B. Whitman (1993). Anabolic pathways in methanogens. In: Ferry, J. G. (Ed.). Methanogenesis. Chapman and Hall, New York. pp 445-472.

Smith, D. C. and A. E. Douglas (1987). The Biology of Symbiosis. E. Arnold, Baltimore.

Smith, D. J. and G. J. C. Underwood (2000). The production of extracellular carbohydrates by estuarine benthic diatoms: the effects of growth phase and light and dark treatment. Journal of Phycology 36: 321-333.

Smith, M. L., J. N. Bruhn and J. B. Anderson (1992). The fungus Armillaria bulbosa is among the largest and oldest living organisms. Nature 356: 428-431.

Snow, B. and D. Snow (1988). Birds and Berries. Poyser, Calton.

Sprent, J. I. (1987). The Ecology of the Nitrogen Cycle. Cambridge University Press, Cambridge.

Staley, J. T., M. P. Bryant, N. Pfennig and J. G. Holt (1989). Bergey's manual of systematic bacteriology. Williams and Wilkins, Baltimore.

Stechmann, A. and T. Cavalier-Smith (2002). Rooting the eukaryote tree by using a derived gene fusion. Science 297: 89-91.

Stracke, S., C. Kistner, S. Yoshida, L. Mulder, S. Sato, T. Kaneko, S. Tabata, N. Sandal, J. Stougaard, K. Szczyglowski and M. Parniske (2002). A plant receptor-like kinase required for both bacterial and fungal symbiosis Nature 417: 959-962.

Stryer, L. (1988). Biochemistry. W. H. Freeman and Co., New York.

Sullivan, M. B., J. B. Waterbury and S. W. Chisholm (2003). Cyanophages infecting the oceanic cyanobacterium Prochlorococcus. Nature 424: 1047-1050.

Taylor, P., T. E. Rummery and D. G. Owen (1979). Reactions of iron monosulfide solids with aqueous hydrogen sulfide up to 160°C. Journal of Inorganic and Nuclear Chemistry 41: 1683-1687.

Tielens, A. G. M., C. Rotte, J. J. van Hellemond and W. Martin (2002). Mitochondria as we don't know them. Trends in Biochemical Sciences 27: 564-572.

van den Berg, H. A. (1998). Multiple Nutrient Limitation in Microbial Ecosystems. PhD Thesis, Vrije Universiteit, Amsterdam.

van den Ent, F., L. A. Amos and J. Lowe (2001). Prokaryotic origin of the actin cytoskeleton. Nature 413: 39-44.

van den Hoek, C., D. G. Mann and H. M. Jahn (1995). Algae; An Introduction to Phycology. Cambridge University Press, Cambridge.

Van Dover, C. L. (2000). The Ecology of Deep-Sea Hydrothermal Vents. Princeton University Press, Princeton.

van Leeuwen, I. M. M. and C. Zonneveld (2001). From exposure to effect: a comparison of modeling approaches to chemical carcinogenesis. Mutations Research 489: 17-45.

van Leeuwen, I. M. M., F. D. L. Kelpin and S. A. L. M. Kooijman (2002). A mathematical model that accounts for the effects of caloric restriction on body weight and longevity. Biogerontology 3: 373-381.

van Wensum, J. (1992). Isopods and Pollutants in Decomposing Leaf Litter. PhD Thesis, Vrije Universeit, Amsterdam.

Vánky, K. (1987). Illustrated Genera of Smut Fungi. Cryptogamic Studies Volume 1. Gustav Fischer Verlag, Stuttgart.

Von Dohlen, C. D., S. Kohler, S. T. Alsop and W. R. McManus (2001). Mealybug β-proteobacterial endosymbionts contain γ-proteobacterial symbionts. Nature 412: 433-436.

Vrede, T. J., D. Dobberfuhl, S. A. L. M. Kooijman and J. J. Elser (2004). The stoichiometry of production - fundamental connections among organism C:N:P stoichiometry, macromolecular composition and growth rate. Ecology 85: 1217-1229.

Wächtershäuser, G. (1988). Pyrite formation, the first energy source for life: A hypothesis. Systematic Applied Microbiology 10: 207 - 210.

Wächtershäuser, G. (1990). Evolution of the 1st metabolic cycles. Proceedings of the National Academy of Sciences of the USA 87: 200-204.

Waddell, T. G., P. Repovic, E. Melendez-Hevia, R. Heinrich and F. Montero (1997). Optimization of glycolysis: A new look at the efficiency of energy coupling. Biochemical Education 25: 204-205.

Woese, C. R. (1979). A proposal concerning the origin of life on the planet earth. Journal of Molecular Evolution 12: 95-100.

Woese, C. R. (2002). On the evolution of cells. Proceedings of the National Academy of Sciences of the USA 99: 8742-8747.

Wood, R. (1999). Reef Evolution. Oxford University Press, Oxford.

Xiong, J., W. M. Fisher, K. Inoue, M. Nakahara and C. E. Bauer (2000). Molecular evidence for the early evolution of photosynthesis. Science 289: 1724-1730.

Zachos, J. C., M. W. Wara, S. Bohaty, M. L. Delaney, M. R. Petrizzo, A. Brill, T. J. Bralower and I. Premoli-Silva (2003). A transient rise in tropical sea surface temperature during the Paeocene-Eocene thermal maximum. Science 302: 1551-1554.

Zubay, G. (2000). Origins of Life on Earth and in the Cosmos. Academic Press, San Diego.

S. A. L. M. Kooijman
Department of Theoretical Biology, Vrije Universiteit Amsterdam

R. Hengeveld
Department of Ecotoxicology, Vrije Universiteit Amsterdam

8

The Founder and Allee Effects in the Patch Occupancy Metapopulation Model

Rampal S. Etienne and Lia Hemerik

ABSTRACT

The problem of ever-increasing habitat fragmentation due to human land use calls for a theoretical framework to study the potential dangers and to find ways of combating these dangers. The metapopulation approach, with the Levins model as its patriarch, provides such a framework. A metapopulation is a collection of populations that live in habitat fragments (called patches). These populations can become extinct, but new populations can be established by dispersing individuals from extant populations. If these colonizations can balance these extinctions, metapopulation persistence is possible. In theoretical literature surprisingly little attention has been paid to the colonization term in the Levins model and its extensions. Specifically, the Allee effect (i.e. reduced probability of colonization due to, e.g., reduced probability of finding a mate, or reduced defence against predators) may play a major role although it has not received appropriate attention. In this paper, we study the colonization term in the Levins model and conclude that it describes the founder effect (i.e. stochastic fluctuations in births and deaths of an establishing population causing colonization to fail). We then incorporate the Allee effect in the colonization term and conclude that previous attempts to do so were erroneous because they ignored some difficulties in the model formulation and interpretation. We devise a phenomenological model for the Allee effect that is consistent in both discrete and continuous time. Although the model with Allee effect shows a fold bifurcation in its deterministic formulation (both in discrete and continuous time), suggesting the possibility of sudden metapopulation extinction when the bifurcation parameter is only changed slightly, the model in its stochastic formulation does not fully support this: the expected occupancy and the expected metapopulation extinction time decrease gradually when the number of patches is moderate.

8.1 INTRODUCTION

As the human population grows, both in number and in its use of natural resources, natural habitat for many organisms is disappearing and becoming fragmented. Not only (rail) roads, but also agricultural, residential and industrial areas fragment previously connected (or even continuous) habitat. Common sense tells us that the answer to habitat fragmentation is

T.A.C. Reydon and L. Hemerik.,(eds.), Current Themes in Theoretical Biology,
203-232.

defragmentation and hence much effort is put into building corridors, of which fauna crossings are just one example. Corridors are conduits connecting two pieces of habitat through an environment of hostile non-habitat. As such, the use of corridors need not be restricted to the animal kingdom; plants can also use them as stepping-stones for their seeds, enabling them to colonize distant habitat. Although corridors may not only act as conduits but also as habitat, filters or even as barriers (Hess and Fischer, 2001), in most cases they are constructed primarily for their conduit function.

Habitat fragmentation and subsequent attempts of habitat defragmentation affect extinction of local populations as well as recolonization of empty but suitable habitat. Because extinction and colonization are the two core ingredients of metapopulation theory, metapopulation theory seems well suited for predicting the consequences of habitat (de)fragmentation. A metapopulation is defined as a population consisting of several more or less loosely connected local populations with colonization and extinction of these local populations analogous to births and deaths of individuals in a population. The father of all metapopulation models is undoubtedly the Levins (1969, 1970) patch occupancy model in which the habitat consists of many distinct patches, which can be either empty or occupied by a population of the species under consideration. An occupied patch can become empty by extinction of the local population and an empty patch can become occupied after colonization by dispersers from extant populations. Because the risk of extinction is spread (Den Boer, 1968), the metapopulation can persist much longer than a local population, if recolonization occurs frequently.

The Levins model has been extended in many ways to study the effects of the extension on metapopulation dynamics. These extensions include the rescue effect (Hanski, 1983; Hanski et al., 1996; Gyllenberg and Hanski, 1997; Etienne, 2000, 2002), the anti-rescue effect (Harding and McNamara, 2002), the patch preference effect (Ray et al., 1991; Etienne, 2000), multiple species interactions (Levins and Culver, 1971; Slatkin, 1974; Sabelis et al., 1991; Nee and May, 1992; Hess, 1994, 1996; Nee et al., 1997; Taneyhill, 2000; Gog et al., 2002; Nagelkerke and Menken, 2002), succession (Amarasekare and Possingham, 2001), heterogeneous habitat (Holt, 1997), the quality of the matrix habitat (Vandermeer and Carvajal, 2001), spatial structure (Hanski and Ovaskainen, 2000; Ovaskainen and Hanski, 2001), and local population dynamics and dynamics of patch formation and destruction (Hastings, 1991; Gyllenberg and Hanski, 1992; Hanski and Zhang, 1993; Hanski and Gyllenberg, 1993; Hastings, 1995; Gyllenberg and Hanski, 1997; Gyllenberg et al., 1997; Keymer et al., 2000). These models are all deterministic. Stochastic versions of the Levins model have also been developed and explored (Day and Possingham, 1995; Gyllenberg and Silvestrov, 1994; Frank and Wissel, 1998; Lande et al., 1998; Ovaskainen, 2001, Etienne and Heesterbeek, 2001; Etienne, 2002; Etienne and Nagelkerke, 2002; Alonso and McKane, 2002).

In all of these models, little change has been made with respect to the colonization process in the Levins model. That is, the colonization rate or the colonization probability (per unit time) has been taken proportional to the number of, on the one hand, empty patches and, on the other hand, the number of dispersers or, when dispersal is not explicitly modelled, the number of occupied patches (which produce dispersers). Yet, colonization of a patch, i.e. the establishment of a new population, may be hampered by the Allee effect (Wissel, 1994), a reduced or even negative growth rate at small population sizes (Allee, 1931), which may affect this proportionality substantially. The Allee effect can be caused by difficulties in, for example, mate finding, food exploitation (e.g. host resistance can only be overcome by sufficient numbers of consumers) and predator avoidance or defence (Allee, 1931; Courchamp *et al.*, 1999; Stephens and Sutherland, 1999; Stephens *et al.*, 1999; McCarthy, 1997) when population size is small. Colonization may also be hampered by sheer stochasticity: even if the growth rate is positive, stochastic fluctuations in birth and death rates may prevent colonization (Goel and Richter-Dyn, 1974). We will call this the founder effect (not to be confused with Mayr's founder effect in evolutionary biology, although the two concepts are related as they both stress the stochastic nature of colonization), and we will show that it is contained in the Levins model, in contrast to a result obtained by Etienne *et al.* (2002a). Sometimes, the founder effect is also considered a type of Allee effect (Lande, 1998; Keitt *et al.*, 2001; Harding and McNamara, 2002), although not unanimously (Stephens *et al.*, 1999), so we prefer to use separate terms for the two effects.

The Allee effect has previously been brought into connection with metapopulations although the subject was often only briefly touched upon (Courchamp *et al.*, 1999; Reed, 1999; Stephens *et al.*, 1999; Stephens and Sutherland, 1999; Cronin and Strong, 1999; Keitt *et al.*, 2001), or the metapopulation consisted of only two patches (Gyllenberg *et al.*, 1996; Amarasekare, 1998a; Gyllenberg *et al.*, 1999), or the study involved simulations rather than an analytical description and treatment of a dynamical model (Berec *et al.*, 2001; Brassil, 2001; Etienne *et al.*, 2002b). Amarasekare (1998b) did present and analyze a Levins-type model with Allee effect, but this Allee effect was active at the metapopulation level. That is, analogous to the Allee effect at the population level that reduces the growth rate of the population, Amarasekare (1998b) assumed that the Allee effect at the metapopulation level reduces the growth rate of the metapopulation. As a possible cause of this Allee effect at the metapopulation level, Amarasekare (1998b) mentioned the Allee effect at the population level, but she did not model this explicitly. Ovaskainen and Hanski (2001) also considered the Allee effect at the metapopulation level, possibly caused by an Allee effect at the population level which they modelled as in Hanski's (1994) incidence function model. However, they did not study this in great detail, as exact (analytical) solutions seemed impossible in the spatial context they focussed on (but see

Ovaskainen *et al.*, 2002). The structured metapopulation model of Gyllenberg *et al.* (1997) and the model of Gotelli and Kelley (1993) allow for incorporation of the Allee effect, but this has not actually been carried out, except for Harding and McNamara (2002) who studied the Allee effect explicitly in the Gotelli and Kelley (1993) model although they modelled this incorrectly, as we will show below.

Thus, the impact of the Allee effect at the population level on metapopulation dynamics has, to our knowledge, never (or at least hardly) been studied analytically in the Levins framework. One reason may be that it seems very simple to perform and the result seems self-evident: a threshold effect. However, this also applies to many models which have been explored analytically, for example, the Levins model with rescue effect (Hanski, 1983; Hanski *et al.*, 1996; Gyllenberg and Hanski, 1997; Etienne, 2000, 2002; Harding and McNamara, 2002), an effect referring to the ability of immigrants to rescue a population from extinction (Brown and Kodric-Brown, 1977). In these cases, more insight was still gained from a mathematical analysis of the model. Moreover, when a model is extended further, results may no longer be self-evident, because the model has become too complicated. How, for example, will the Allee effect influence metapopulation dynamics when modelled in conjunction with interspecific competition or predator-prey interactions (as in the references mentioned above)? It is important to have a thorough understanding of the impact of the Allee effect when the model is still relatively easily tractable in order to properly evaluate the outcomes in more complicated models. Therefore, we will present a Levins-type model with a population-level Allee effect.

We will start with a review of the Levins model and the more general patch occupancy model, of which it is a special case. We will present deterministic and stochastic formulations of this patch occupancy model, both in discrete and continuous time. The occupancy model enables incorporation of the founder effect and the Allee effect. We will show that this is not straightforward by pointing out where previous attempts (notably Etienne *et al.*, 2002a; Harding and McNamara, 2002) have failed. We will study the dynamics of the resulting models, particularly pertaining to metapopulation persistence, and discuss some implications for conservation. We will end with a summary and discussion of our findings.

8.2 THE LEVINS MODEL

The classical Levins model is generally written in terms of the following differential equation

$$\frac{dp}{dt} = cp(1-p) - mp \tag{8.1}$$

where p is the fraction of occupied patches, c is the colonization rate of empty patches, and m is the extinction rate of occupied patches. Note that the colonization term $cp(1-p)$ is proportional to both the occupied patches (p) and the empty patches ($1-p$), as we mentioned above.

Defining $\beta := c/m$, (8.1) has the nontrivial equilibrium,

$$p^* = 1 - \frac{1}{\beta} =: K. \tag{8.2}$$

We call this K because of its interpretation as the metapopulation carrying capacity (Amarasekare, 1998b). This equilibrium is stable as long as $\beta > 1$ (if $\beta \le 1$ the trivial equilibrium $p^* = 0$ is stable).

To enable comparison with its stochastic version (see below) the Levins model has also been written in terms of the number of occupied patches (Etienne, 2002; Etienne and Nagelkerke, 2002; Harding and McNamara, 2002),

$$\frac{dn}{dt} = c'n(N-n) - mn \tag{8.3}$$

where n is the number of occupied patches and N is the total number of patches in the metapopulation, so $p = n/N$. The colonization parameters c' and c are related to one another by $c = c'N$. The non-trivial equilibrium is, obviously, $n^* = N - m/c'$.

8.3 THE PATCH OCCUPANCY MODEL

The Levins model is a special case of a family of models, called patch occupancy models. A patch occupancy model assumes that patches can be either occupied or empty and describes the dynamics of the number of occupied patches. There are deterministic and stochastic patch occupancy models, formulated in either continuous or discrete time. We will describe each of the four possible combinations. Because the Levins model is a patch occupancy model, there are also stochastic and discrete-time analogues of the Levins model. Treating the Levins model as a special case of a more general model clarifies the interpretation of the Levins model which may otherwise lead to inconsistencies, as we will see below.

The deterministic model in continuous time

The deterministic continuous-time patch occupancy model was presented by Gotelli and Kelley (1993) as

$$\frac{dp}{dt} = C(p)(1-p) - M(p)p \tag{8.4}$$

with $C(p) = cp$ and $M(p) = m$. Again p is the fraction of occupied patches. In (8.4) it is clear that p can be interpreted as the probability of a patch being occupied (Etienne, 2002; Etienne and Nagelkerke, 2002), although this was not explicitly mentioned by Gotelli and Kelley (1993). If the patch is empty (probability $1 - p$), then it becomes occupied at rate $C(p)$ and when it is occupied (probability p), it becomes empty at rate $M(p)$.

When written in terms of the number of occupied patches, (8.4) becomes (Etienne, 2002; Harding and McNamara, 2002)

$$\frac{dn}{dt} = C(n)(N - n) - M(n)n. \tag{8.5}$$

The Levins model has $C(n) = c'n$ and $M(n) = m$.

The stochastic model in continuous time

The general formulation of the stochastic continuous-time patch occupancy model describes the probability P_n of n patches (out of N patches) being occupied, for all $0 \leq n \leq N$:

$$\frac{dP_n}{dt} = C(n-1)\big(N - (n-1)\big)P_{n-1}$$
$$- \big[M(n)n + C(n)(N - n)\big]P_n + M(n)(n+1)P_{n+1} \tag{8.6}$$

with $P_n = 0$ for $n < 0$ and $n > N$.

The model described by (8.6) is a Markov model (Frank and Wissel, 1998; Ovaskainen, 2001; Etienne, 2002; Etienne and Nagelkerke, 2002), so properties of Markovian models can be used to analyse the model. For example, when the model is written in matrix notation, the resulting matrix has the property that its second largest (in absolute value) eigenvalue is a measure of the expected time to extinction of the metapopulation when started in the quasi-stationary state. This quasi-stationary state is the left eigenvector corresponding to this second largest eigenvalue. The elements $i = 2,...,N$ of this eigenvector give the probability of $i - 1$ patches being occupied, conditional on non-extinction of the metapopulation. The quasi-stationary state is therefore a probability distribution over all possible states; it is the probability distribution reached when the system has been undisturbed for some time, and can thus be considered a pseudo-equilibrium (it cannot be a real equilibrium, because the only real equilibrium is metapopulation extinction).

Formula (8.6) is also a special case of the stochastic birth-death model (Goel and Richter-Dyn, 1974),

$$\frac{dP_n}{dt} = b_{n-1}P_{n-1} - \big[b_n + d_n\big]P_n + d_nP_{n+1} \tag{8.7}$$

where b_n and d_n are the birth and death rates, respectively. With $b_n = C(n)(N - n)$ and $d_n = M(n)\,n$ we retrieve (8.6). Being a birth-death model,

(8.6) possesses the convenient property that the expected time to absorption (i.e. extinction), when starting with n_0 patches occupied, is given by (Goel and Richter-Dyn, 1974):

$$T_{ext}(n_0) = \sum_{i=1}^{n_0} \frac{1}{d_i} + \sum_{j=i+1}^{N} \left(\frac{1}{d_j} \prod_{k=i}^{j-1} \frac{b_k}{d_k} \right) \tag{8.8}$$

This formula provides an alternative way of calculating the expected time to metapopulation extinction, when starting in the quasi-stationary state (Etienne and Nagelkerke, 2002).

The deterministic continuous-time model is a limiting case of the stochastic continuous-time model. We will not go into the appropriate limit here, but refer to Etienne (2002) for more details.

The stochastic continuous-time analogue of the Levins model is obtained from the stochastic continuous-time occupancy model by setting $C(n) = c'n$ and $M(n) = m$.

The deterministic model in discrete time

The discrete-time analogue of (8.4) is given by the difference equation

$$n_{t+\Delta t} = n_t + C(n_t, \Delta t)(N - n_t) - M(n_t, \Delta t)n_t \tag{8.9}$$

where n_t is the number of occupied patches at time t, $M(n_t, \Delta t)$ is the probability of extinction of a local population in one time-step of length Δt, and $C(n_t, \Delta t)$ is the probability that an empty patch will be colonized. We will give some examples of $C(n_t, \Delta t)$ below.

If we assume that

$$\lim_{\Delta t \downarrow 0} \frac{C(n_t, \Delta t)}{\Delta t} = C(n) \tag{8.10a}$$

$$\lim_{\Delta t \downarrow 0} \frac{M(n_t, \Delta t)}{\Delta t} = M(n) \tag{8.10b}$$

then (8.9) reduces to the continuous-time model

$$
\begin{aligned}
\frac{dn}{dt} &= \lim_{\Delta t \downarrow 0} \frac{n_{t+\Delta t} - n_t}{\Delta t} = \\
&= \lim_{\Delta t \downarrow 0} \frac{C(n_t, \Delta t)(N - n_t) - M(n_t, \Delta t)n_t}{\Delta t} = \\
&= (N - n_t)\lim_{\Delta t \downarrow 0} \frac{C(n_t, \Delta t)}{\Delta t} - n_t \lim_{\Delta t \downarrow 0} \frac{M(n_t, \Delta t)}{\Delta t} = \\
&= C(n)(N - n) - M(n)n
\end{aligned}
\tag{8.11}
$$

where in the last expression we have dropped the subscript t to indicate that we are dealing with a continuous-time formulation. We use this notation throughout this paper.

There are usually several functions for C and M that satisfy (8.8.10a) and (8.8.10b), because they only need to behave similarly near $\Delta t = 0$. Therefore, there are several discrete-time analogues of the continuous-time patch occupancy model, and hence there are several discrete-time analogues of the original (deterministic continuous-time) Levins model as well.

The stochastic model in discrete time

In discrete time multiple extinctions and colonizations per time step are plausible. Suppose that the probability of a local population of becoming extinct in one time-step of length Δt is $M(n, \Delta t)$ and that the probability of an empty patch of becoming occupied is $C(n, \Delta t)$ if there are n occupied patches. This is fully analogous to the deterministic model in discrete time. The transition probability P_{kl} to move from k occupied patches to l occupied patches in one time-step is then given by (see Gyllenberg and Silvestrov, 1994)

$$P_{kl} = \sum_{i=0}^{\min(k,l)} \binom{k}{i} M(k,\Delta t)^{k-i} \cdot$$

$$\left[1 - M(k,\Delta t)\right]^{i} \binom{N-i}{l-i} C(k,\Delta t)^{l-i} \left[1 - C(k,\Delta t)\right]^{N-l}. \tag{8.12}$$

The P_{kl} form a matrix, of which the second largest eigenvalue is a measure of the metapopulation extinction time, similar to the continuous-time model. In (8.12), extinctions and colonizations can occur simultaneously. There is also a version where there is an extinction phase followed by a colonization phase (Akçakaya and Ginzburg, 1991; Day and Possingham, 1995). We obtain the corresponding formula for P_{kl} by replacing $C(k, \Delta t)$ in (8.12) by $C(i, \Delta t)$.

We refrain from describing the transition from the stochastic discrete-time model to the stochastic continuous-time model, because this is more complicated and because we will not be using it in this paper.

8.4 THE FOUNDER EFFECT

In the introduction we defined the founder effect as the effect that colonization may be hampered by sheer stochasticity: even if the growth rate is positive, stochastic fluctuations in birth and death rates may prevent colonization. We will first incorporate the founder effect into the deterministic discrete-time patch occupancy model and then take the appropriate limit to obtain the continuous-time model. Subsequently, we will present a direct way of incorporating the founder effect in the continuous-time model. This gives a different model, but this model also has a different interpretation. We conclude the section with a more detailed model of the founder effect where stochasticity is not only present in the establishment of a new population, but already in reaching the destination patch.

The deterministic model in discrete time

In the Levins model the mechanism of colonization is not modelled explicitly. Here, we attempt to find a more mechanistic basis for the colonization term in the Levins model. This is most comprehensibly achieved in the discrete-time model (8.9). Colonization may be described by a cumulative-hit model. In this model, arrival of a disperser does not necessarily imply colonization, but the more dispersers arrive during the time interval Δt, the larger the probability of colonization is. The probability $C(n_t, \Delta t)$ of an empty patch being colonized during the time interval Δt thus increases with Δt, because the number of dispersers arriving at the patch increases with Δt.

Assume that each patch produces $L(\Delta t)$ dispersers that immigrate into any patch, empty or occupied with probability $p_d(\Delta t)$ (d of dispersal). Hence, each patch receives

$$I(n_t, \Delta t) = \frac{L(\Delta t) p_d(\Delta t) n_t}{N} \qquad (8.13)$$

dispersers during the time interval Δt; note that dispersers from all occupied patches are added together. We further assume that these dispersers arrive simultaneously, which is fairly realistic for species with a short dispersal period, and that they reproduce asexually. Local population dynamics are governed by a birth-death model (Goel and Richter-Dyn, 1974). The colonization probability, starting from $I(n_t, \Delta t)$ immigrants, is given by

$$C(n_t, \Delta t) = 1 - \left(\frac{1}{R_0}\right)^{I(n_t, \Delta t)}, \quad R_0 > 1 \qquad (8.14)$$

where R_0 is the basic reproduction number of the population, that is, the expected number of off-spring of each founder of the colony (Goel and Richter-Dyn, 1974). Because $C(n_t, \Delta t) < 1$, the arrival of dispersers at an empty patch does not guarantee colonization, even if $R_0 > 1$, although unsuccessful colonization is very unlikely if there are many dispersers. This is the founder effect or establishment effect (Lande et al., 1998).

Let us now insert (8.14) into (8.9). Furthermore, we assume that the extinction probability does not depend on the number of occupied patches, so $M(n_t, \Delta t) = M(\Delta t)$. We will also assume that condition (8.8.10b) holds. Because our interest in this paper lies in the colonization process, this assumption will be used throughout the rest of this paper. Because

$$\frac{d^2}{dn_t^2} C(n_t, \Delta t)(N - n_t) < 0 \qquad (8.15)$$

for all $n_t < N$, the condition for a stable equilibrium is found to be (see also Figure 8.1):

$$\frac{L(\Delta t)p_d(\Delta t)\ln R_0}{M(\Delta t)} > 1. \tag{8.16}$$

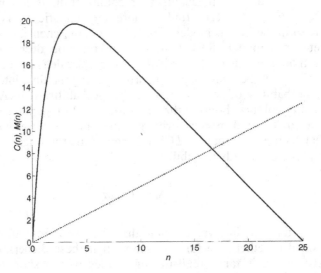

Figure 8.1. The contributions of colonization ($C(n_t, \Delta t)(N - n_t)$, solid curve) and extinction ($M(\Delta t)n_t$, dotted curve) in the deterministic discrete-time Levins model, that is, (8.9) with $C(n_t, \Delta t)$ given by (8.17) and $M(n_t, \Delta t) = M(\Delta t)$. The equilibria are at the intersection points of these curves. Because the contribution of colonization always has a negative second derivative, a non-trivial equilibrium can only exist if the colonization contribution has a larger slope at the origin than the extinction contribution. In this figure we used the numerical values $N = 25$, $s(\Delta t) = 0.5$ and $M(\Delta t) = 0.5$.

We would like to point out here that a similar condition would have been obtained without the birth-death model, and even without the cumulative-hit hypothesis; a one-hit model (arrival of at least one disperser at an empty patch entails colonization) may also apply. Suppose that during a time interval of length Δt each occupied patch can colonize an empty patch with probability $s(\Delta t)$. This probability, which by definition does not depend on n_t, can represent the probability that at least one disperser will reach the empty patch (which is sufficient for colonization under the one-hit hypothesis) or the probability that the dispersers that arrive at the empty patch establish a population (as we assumed above), or both. The colonization probability is given by

$$C(n_t, \Delta t) = 1 - \left(1 - s(\Delta t)\right)^{n_t} \tag{8.17}$$

and this is mathematically equivalent to (8.17) if

$$s(\Delta t) = 1 - \left(\frac{1}{R_0}\right)^{\frac{L(\Delta t)p_d(\Delta t)}{N}}. \tag{8.18}$$

Note that indeed $s(\Delta t)$ does not depend on n_t. With (8.17), the condition for a stable nontrivial equilibrium reads

$$-\frac{N\ln(1 - s(\Delta t))}{M(\Delta t)} > 1. \tag{8.19}$$

The deterministic model: from discrete time to continuous time

We remarked above that under certain conditions the discrete-time Levins model reduces to the classical continuous-time Levins model. The condition for the colonization term, (8.8.10a), where $C(n) = c'n$ and $M(n) = m$ for the Levins model, translates to the following expression in our model with (8.17):

$$\lim_{\Delta t \downarrow 0} \frac{s(\Delta t)}{\Delta t} = c' \tag{8.20}$$

where c' is a constant. For the cumulative-hit model, this in turn translates to

$$\lim_{\Delta t \downarrow 0} \frac{I(n_t, \Delta t)}{\Delta t} = c'' \tag{8.21}$$

with c'' another constant. With (8.20) condition (8.8.10a) is indeed satisfied, because

$$\lim_{\Delta t \downarrow 0} \frac{C(n_t, \Delta t)}{\Delta t} = \lim_{\Delta t \downarrow 0} \frac{1 - \left(1 - s(\Delta t)\right)^{n_t}}{\Delta t}$$

$$= \lim_{\Delta t \downarrow 0} \frac{n_t\left(1 - s(\Delta t)\right)^{n_t - 1} \frac{ds(x)}{dx}\Big|_{x=\Delta t}}{1} = c'n_t \tag{8.22}$$

Thus the original Levins model is recovered. Hence, the colonization term in the Levins model describes the founder effect.

The deterministic model in continuous time

Let us now look at an attempt to incorporate the founder effect in the deterministic continuous-time model directly (Etienne *et al.*, 2002a). We adopt the interpretation of c' as the product of the rate c_{out} at which dispersers are produced by an occupied patch and the probability c_{in} with which a disperser successfully colonizes a patch (Frank and Wissel, 1998; Etienne and Heesterbeek, 2000; Ovaskainen and Hanski, 2001). The founder effect must evidently be inserted into c_{in}. Suppose, for instance, that $c_{in} = c_{out}n/(c_{out}n + a)$ for some parameter a, a formula that is similar to (8.14) and seems a reasonable choice to model the founder effect phenomenologically (Etienne *et al.*, 2002a). It says that c_{in} is a saturating function of the number of dispersers produced by all occupied patches per unit of time, $c_{out}n$. Note that this is different from the assumptions that led to (8.17), not just because a different saturating function is taken, but particularly because the argument of the function is the rate of dispersers produced instead of a number within a time window. The model becomes (Etienne *et al.*, 2002a)

$$\frac{dn}{dt} = c_{out}n\frac{c_{out}n}{c_{out}n+a}\left(N-n\right) - mn \qquad (8.23)$$

which shows entirely different dynamics than the Levins model (8.3). In fact, it is mathematically similar to the Levins model with Allee effect that we will construct below.

The difficulty with (8.23) is that the use of saturating functions in continuous-time models is tricky, because they often seem to rely on discrete-time considerations, as we will show below for the Allee effect. In the current case, it is unclear what the mechanistic basis is for the saturating function for c_{in}, and therefore it is unclear whether it really describes the founder effect as we understand it. We expect c_{in} to become constant for large $c_{out}n$, because for large $c_{out}n$ we no longer expect to find a founder effect, but it is not evident what the behavior for small $c_{out}n$ should be. Our expressions (8.14) and (8.17) are less phenomenological than the expression for c_{in}, because they are based on probabilistic arguments of whether a set of dispersers can establish a population.

In summary, the founder effect is contained in the Levins model as our transition from discrete to continuous time has shown, and the model (8.23) does not describe the founder effect (but the Allee effect, see below).

Stochasticity in dispersal: colonization average vs. average colonization in the founder effect

Until now we have assumed that a fixed number of dispersers, $L(\Delta t)p_d(\Delta t)n_t/N$, arrives at an empty patch during the time interval Δt upon which colonization may or may not occur. Hence, the number of arriving dispersers is modelled deterministically. If, instead, we take the number of dispersers leaving the occupied patch to be a fixed number, say $L(\Delta t)$, and if we let the probability of reaching a patch be governed by a binomial distribution with parameter $p_d(\Delta t)$, then we only have *on average*

$$\bar{I}(n_,\Delta t) = \frac{L(\Delta t)p_d(\Delta t)n_t}{N}. \qquad (8.24)$$

By inserting this expression into (8.14), we would effectively use the colonization probability of the average number of arriving dispersers:

$$C(\bar{I}(n_t, \Delta t)) = 1 - \left(\frac{1}{R_0}\right)^{\frac{L(\Delta t)p_d(\Delta t)n_t}{N}}. \tag{8.25}$$

This is incorrect. We should use the average colonization probability instead:

$$\overline{C(I(n_t, \Delta t))} = \sum_{j=0}^{\frac{L(\Delta t)n_t}{N}} \left[1 - \left(\frac{1}{R_0}\right)^j\right] \left(\frac{\frac{L(\Delta t)n_t}{N}}{j}\right) [p_d(\Delta t)]^j [1 - p_d(\Delta t)]^{\frac{L(\Delta t)n_t}{N} - j}$$

$$= 1 - \left[1 - p_d(\Delta t)\left(1 - \frac{1}{R_0}\right)\right]^{\frac{L(\Delta t)n_t}{N}}. \tag{8.26}$$

Equations (8.25) and (8.26) are not equal. Figure 8.2 shows them for $L(\Delta t)n_t/N = 100$ and $R_0 = 3$. Using (8.25) therefore leads to a higher colonization probability.

In the following sections, we will adhere to the original model with a deterministic (fixed) number of dispersers arriving at an empty patch, because it is qualitatively equivalent to, but much simpler than the more detailed model where the number of dispersers arriving at an empty patch is a stochastic variable. As for quantitative differences, we checked these for several parameter settings and found that they are never very large.

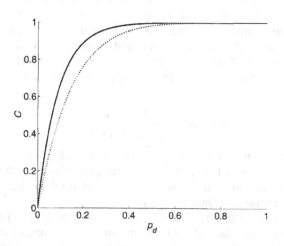

Figure 8.2. The colonization function of the average number of immigrants (solid curve) and the average colonization function of the number of immigrants (dotted line) plotted against the probability of successful dispersal p_d. Taking the average at the level of arriving immigrants thus leads to an overestimate of the colonization function. Numerical values of the other parameters are $L(\Delta t)n_t/N = 100$ and $R_0 = 3$.

8.5 THE ALLEE EFFECT

The Allee effect differs from the founder effect in that the Allee effect denotes a reduced or negative *deterministic* growth rate at small population sizes, whereas the founder effect is a purely *stochastic* phenomenon, most active at small population sizes, that occurs even with an unreduced positive growth rate. We will incorporate the Allee effect (at the population level, not at the metapopulation level as in Amarasekare, 1998b) in the deterministic discrete-time model, and then take the appropriate limit to get the analogous continuous-time model. Because this yields a rather undesirable result, we will then start from a continuous-time formulation and convert this back to discrete-time. In all cases we will study metapopulation dynamics. To gain insight into the threshold phenomenon exhibited in the deterministic models, we move to a stochastic formulation. We end by examining the Allee effect in conjunction with the patch preference effect, because we think that these will often occur together, and because they may represent opposing forces making the outcome less obvious.

The deterministic model in discrete time

The founder effect typically represents the case where the colonization probability $C(n_t, \Delta t)$ is linear in the number of occupied patches n_t when $s(\Delta t)$ is small. This means that contributions from different patches are simply additive. When the Allee effect is active, the colonization probability behaves differently. There may be a minimum number of immigrants necessary for the population to grow and thus for colonization to occur or $C(n_t, \Delta t)$ may initially be quadratic in the number of immigrants (or the number of occupied patches producing dispersers). Instead of a detailed model of colonization for a population suffering from the Allee effect, we consider a more phenomenological model, again in discrete time because of its easier interpretation:

$$C(n_t, \Delta t) = \frac{I^2(n_t, \Delta t)}{I^2(n_t, \Delta t) + y^2} \qquad (8.27)$$

where y is some constant, measuring the strength of the Allee effect. Thus, colonization as a function of the number of immigrants has a sigmoidal shape. Formula (8.27) is not just an ad hoc choice; it is used in the well-known incidence function model of Hanski (1994).

Let us again assume that the number of immigrants $I(n_t, \Delta t)$ entering an empty patch during time interval Δt is determined by the number of dispersers that are produced by each occupied patch, $L(\Delta t)$, and by the probability that each disperser reaches the empty patch, that is, we use (8.13). We can rewrite (8.27) as

$$C(n_t, \Delta t) = \frac{n_t^2}{n_t^2 + y'^2(\Delta t)} \tag{8.28}$$

where $y'(\Delta t) = yN/(L(\Delta t)p_d(\Delta t))$.

Inserting this into (8.9) leads to the following equation for the equilibrium n^*:

$$\frac{n^{*2}}{n^{*2} + y'^2(\Delta t)}(N - n^*) = M(\Delta t)n^*. \tag{8.29}$$

This results in three equilibria, dropping the dependence on Δt for notational convenience:

$$n_0^* = 0$$

$$n_-^* = \frac{N}{1+M}\left(\frac{1}{2} - \frac{1}{2}\sqrt{1 - 4\frac{My'^2(1+M)}{N^2}}\right)$$

$$= \frac{N}{1+M}\left(\frac{1}{2} - \frac{1}{2}\sqrt{1 - 4My''^2(1+M)}\right) \tag{8.30}$$

$$n_+^* = \frac{N}{1+M}\left(\frac{1}{2} + \frac{1}{2}\sqrt{1 - 4\frac{My'^2(1+M)}{N^2}}\right)$$

$$= \frac{N}{1+M}\left(\frac{1}{2} + \frac{1}{2}\sqrt{1 - 4My''^2(1+M)}\right)$$

where we have defined $y'' = y'/N = y/(Lp_d)$. Of these solutions, n_-^* is unstable and forms the separatrix between the other two, stable, solutions. This means that, when the system is started below the separatrix, the metapopulation will become extinct, and when started above the separatrix, it will converge to the nontrivial equilibrium n_+^*. See Figure 8.3.

Figure 8.3A shows the catastrophic consequences of increased dispersal resistance due to (rail) roads and the like (increased dispersal resistance means a reduction of the probability of reaching a patch p_d and hence an increase in y''). If this dispersal resistance grows gradually (for example, because there is a gradual increase of the amount of cars using the road), then p^* may seem to be hardly affected, until suddenly the metapopulation disappears (in Figure 8.3A this happens for $y'' \approx 2.88$). Figure 8.3B shows similar behaviour for increases in the probability of local extinction, for example due to habitat deterioration.

Harding and McNamara (2002) also used (8.27), but they did so in the continuous-time model (8.5), that is, they set $C(n) = \alpha^2(n)/(\alpha^2(n) + y^2)$, following Hanski (1994). However, Hanski's (1994) model is a discrete-time model, and $C(n)$ is the probability of colonization per time-step, whereas a continuous-time model requires a probability *rate*. They do define α as the immigration rate, which is correctly used in the Levins model where they have $C(n) \sim \alpha(n)$, but this does not make $C(n)$ a rate in this case. Specifically, $C(n)$ cannot become greater than unity in this formulation (because it is in fact a probability), but the colonization rate may surely become much greater than unity; it usually becomes infinite when the immigration rate becomes infinite. This is not just a matter of a scaling factor. Although Harding and McNamara (2002) may have omitted a constant to simplify their expressions, addition of a constant, that is, using $C(n) = k\alpha^2(n)/(\alpha^2(n) + y^2)$ instead where k is a constant with the dimension of rate, does not resolve the issue. The colonization rate is then explicitly allowed to become larger than unity, but an infinite immigration rate still does not result in an infinite colonization rate. Of course, one can have a model where an infinite immigration rate does not automatically correspond to an infinite colonization rate, but then one is trying to model a different phenomenon, not the Allee effect. Thus, apparently it is not straightforward to incorporate the Allee effect into the continuous-time Levins model as it seems, so our attempt seems well justified.

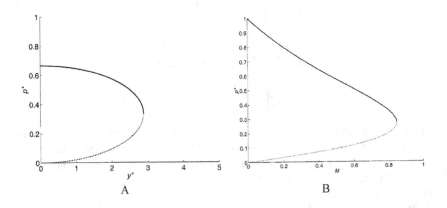

Figure 8.3. Bifurcation diagrams of the discrete-time model (8.9) with Allee effect as in (8.27) and $M(n_t, \Delta t) = M(\Delta t) = M$. The solid curves represent the stable equilibria, the dotted line represents the unstable equilibrium. A fold bifurcation occurs where the unstable and stable branches meet. A: The bifurcation parameter is y''; $M = 0.5$. B: The bifurcation parameter is M; $y'' = 0.4$.

The deterministic model: from discrete to continuous time and back

Above we noted that the Levins model with founder effect is just the Levins model if (8.21) is satisfied. Now, let us examine what happens if we use this same condition in the Levins model with Allee effect. Using (8.27):

$$\lim_{\Delta t \downarrow 0} \frac{C(n_t, \Delta t)}{\Delta t} = \lim_{\Delta t \downarrow 0} \frac{\dfrac{I^2(n_t, \Delta t)}{I^2(n_t, \Delta t) + y^2}}{\Delta t} =$$

$$\lim_{\Delta t \downarrow 0} \frac{\left(I^2(n_t, \Delta t) + y^2\right) 2 I(n_t, \Delta t) \left.\dfrac{dI(x)}{dx}\right|_{x=\Delta t} - I^2(n_t, \Delta t) 2 I(n_t, \Delta t) \left.\dfrac{dI(x)}{dx}\right|_{x=\Delta t}}{\left(I^2(n_t, \Delta t) + y^2\right)^2} = 0 \qquad (8.31)$$

because $I(n_t, 0) = 0$. Hence, the conversion to a continuous-time model leads to a model in which the metapopulation cannot persist!

This is certainly an undesirable result, so the suggestion presents itself that we are dealing with an artefact inherent to our choice of (8.27). To examine this in more detail, we take a little detour. We start with the colonization term of the continuous-time Levins model in the Harding and McNamara (2002) framework, that is,

$$C(n) = k \alpha n \qquad (8.32)$$

where k is a positive constant and α the immigration rate, with the metapopulation dynamics governed by (8.5), so in effect we have $c' = k\alpha$. This $C(n)$ represents the colonization probability per unit time. Consider a patch that is empty at $t = 0$ and suppose that n patches are occupied at time $t = 0$ and that local extinction is impossible. The probability p that the empty patch is colonized (and hence occupied) at time t is described by

$$\frac{dp}{dt} = C(n)(1 - p) = k \alpha n (1 - p). \qquad (8.33)$$

The solution is readily found to be $p(t) = 1 - \exp(-k \alpha n t)$. Hence, after a time interval Δt the probability that the patch is occupied is $p(\Delta t) = 1 - \exp(-k \alpha n \Delta t)$. We can use this as the colonization probability in the discrete-time model (8.9), that is,

$$C(n_t, \Delta t) = 1 - \exp(-k I(n_t, \Delta t)) \qquad (8.34)$$

with $I(n_t, \Delta t) = n \alpha \Delta t$. This equation is mathematically equivalent to (8.14) with $R_0 = \exp(k)$. So, the continuous-time model with linear colonization rate corresponds to the discrete-time model with colonization probability that approaches its maximum of unity at an exponentially decreasing rate. The question now arises whether we can find a colonization formula in the continuous-time model that corresponds to a colonization probability in the discrete-time model with the sigmoidal shape that is so typical of the Allee effect. Evidently, (8.27) is surely not the only function with a sigmoidal shape.

Of this colonization formula in the continuous-time model we require that it is linear in the immigration rate α for large α and that it is quadratic in the immigration rate for small α. Perhaps the simplest formula satisfying these requirements is

$$C(n) = k \frac{(\alpha n)^2}{\alpha n + \alpha_0} \tag{8.35}$$

where α_0 measures the strength of the Allee effect; for $\alpha_0 = 0$ we obtain the Levins model colonization term $C(n) = k\alpha n$. Using an argument similar to the one making the transition from (8.32) to (8.34), we find that equation (8.35) corresponds to a colonization probability in the discrete-time model

$$C(n_t, \Delta t) = 1 - \exp\left(-k \frac{(\alpha n_t)^2}{\alpha n_t + \alpha_0} \Delta t\right). \tag{8.36}$$

The colonization rates and corresponding colonization probabilities are shown in Figure 8.4. We stress that although (8.35) is not a mechanistically derived model, its parameters have a biological meaning, which may be substituted by expressions that *are* derived from mechanistic submodels. Below, we will present an example of this.

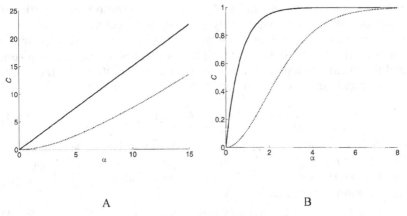

A   B

Figure 8.4. Colonization as a function of the immigration rate α for the Levins model with founder effect (equations (8.32) and (8.34), solid curves) and with Allee effect (equations (8.35) and (8.36), dotted curves) for continuous time (A) and for discrete time (B). Parameter values are $k = 1.5$ and $\alpha_0 = 10$. We further set $n = 1$.

The equilibrium number of patches for the discrete-time model with (8.36) is given by the solution of

$$\left[1 - \exp\left(-k \frac{(\alpha n^*)^2}{\alpha n^* + \alpha_0} \Delta t\right)\right](N - n^*) = Mn^*. \tag{8.37}$$

In Figure 8.5 we have plotted the colonization and extinction terms. It can easily be seen that, in contrast to Figure 8.1, there may be two nontrivial equilibria, of which the larger one is stable. They coincide if

$$\frac{d}{dn}\left[1-\exp\left(-k\frac{(\alpha n)^2}{\alpha n+\alpha_0}\Delta t\right)\right](N-n)\Bigg|_{n=n^*} = M \qquad (8.38)$$

which is again a fold bifurcation.

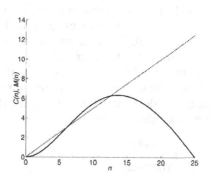

Figure 8.5. The colonization (solid curve) and extinction functions (dotted curve) in the deterministic discrete-time Levins model (8.9) with (8.36) and $M(n_t, \Delta t) = M(\Delta t)$. The equilibria are at the intersection points of these curves. There can be two non-trivial equilibria, the larger one of which is stable. In this figure we used the numerical values $N = 25$, $\alpha_0 = 100$, $k = 0.5$, $\Delta = 1$, $\Delta t = 1$, and $M(\Delta t) = 0.5$.

Thus, this model behaves qualitatively similarly to the solution to (8.29). We conclude that (8.35) and (8.36) provide a consistent way to incorporate the Allee effect into the continuous-time and discrete-time Levins model respectively. We think that (8.36) is also a better formula than (8.27), because (8.36) does not lead to undesirable results when converted to its continuous-time analogue and it can be reduced to (8.34) by setting $\alpha_0 = 0$.

There is one subtlety in (8.36) that needs some attention. If we rewrite (8.36) in terms of the number of immigrants $I(n_t, \Delta t) = \alpha n_t \Delta t$, we obtain

$$C(n_t, \Delta t) = 1-\exp\left(-k\frac{I^2(n_t, \Delta t)}{I(n_t, \Delta t)+I_0(\Delta t)}\right) \qquad (8.39)$$

where $I_0(\Delta t) = \alpha_0 \Delta t$. Hence, the strength of the Allee effect as measured by $I_0(\Delta t)$ depends on Δt. A longer time interval Δt will therefore have a positive and a negative effect on the colonization probability: the number of arriving dispersers increases, but so does $I_0(\Delta t)$. The net increase is positive and linear in Δt for small Δt, as it should be in order to obey (8.8.10a).

When we compare the Allee effect model, (8.35), to the colonization term in (8.23), which was used by Etienne *et al.* (2002a) for the founder effect, we see that the colonization terms are identical, apart from the constant k. So, the results obtained by Etienne *et al.* (2002a) actually apply to the Levins model with (sigmoidal) Allee effect.

The stochastic model in continuous time

The bifurcation in the deterministic models can be used as a warning that small changes in parameters (such as m) may cause the nontrivial equilibrium to disappear suddenly, and thus metapopulation extinction. If these changes are really so abrupt, it is interesting to investigate if there are also abrupt changes in the stochastic model. To examine this, we will consider the continuous-time model (8.6) with Allee effect as modelled in (8.35). We choose the continuous-time model, merely because it allows for explicit analytical solutions. Results obtained with the discrete-time model are similar.

But let us first look at the deterministic counterpart, (8.5) and (8.35), for comparison. The equilibria are the solutions of

$$k\frac{\left(\alpha n^*\right)^2}{\alpha n^* + \alpha_0}(N - n^*) = mn^* \tag{8.40}$$

which are readily found to be (see also Etienne *et al.* 2002a)

$$n_0^* = 0$$

$$n_-^* = \frac{1}{2}\left(N - \frac{m}{c'}\right)\left(1 - \sqrt{1 - 4\frac{\dfrac{m}{c'}\dfrac{\alpha_0}{\alpha}}{\left(N - \dfrac{m}{c'}\right)^2}}\right) \tag{8.41}$$

$$n_+^* = \frac{1}{2}\left(N - \frac{m}{c'}\right)\left(1 + \sqrt{1 - 4\frac{\dfrac{m}{c'}\dfrac{\alpha_0}{\alpha}}{\left(N - \dfrac{m}{c'}\right)^2}}\right)$$

where $c' := k\alpha$ as before.

Returning to the stochastic model, for a system starting in the quasi-stationary state the expected time to metapopulation extinction and the expected occupancy conditional on non-extinction can be computed using the eigenvalues and eigenvectors of the matrix corresponding to (8.6), as we mentioned above. For systems starting in another state, e.g. all patches occupied, the expected metapopulation extinction time and the expected occupancy at some specified time can be calculated using (8.8) and the solution to (8.6) respectively. The mean occupancy at time $t = 10$ of a metapopulation with all patches occupied at time $t = 0$ is plotted in Figure 8.6A, together with the deterministic non-trivial equilibria. The results for the Levins model without Allee effect are also shown. We see that there is no abrupt change in the expected occupancy, although it does decrease faster than in the model without Allee effect. The expected time to metapopulation extinction, plotted in Figure 8.6B, behaves similarly. Neither does it make a difference if we use the expected occupancies and extinction times for systems starting in the quasi-stationary state or any other state (results not shown). As N increases, the expected occupancy follows the deterministic equilibrium more closely and the expected extinction time increases more steeply with increasing immigration rate, but the bifurcation point is found at lower immigration rates. We therefore conclude that the Allee effect does have a detrimental effect on metapopulation persistence, but small parameter changes will not lead to sudden extinction when the number of patches is moderate. Only when there are many patches, the bifurcation of the deterministic model is reflected in the stochastic quantities of expected occupancy and extinction time, but the inviability region is then much smaller (Figure 8.7).

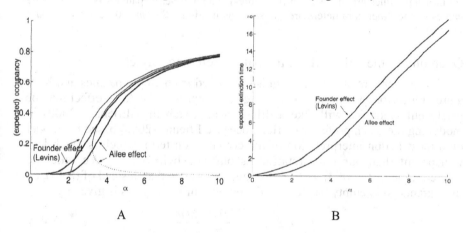

A   B

Figure 8.6. The mean occupancy at time $t = 10$ starting with all patches occupied (A) and the expected time to metapopulation extinction (B) for the Levins model with and without the Allee effect as a function of the immigration rate α. In A the deterministic equilibrium values are also shown (thin lines). The values of the other parameters are $m = 0.52$, $N = 25$, $k = 0.01$ and $\alpha_0 = 10$.

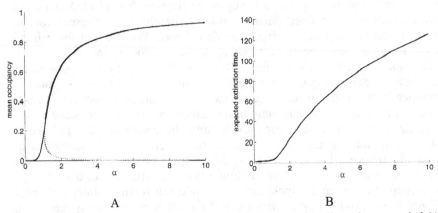

A   B

Figure 8.7. The mean occupancy at time $t = 10$ starting with all patches occupied (A) and the expected time to metapopulation extinction (B) for the Levins model with Allee effect as a function of the immigration rate α. In A the deterministic equilibrium values are also shown (thin lines). The number of patches is $N = 75$. The values of the other parameters are the same as in Figure 8.6: $m = 0.52$, $k = 0.01$ and $\alpha_0 = 10$.

Overcoming the Allee effect: the patch preference effect

If dispersers are able to distinguish occupied from empty patches and have a preference for empty patches, the negative impact of the Allee effect may be partly nullified. We will take a different approach than Etienne (2000) in modelling the patch preference effect, because Etienne (2000) based his model on a mass action interpretation of the colonization term, which is much less transparent than our interpretation as outlined below (8.4). We take the discrete-time model (8.9) with (8.36), but we assume that, instead of (8.13), all immigrants go to empty patches, so the number of immigrants is given by

$$I'(n_t, \Delta t) = \frac{L(\Delta t) p_d(\Delta t) n_t}{N - n_t}. \tag{8.42}$$

With (8.42) the colonization probability becomes

$$C(n_t, \Delta t) = 1 - \exp\left(-k \frac{I'^2(n_t, \Delta t)}{I'(n_t, \Delta t) + I_0(\Delta t)}\right) =$$

$$= 1 - \exp\left(-kL(\Delta t) p_d(\Delta t) \frac{\left(\dfrac{n_t}{N - n_t}\right)^2}{\dfrac{n_t}{N - n_t} + \dfrac{I_0(\Delta t)}{L(\Delta t) p_d(\Delta t)}}\right). \tag{8.43}$$

The equilibria are the solutions of

$$\left[1-\exp\left(-k'\frac{\left(\dfrac{n^*}{N-n^*}\right)^2}{\dfrac{n^*}{N-n^*}+y''}\right)\right](N-n^*)=Mn^* \qquad (8.44)$$

where we defined $k':=kL(\Delta t)p_d(\Delta t)$, $y'':=I_0(\Delta t)/(L(\Delta t)p_d(\Delta t))$ and $M=M(\Delta t)$. There is no explicit expression for the solutions, except for $n_0^*=0$ which is a stable equilibrium, but we do have explicit expressions for the bifurcation parameters y'' and M:

$$y''=-\frac{n^*}{N-n^*}\left(1+\frac{k'n^*}{(N-n^*)\ln\left(\dfrac{N-n^*(1+M)}{N-n^*}\right)}\right) \qquad (8.45)$$

$$M=\left[1-\exp\left(-k'\frac{\left(\dfrac{n^*}{N-n^*}\right)^2}{\dfrac{n^*}{N-n^*}+y''}\right)\right]\frac{N-n^*}{n^*}. \qquad (8.46)$$

The bifurcation diagrams are shown in Figure 8.8.

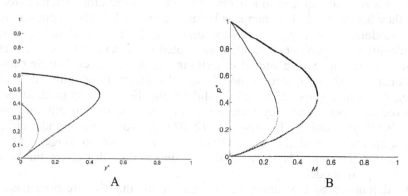

Figure 8.8. Bifurcation diagrams of the discrete-time model (8.9) with Allee effect and patch reference effect as in (8.43) and $M(n_t,\Delta t)=M(\Delta t)=M$ (thick lines) and Allee effect only (thin lines). The solid curves represent the stable equilibria, the dotted line represents the unstable equilibrium. A fold bifurcation occurs where the unstable and stable branches meet. A. The bifurcation parameter is y''; $M=0.5$. B. The bifurcation parameter is M; $y''=0.4$. In both panels $k'=1$.

As expected, the equilibrium fraction of occupied patches in the model with Allee effect and patch preference effect is higher than the equilibrium fraction in the model with Allee effect but without patch preference effect. It is even higher than the model without Allee effect and patch preference effect (for which $y'' = 0$) for all values of y'' that allow a non-trivial equilibrium. In this sense we can say that patch preference can overcome the Allee effect. In other words: aggregation overcomes the Allee effect.

8.6 DISCUSSION

In this paper we have explicitly shown that the founder effect is contained in the Levins model. We also demonstrated how to consistently incorporate the Allee effect (acting at the population level) into the Levins model. We moved between discrete-time and continuous-time, deterministic and stochastic formulations of the metapopulation model to see how these formulations are related and to cast different lights on the founder and Allee effects. In these formulations, we prefer the interpretation of the patch occupancy as the probability of being occupied. This is equivalent to the patch occupancy in the mean-field approximation, but has the advantage that it can be consistently extended to cases in which this approximation no longer applies.

In our study of the founder and Allee effects we stressed the difference between the colonization term in the continuous-time and discrete-time formulations of the models, which, in spite of its apparent triviality, has led to some misunderstanding. The direct attempts of Etienne *et al.* (2002a) and Harding and McNamara (2002) to model the founder and Allee effects in a continuous-time model lead to models that describe something different from what they had in mind. This happens because they used, in their continuous-time models, saturating functions that are based on discrete-time considerations. Etienne *et al.* (2002a) attempted to model the founder effect, but they ended up with a model describing the Allee effect. Harding and McNamara (2002) attempted to model the Allee effect in a continuous-time model, but their results are actually valid for the discrete-time model (8.9). They did not notice that they were erroneous, because their results did not contradict their intuitions; Etienne *et al.* (2002a) considered the founder effect to be a special case of the Allee effect (with which we no longer agree), whereas Harding and McNamara (2002) still studied the Allee effect, but for a different type of model.

We demonstrated that under certain conditions the discrete-time model reduces to the continuous-time model when the appropriate limit is taken. The colonization function in the birth-death model (Goel and Richter-Dyn, 1974), representing the founder effect, satisfies these conditions, and hence the Levins model can be claimed to incorporate the founder effect, as has always been assumed. However, these conditions are not satisfied by the widely applied Hanski (1994) sigmoidal formulation considered to represent the Allee effect. On the contrary, this formulation leads to metapopulation extinction in all cases when transformed to the continuous-time model. We formulated a new function to describe the Allee effect that has the correct properties in both the discrete-time and the continuous-time formulations, and has the advantage over the function of Hanski (1994) that it reduces to the founder effect function when the parameter measuring the Allee effect is set to 0. This comes at a price, though: the Allee effect is no longer associated solely with a threshold of the number of immigrants arriving in a certain time interval, but this threshold also depends on the length of the time interval. This is not biologically unreasonable: colonization is less successful when the time interval is longer while the number of immigrants remains the same, because in this case immigrants arrive at the patch in smaller groups less simultaneously, making the Allee effect potentially stronger.

Our new function for the Allee effect is, just as the one based on Hanski (1994), a purely phenomenological model of the Allee effect: no mechanistic basis has been provided. Incorporation of a phenomenological submodel in a mechanistic model need not be a problem at all, as this has been common practice with the rescue effect as well (Hanski, 1983; Hanski *et al.*, 1996); in a broader population modeling context, the logistic growth model provides a well-known example. Still, for the rescue effect the underlying mechanism has been studied in more detail (Gyllenberg and Hanski, 1997; Etienne, 2002), so a more mechanistic approach is also recommended for the Allee effect.

Using our new function in the stochastic and deterministic formulations, we showed that the fold bifurcation in the deterministic model should not be given the interpretation that a small change in parameter values may cause an abrupt change in metapopulation persistence, that is, sudden metapopulation extinction, if the network contains a moderate number of patches. This is readily explained. In the deterministic model the equilibrium suddenly (dis)appears when the bifurcation parameter passes the bifurcation point, but the stability of the equilibrium changes continuously, vanishing at the bifurcation point (because the stable and unstable branch meet at the fold bifurcation). Hence, the attracting force of the non-trivial equilibrium changes gradually. Only when the number of patches is large, are there sharp changes in the attracting force, so that the stochastic model increasingly resembles the deterministic model as the number of patches increase. This is indeed what we observed.

To reduce the impact of the Allee effect, one need not increase the immigration rate *per se*. We showed that increasing patch preference for empty patches also diminishes the Allee effect. Preference for empty patches can be a natural phenomenon, for example it is plausible for territorial species. But preference for empty patches can also be enforced by (temporarily) closing corridors connecting occupied patches, thus redirecting migration to empty patches. However, preference for empty patches may also decrease the benefits of the rescue effect (Etienne, 2000). The net result for metapopulation persistence depends on the relative strengths of the effects. This deserves further study.

We have looked at the founder and Allee effects in the simple patch occupancy model that ignores population size. It seems more appropriate to study the Allee effect with an individual-based model or a model where population sizes are explicitly taken into account, instead of in a patch occupancy model where individuals are only implicit. These individual-based or size-structured metapopulation models are, however, usually much more complicated and thus more difficult to comprehend. A simple model as the patch occupancy model provides a strong metaphor with which predictions can be obtained fairly quickly. These predictions can (and must) then be tested in more complicated models, and if the predictions of these models are different, one must critically examine the structure of these models to find out why. In this way, one can systematically gain more insight into what processes are relevant. This is the primary value of simple models. For the very same reason, the rescue effect, the patch preference effect and the like have also been studied in the simple patch occupancy model (see the introduction for references). The secondary value of simple models is that they can be extended to more realistic (and therefore usually more complicated) models. It is then of great importance that the mathematical description of the simple model exactly corresponds with one's assumptions. If, for example, the models of Etienne *et al.* (2002a) and Harding and McNamara (2002) were extended to include competition or predator-prey interactions (see the introduction for references on how to do so), the results may no longer correspond with their initial assumptions.

Analysis of the dynamics (or rather the equilibria) of the patch occupancy model with founder and Allee effect did not yield results that have not been obtained earlier with more complicated models. However, this does not make our model superfluous. On the contrary, it means that for many purposes, our simple model suffices as a description of the founder and Allee effects.

REFERENCES

Akçakaya, H.R. and L.R. Ginzburg (1991). Ecological risk analysis for single and multiple populations. In: Seitz, A. and V. Loeschcke (Eds). Species conservation: a population-biological approach. Birkhäuser Verlag, Basel, Switzerland. pp. 73-85.

Allee, W.C. (1931). Animal aggregations, a study in general sociology. University of Chicago Press, Chicago, USA.

Alonso, D. and A. McKane (2002). Extinction dynamics in mainland-island metapopulations: an N-patch stochastic model. Bulletin of Mathematical Biology 64: 913-958.

Amarasekare, P. (1998a). Interactions between local dynamics and dispersal: insights from single species models. Theoretical Population Biology 53: 44-59.

Amarasekare, P. (1998b). Allee effects in metapopulation dynamics. American Naturalist 152: 298-302.

Amarasekare, P. and H.P. Possingham (2001). Patch dynamics and metapopulation theory: the case of successional species. Journal of Theoretical Biology 209: 333-344.

Berec, L., D.S. Boukal and M. Berec (2001). Linking the Allee effect, sexual reproduction, and temperature-dependent sex determination via spatial dynamics. American Naturalist 157: 217-230.

Brassil, C.E. (2001). Mean time to extinction of a metapopulation with an Allee effect. Ecological Modelling 143: 9-16.

Brown, J.H. and A. Kodric-Brown (1977). Turnover rate in insular biogeography: effect of immigration on extinction. Ecology 58: 445-449.

Courchamp, F., T. Clutton-Brock and B. Grenfell (1999). Inverse density dependence and the Allee effect. Trends in Ecology and Evolution 14: 405-410.

Cronin, J.T. and D.R. Strong (1999). Dispersal-dependent oviposition and the aggregation of parasitism. American Naturalist 154: 23-36.

Day, J.R. and H.P. Possingham (1995). A stochastic metapopulation model with variability in patch size and position. Theoretical Population Biology 48: 333-360.

Den Boer, P.J. (1968). Spreading of risk and stabilization of animal numbers. Acta Biotheoretica 18: 165-194.

Etienne, R.S. (2000). Local populations of different sizes, mechanistic rescue effect and patch preference in the Levins metapopulation model. Bulletin of Mathematical Biology 62: 943-958.

Etienne, R.S. (2002). A scrutiny of the Levins metapopulation model. Comments on Theoretical Biology 7: 257-281.

Etienne, R.S. and J.A.P. Heesterbeek (2000). On optimal size and number of reserves for metapopulation persistence. Journal of Theoretical Biology: 203: 33-50.

Etienne, R.S. and J.A.P. Heesterbeek (2001). Rules of thumb for conservation of metapopulations based on a stochastic winking-patch model. American Naturalist 158: 389-407.

Etienne, R.S. and C.J. Nagelkerke (2002). Non-equilibria in small metapopulations: comparing the deterministic Levins model with its stochastic counterpart. Journal of Theoretical Biology 219: 463-478.

Etienne, R.S., M. Lof and L. Hemerik (2002a). The Allee effect in metapopulation dynamics revisited. In: Etienne, R.S. Striking the metapopulation balance. Mathematical models and methods meet metapopulation management. pp. 71-78. PhD Thesis, Wageningen University, Wageningen, The Netherlands.

Etienne, R.S , B. Wertheim, L. Hemerik, P. Schneider and J.A. Powell (2002b). The interaction between dispersal, the Allee effect and scramble competition affects population dynamics. Ecological Modelling 148: 153-168.

Frank, K. and C. Wissel (1998). Spatial aspects of metapopulation survival - from model results to rules of thumb for landscape management. Landscape Ecology 13: 363-379.

Goel, N.S. and N. Richter-Dyn (1974). Stochastic models in biology. Academic Press, New York, NY.

Gog, J., R. Woodroffe and J. Swinton (2002). Disease in endangered metapopulations: the importance of alternative hosts. Proceedings of the Royal Society of London B 269: 671-676.

Gotelli, N.J. and W.G. Kelley (1993). A general model of metapopulation dynamics. Oikos 68: 36-44.

Gyllenberg, M. and I. Hanski (1992). Single-species metapopulation dynamics: a structured model. Theoretical Population Biology 42: 35-61.

Gyllenberg, M. and I. Hanski (1997). Habitat deterioration, habitat destruction, and metapopulation persistence in a heterogenous landscape. Theoretical Population Biology 52: 198-215.

Gyllenberg, M. and D.S. Silvestrov (1994). Quasi-stationary distributions of a stochastic meta-population model. Journal of Mathematical Biology 33: 35-70.

Gyllenberg, M., A.V. Osipov and G. Söderbacks (1996). Bifurcation analysis of a metapopulation model with sources and sinks. Journal of Nonlinear Science 6: 329-366.

Gyllenberg, M., I. Hanski and A. Hastings (1997). Structured metapopulation models. In: Hanski, I.A. and M.E. Gilpin (Eds). Metapopulation biology: ecology, genetics, and evolution. Academic Press, San Diego, CA. pp. 93-122.

Gyllenberg, M., J. Hemminki and T. Tammaru. (1999). Allee effects can both conserve and create spatial heterogeneity in population densities. Theoretical Population Biology 56: 231-242.

Hanski, I. (1983). Coexistence of competitors in patchy environment. Ecology 64: 493-500.

Hanski, I. (1994). A practical model of metapopulation dynamics. Journal of Animal Ecology 63: 151-162.

Hanski, I. (1999). Metapopulation ecology. Oxford University Press, Oxford, U.K..

Hanski, I. and M. Gyllenberg (1993). Two general metapopulation models and the core-satellite species hypothesis. American Naturalist 142: 17-41.

Hanski, I. and D.-Y. Zhang (1993). Migration, metapopulation dynamics and fugitive co-existence. Journal of Theoretical Biology 163: 491-504.

Hanski, I. and O. Ovaskainen (2000). The metapopulation capacity of a fragmented landscape. Nature 404: 755-758.

Hanski, I., A. Moilanen and M. Gyllenberg (1996). Minimum viable metapopulation size. American Naturalist 147: 527-541.

Harding, K.C. and J.M. McNamara (2002). A unifying framework for metapopulation dynamics. American Naturalist 160: 173-185.

Hastings, A. (1991). Structured models of metapopulation dynamics. Biological Journal of the Linnean Society 42: 57-70.

Hastings, A. (1995). A metapopulation model with population jumps of varying sizes. Mathematical Biosciences 128: 285-298.

Hess, G.R. (1994). Conservation corridors and contagious disease: a cautionary note. Conservation Biology 8: 256-262.

Hess, G.R. (1996). Disease in metapopulation models: implications for conservation. Ecology 77: 1617-1632.

Hess, G.R. and R.A. Fischer (2001). Communicating clearly about conservation corridors. Landscape and Urban Planning 55: 195-208.

Holt, R.D. (1997). From metapopulation dynamics to community structure: some consequences of spatial heterogeneity. In: Hanski, I.A. and M.E. Gilpin (Eds). Metapopulation biology: ecology, genetics, and evolution. Academic Press, San Diego, CA. pp. 149-164.

Keitt, T.H., M.A. Lewis and R.D. Holt (2001). Allee effects, invasion pinning, and species' borders. American Naturalist 157: 203-216.

Keymer, J.E., P.A. Marquet, J.X. Velasco-Hernández and S.A. Levin (2000). Extinction thresholds and metapopulation persistence in dynamic landscapes. American Naturalist 156: 478-494.

Lande, R. (1998). Demographic stochasticity and Allee effect on a scale with isotropic noise. Oikos 83: 353-358.

Lande, R., S. Engen and B-E. Saether (1998). Extinction times in finite meta-population models with stochastic local dynamics. Oikos 83: 383-389.

Levins, R. (1969). Some demographic and genetic consequences of environmental heterogeneity for biological control. Bulletin of the Entomological Society of America 15: 237-240.

Levins, R. (1970). Extinction. In: Gertenhaber, M. (Ed.). Some mathematical problems in biology. American Mathematical Society, Providence, RI. pp. 75-107.

Levins, R. and D. Culver (1971). Regional coexistence of species and competition between rare species. Proceedings of the National Academy of Science of the USA 68: 1246-1248.

McCarthy, M.A. (1997). The Allee effect, finding mates and theoretical models. Ecological Modelling 103: 99-102.

Nagelkerke, C.J. and S.B.J. Menken (2002). Local vs. global power. Coexistence of specialist and generalist metapopulations. Manuscript.

Nee, S. and R.M. May. (1992). Dynamics of metapopulations: habitat destruction and competitive coexistence. Journal of Animal Ecology 61: 37-40.

Nee, S., R.M. May and M.P. Hassell (1997). Two-species metapopulation models. In: Hanski, I.A. and M.E. Gilpin (Eds). Metapopulation biology: ecology, genetics, and evolution. Academic Press, San Diego, CA. pp. 123-147.

Ovaskainen, O. (2001). The quasi-stationary distribution of the stochastic logistic model. Journal of Applied Probability 38: 898-907.

Ovaskainen, O. and I. Hanski (2001). Spatially structured metapopulation models: global and local assessment of metapopulation capacity. Theoretical Population Biology 60: 281-302.

Ovaskainen, O., K. Sato, J. Bascompte and I. Hanski (2002). Metapopulation models for extinction threshold in spatially correlated landscapes. Journal of Theoretical Biology 215: 95-108.

Ray, C., M. Gilpin and A.T. Smith. (1991). The effect of conspecific attraction on metapopulation dynamics. Biological Journal of the Linnean Society 42: 123-134.

Reed, J.M. (1999). The role of behavior in recent avian extinctions and endangerments. Conservation Biology 13: 232-241.

Sabelis, M., O. Diekmann and V.A.A. Jansen (1991). Metapopulation persistence despite local extinction: predator-prey patch models of the Lotka-Volterra type. Biological Journal of the Linnean Society 42: 267-283.

Slatkin, M. (1974). Competition and regional coexistence. Ecology 55: 128-134.

Stephens, P.A. and W.J. Sutherland (1999). Consequences of the Allee effect for behaviour, ecology and conservation. Trends in Ecology and Evolution 14: 401-405.

Stephens, P.A., W.J. Sutherland and R.P. Freckleton (1999). What is the Allee effect? Oikos 87: 185-190.

Taneyhill, D.E. (2000). Metapopulation dynamics of multiple species: the geometry of competition in a fragmented habitat. Ecological Monographs 70: 495-516.

Vandermeer, J. and R. Carvajal (2001). Metapopulation dynamics and the quality of the matrix. American Naturalist 158: 211-220.

Wissel, C. (1994). Stochastic extinction models discrete in time. Ecological Modelling 75: 183-192.

Rampal S. Etienne

Community and Conservation Ecology Group, University of Groningen

Lia Hemerik

Biometris, Wageningen University and Research Centre

9

Balancing Statistics and Ecology: Lumping Experimental Data for Model Selection

Nelly van der Hoeven, Lia Hemerik and Patrick A. Jansen

ABSTRACT

Ecological experiments often accumulate data by carrying out many replicate trials, each containing a limited number of observations, which are then pooled and analysed in the search for a pattern. Replicating trials may be the only way to obtain sufficient data, yet lumping disregards the possibility of differences in experimental conditions influencing the overall pattern. This paper discusses how to deal with this dilemma in model selection. Three methods of model selection are introduced: likelihood-ratio testing, the Akaike Information Criterion (AIC) with or without small-sample correction and the Bayesian Information Criterion (BIC). Subsequently, we apply the AICc method to an example on size-dependent seed dispersal by scatterhoarding rodents.

The example involves binary data on the selection and removal of *Carapa procera* (Meliaceae) seeds by scatterhoarding rodents in replicate trials during years of different ambient seed abundance. The question is whether there is an optimum size for seeds to be removed and dispersed by the rodents. We fit five models, varying from no effect of seed mass to an optimum seed mass. We show that lumping the data produces the expected pattern but gives a poor fit compared to analyses in which grouping levels are taken into account. Three methods of grouping were used: per group a fixed parameter value; per group a randomly drawn parameter value; and some parameters fixed per group and others constant for all groups. Model fitting with some parameters fixed for all groups, and others depending on the trial give the best fit. The general pattern is however rather weak.

We explore how far models must differ in order to be able to discriminate between them, using the minimum Kullback-Leibler distance as a measure for the difference. We then show by simulation that the differences are too small to discriminate at all between the five models tested at the level of replicate trials.

We recommend a combined approach in which the level of lumping trials is chosen by the amount of variation explained in comparison to an analysis at the trial level. It is shown that combining data from different trials only leads to an increase in the probability of identifying the correct model with the AIC criterion if the distance of all simpler (=less extended models) to the simulated model is sufficiently large in each

T.A.C. Reydon and L. Hemerik.,(eds.), Current Themes in Theoretical Biology,
233-265.

trial. Otherwise, increasing the number of replicate trials might even lead to a decrease in the power of the AIC.

Keywords: AIC, *Carapa procera*, Kullback-Leibler distance, Likelihood-Ratio test, model selection, *Myoprocta acouchy*, non-central chi-square distribution, power, Red acouchy, scatterhoarding, seed dispersal, seed size.

9.1 INTRODUCTION

It is quite common in ecology to have several candidate models for describing ecological observations (Hilborn and Mangel, 1997). In some cases, models are based on different assumptions about the underlying mechanism, whereas in others, models are used to describe the relationship between factors. Both cases however, require the identification of the model best conforming to the observations.

Several criteria exist to determine which model fits best, for instance likelihood-ratio (LR) testing, the AIC (Akaike Information Criterion) and the BIC (Bayesian Information Criterion) (see Burnham and Anderson, 2002; Hilborn and Mangel, 1997; Linhart and Zucchini, 1986; Borowiak, 1989 for extensive reviews of model discrimination methods). After an initial comparison of the three methods (LR, BIC and AIC) we focus in this paper on the AIC that treats all models as equivalent and allows comparison of nested and non-nested models. Thus, the AIC assumes that each model can be the true model and none of the models is preferred.

Ecological experiments often accumulate data for model fitting by carrying out several independent trials, each containing a limited number of observations, which are then pooled and analysed for a pattern. Replicating trials may be the only way to obtain sufficient data, yet lumping is not a priori admissible. If conditions between trials differ, simply lumping all trials is even a priori inadmissible. Such situations require the model be fitted to the data of each trial separately, each with different model parameters. This will, however, affect the ability to distinguish between models (the identifiability), and the possibility to derive general conclusions from the properties of the best fitting model. We consider a model identifiable if the probability of being the best-fitting on its own simulated data exceeds 80%. An alternative approach is to assume that the parameters in each trial are independent drawings from some probability distribution.

This paper explores the consequences of data lumping for model selection using data on seed selection by scatterhoarding rodents as an example. The question to be answered is whether there is an optimum size for seeds to be selected and dispersed by these rodents. In our example, it is biologically unrealistic as well as technically difficult to provide a single animal with

>1000 marked seeds at a given time, while it is ecologically desirable to consider selection by different individuals. The only way to detect a trend was to carry out many independent replicate trials with small batches of seeds, spaced apart in time and space, and involving different individual rodents. The challenge is to balance statistical requirements with ecological feasibility.

We start with the description of three methods for model selection (Section 2). In Section 3, we apply two of these methods to a data set on seed dispersal by scatterhoarding rodents. Next, we have fitted the same models to simulated data in order to obtain an impression of the identifiability of the chosen models for certain combinations of parameter values, that is which percentage of the simulation runs are classified correctly (Section 4). Finally, conclusions of the model fitting both the experimental data and the simulated models are given and discussed (Section 5).

9.2 METHODS FOR MODEL DISCRIMINATION

Hypothesis testing

Models that are to be compared are often nested: one model (the nested model) is a special case of another, more complex model with one or more of the parameters of the complex model fixed. For example, the linear model $y = a + bx$ is a special case of the quadratic model $y = a + bx + cx^2$ with $c = 0$. If these models are compared with the usual hypothesis testing method, the null hypothesis is that the simplest model is true, unless the observed data are much more likely under the more complex model. A general method to test the simple model against the more complex is the Likelihood-Ratio test (LR test). This test compares the ratio of the maximum likelihood (ML) for the two models to a critical value. Instead of the ratio between the ML's the difference between the log of both ML's can be used. Twice the difference between these maximized log-likelihoods is approximately χ^2 distributed. This means that for large numbers of observations the α-critical value for 2×(the difference in maximized log-likelihoods) is approximately $\chi^2_{\alpha,\nu}$ with ν the difference in the number of parameters of the extended (k_2) and the more simple model (k_1), so $\nu = k_2 - k_1$. For a small number of observations, the χ^2 approximation may not hold.

So, in general let L_1 and L_2 be the maximum of the likelihood function for the simple and the extended model. Then, for large numbers of observations

$$T = 2 \times (\ln(L_2) - \ln(L_1)) \to \chi^2_\nu. \tag{9.1}$$

The AIC: finding the model giving the best approximation

One approach to discriminate between models, described by Akaike (1974), is to assume that there is some - unknown - "real" model, and that the model having the minimum distance to that unknown real model is the best approximation. It uses the so-called Kullback-Leibler (K-L) distance (Kullback and Leibler, 1951) as a measure of the distance between models. For continuous models, the K-L distance of the approximate model g with parameter θ to the real model f is

$$I(f, g_\theta) = \int f(y) \ln\left(\frac{f(y)}{g(y \mid \theta)}\right) dy. \tag{9.2}$$

This distance is related to the information lost by using model g with parameter θ instead of the real model f. It indicates how good model g with parameter θ approximates model f. Note that $I(f,g) \neq I(g,f)$, that is, the K-L distance is not commutative and therefore is not a real distance.

For discrete models with k possible outcomes y_i ($i = 1, \ldots, k$), the K-L distance can be written as

$$I(f, g_\theta) = \sum_{i=1}^{k} p(y_i \mid f) \ln\left(\frac{p(y_i \mid f)}{p(y_i \mid g_\theta)}\right). \tag{9.3}$$

In general, the real model, f, will be unknown. Fortunately, when two models, g_1 and g_2 have to be compared, the difference $I(f,g_1) - I(f,g_2)$ does not depend on the real model f. Using this, Akaike (1974) developed the AIC (An Information Criterion, better known as Akaike's Information Criterion) which is defined as

$$AIC = 2[k - \ln(L)] \tag{9.4}$$

where k denotes the number of estimated parameters and L is the maximum of the likelihood function. The model with the minimum AIC is considered to be the best fitting model. This approach allows a simple ranking of the models and is also appropriate for comparing non-nested alternatives. Using the AIC, Model 2 is preferred above Model 1 if $AIC_1 - AIC_2 > 0$, so if

$$T = 2[\ln(L_2) - \ln(L_1)] > 2(k_2 - k_1) = 2\nu. \tag{9.5}$$

A correction term should be added to the AIC if the number of parameters, k, is large, or the number of observations, n, is small. There is no universal best correction term, but the corrected AIC, $AICc$ as given by Hurvich and Tsai (1989),

$$AICc = AIC + C(k,n) = 2[k - \ln(L)] + \frac{2k(k+1)}{n - (k+1)}, \tag{9.6}$$

performs reasonably well for most models (Burnham and Anderson, 2002).

Using the *AICc*, Model 2 is preferred over Model 1 if

$$T > 2(k_2 - k_1)\left(1 + \frac{n(1 + k_1 + k_2) - (k_1 + 1)(k_2 + 1)}{(n - (k_1 + 1))(n - (k_2 + 1))}\right) \qquad (9.7)$$
$$= 2\,v(1 + \text{correction term}).$$

The correction term only depends on the number of parameters in both models (k_1, k_2) and the number of observations (n).

The BIC: finding the true model within a set of models

There may be a reason to believe a priori that one of the models in a set of models is true. The BIC described by Schwarz (1978) is a selection criterion for identifying such a true model with an as large as possible probability. The BIC is also based on twice the log ML's, and uses a correction term increasing with the number of observations,

$$BIC = k\,\ln(n) - 2\ln(L) \qquad (9.8)$$

The model with the minimum *BIC* is considered to be the best fitting model. Using the *BIC*, Model 2 is preferred above Model 1 if $BIC_1 - BIC_2 > 0$, so if

$$T = 2[\ln(L_2) - \ln(L_1)] > (k_2 - k_1)\ln(n) = v\ln(n). \qquad (9.9)$$

The *BIC* is a consistent estimator for the model type: if the number of observations becomes very large, the probability that the correct model is identified increases to 1. It should be noted however, that to meet the condition "very large" extremely large sample sizes are indeed required. For instance, identifying the correct model with high probability requires a very large number of observations. Umbach and Wilcox (1996), for example, needed as much as 125,000 simulated observations to reach a power of 0.79.

Comparison between the three methods

LR, AIC, AICc and BIC use the same test statistic T to find the best approximate model. If the extended model has one extra parameter ($v = 1$), the χ^2 approximation for the LR test criterion at $\alpha = 5\%$ leads to rejection of the more simple model if $T > 3.84$. The threshold value for T increases with an increasing degree of freedom (see Figure 9.1). The AIC considers all models equivalent and for $v = 1$ chooses the more extended model if $T > 2$. The threshold value for T increases linearly with higher values of v (see Figure 9.1). For a difference of one parameter, the AIC criterion will choose the extended model with (approximately) probability 0.16 if the simple model is true.

The critical value for T in the AICc criterion is more complex. If the simple model has two parameters and the extended model three, the AICc criterion chooses the more extended model if $T > (2n(n-1))/((n-3)(n-4))$. The critical value for T in a trial of only five observations ($n = 5$), for example, is 20. The AICc criterion becomes less strict for the extended model with increasing n, and for $n > 12$, the AICc criterion is less strict than the LR one. Figure 9.1 shows the threshold values of the *AICc* for T with $n = 25$ or 100 and a 2-parameters simplest model. In this figure, the BIC criterion is given for the same numbers of observations.

In contrast to the AICc, the BIC criterion becomes stricter for the extended model as the number of observations increases (Figure 9.1). If the difference in the number of parameters is one ($v = 1$), the *AIC* and *BIC* are almost identical for $n = 8$, and the results of the BIC criterion and the χ^2 approximation of the LR test are about the same for $n = 47$. The preference for the more parsimonious model with increasing difference in number of parameters increases faster for the AIC, the AICc and the BIC than for the LR test (Burnham and Anderson, 2002).

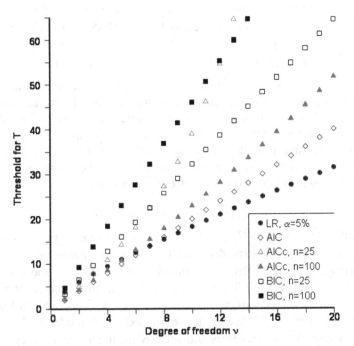

Figure 9.1. The threshold value for the test statistic T as a function of the difference in the number of parameters (degrees of freedom v) of the two compared models. For the AICc the number of parameters in the simpler model is set at 2.

The differences in critical value for model selection illustrate a fundamental difference between the three methods. Classical hypothesis testing with a likelihood-ratio test assumes that the most simple model is true unless the observed values are very unlikely (probability less than α, with $\alpha = 0.05$ as most common choice). Using the AIC or AICc criterion, it is assumed that none of the models is true, but it is tried to minimize the (K-L) distance to the real, unknown, model in order to choose the model giving the best prediction for new data sets. Using the BIC criterion, it is assumed that one of the models is true, and the probability of choosing that true model is maximized. Note that only the difference between the log-likelihoods is of interest in each method. Therefore, all *AIC, AICc* and *BIC* can be decreased by a constant. The smallest *AIC (AICc, BIC)* is often subtracted from all *AIC (AICc, BIC)* values, making the smallest *AIC (AICc, BIC)* 0.

The Kullback-Leibler distance between models

Choosing among models requires quantification of the difference between them. The fundamental distance measure for the AIC is the Kullback-Leibler (K-L) distance. The K-L distance between models can be determined for one realization, but also cumulative for a combination of n observations, which is of interest for model selection. Then, the K-L distance between the models is the expectation, given the extended model, of the difference between the simpler model and the more extended model in the log-likelihoods for n combined realizations. It depends, among others, on the values of the independent variables in the observations.

In an example, we will show how the K-L distance between models can be determined if n realizations of the model are observed. We assume that some discrete variable, y, is observed, and that y can have m different realizations, $w_1, w_2, ..., w_m$. The probability to attain w_j, can be described by some model and depends on the independent variable, x, and a model parameter. We consider two models f and g with parameters φ and θ, respectively, and n independent observations. Thus, for a certain x and θ, model g gives the probability that the realization for the observed variable y is w_j. This probability is written as $P_g(y = w_j \mid x, \theta)$. Suppose that n independent discrete observations are obtained, each with its own value for x and all with the same set of m different possible realizations. Then, the K-L distance of model g for all n observations together (g^n) to model f for the same combination (f^n) is

$$I^n(f,g) = \sum_{i=1}^{n} \sum_{j=1}^{m} P_f(y = w_j \mid x_i, \varphi) \ln\left(\frac{P_f(y = w_j \mid x_i, \varphi)}{P_g(y = w_j \mid x_i, \theta)}\right). \quad (9.10)$$

The K-L distance can be calculated for fixed parameters φ and θ and a specific set of independent variables $x = (x_1, x_2, ..., x_n)^T$. So if f^n is completely defined, i.e. parameter φ and the independent variable $\vec{x}$ are fixed, then the K-L distance of g'' for each possible value of its parameter θ can be calculated. Thus, the minimum K-L distance of g'' to this specific version of f'' can be determined.

The minimum K-L distance gives an indication of how easily model g will be preferred over model f if this specific version of model f is true. If none of the models is true, as the AIC criterion assumes, the minimum distance of model g to the best approximating version of model f can be used as an indication of how easily model g will be preferred over model f. Note that the minimum K-L distance is 0 if model f is nested in model g.

The term

$$\sum_{j=1}^{m} P_f(y = w_j \mid x_i, \varphi) \ln\left(\frac{P_f(y = w_j \mid x_i, \varphi)}{P_g(y = w_j \mid x_i, \theta)}\right) \qquad (9.11)$$

in equation (9.10) depends on the values of x_i. Adding an extra data point will lead to an increase in the K-L distance depending on the position of the independent variable in that data point. However, if the models f and g are reasonably smooth and the frequency distribution of the independent variables is (nearly) unaffected by addition of extra data points, $I(f,g)$ will increase nearly proportional to the number of observations (see e.g. Linhart and Zucchini, 1986). In other words, the minimum distance of g'' to f'' becomes proportional to n for large n if the distribution of the independent variable does not depend on the number of observations. This result is clearly only intended for large samples. For the first few data points, it might easily be possible to estimate parameter θ of g so that the probability $P_g(y \mid x_i, \theta) = P_f(y \mid x_i, \varphi)$ in the few data points x_i. This will generally be true for linear models if the number of data points does not exceed the number of parameters of g.

9.3 EXAMPLE: SEED SIZE DISCRIMINATION BY SCATTERHOARDING RODENTS

Methods

Ecological background

The dispersal phase is one of the most critical phases in plant life history. Plants have evolved a wide variety of mechanisms to have their seeds dispersed. Many nut-bearing tree species depend on scatterhoarding birds or rodents for dispersal. Such animals bury seeds as food supplies in numerous spatially scattered caches in the soil surface. This behaviour provides effective dispersal because some seeds are left to germinate and establish seedlings (Vander Wall, 1990). Non-scatterhoarded seeds, in contrast, probably die underneath the parent tree due to fungi, invertebrates and non-hoarding mammals (Jansen, 2003).

The benefits of scatterhoarding have given rise to the idea that the production of large, nutritious seeds in nut-bearing tree species has evolved in response to feeding preferences of scatterhoarding animals (Smith and Reichman, 1984). Large seeds are more nutritious and may therefore be more suitable for hoarding than smaller seeds. Indeed, several studies have shown that scatterhoarding animals disperse large seeds further than small ones (e.g. Hallwachs, 1994; Jansen *et al.*, 2002; Vander Wall, 2003). However, there must be a point beyond which seeds become too large to efficiently be handled by a given animal taking into account its limited body mass and mouth width. Therefore, there should be an optimum seed size for dispersal by a given scatterhoarding animal (Jansen *et al.*, 2002).

Data

Jansen (2003) experimentally studied the effect of seed size on dispersal by scatterhoarding rodents in the Nouragues rainforest reserve in French Guyana, South America ($4°02$'N and $52°42$'W). During five consecutive years (1996-2000), numerous cafeteria plots were laid out in the territories of Red acouchy (*Myoprocta acouchy*), a cavi-like scatterhoarding rodent. Each plot contained 25 (1996-1997) or 49 (1998-2000) individually marked seeds of the canopy tree *Carapa procera* (Meliaceae), numbers that agree with the approximate daily production by average individuals of this species. Seed batches were assembled as to have seed mass within plots ranging from 3 to 60g, offering acouchies a wide choice. Seed removal from the experimental plots was monitored at days 1, 2, 4, 8, 16, 32, 64 and 128 after the start of the experiment. Moreover, the plots were also continuously monitored on video during the first day or first few days. Seeds that were eaten on the plot were included in the removed seeds, with the annotation of being eaten. See Jansen (2003) for further details.

The data set used in this paper consists of 66 plots (trials) with complete data on seed masses and seed removal. The structure of this data set allows us to apply our model selection methods at four different levels: (1) all trials lumped; (2) trials grouped in years of poor and rich fruiting; (3) trials grouped by year; and (4) individual trials. Moreover, there was variation among plots within and between years. Plots were laid out at different sites, under different forest conditions and in different rodent territories. Moreover, years differed in fruit availability. Seeds were abundant during the even years and seeds were scarce during the odd years. This distinction is important, because feeding preferences are more pronounced under seed abundance, allowing animals to be more choosy, than under conditions of scarcity (Jansen *et al.*, 2002).

The models

We modelled the probability of seed removal as a function of seed mass using a hierarchical set of models (Huisman *et al.*, 1993),

Model I: $$p(x) = \frac{1}{1+e^a},$$ (9.12a)

Model II: $$p(x) = \frac{1}{1+e^{a+bx}},$$ (9.12b)

Model III: $$p(x) = \frac{1}{1+e^{a+bx}} \cdot \frac{1}{1+e^c},$$ (9.12c)

Model IV: $$p(x) = \frac{1}{1+e^{a+bx}} \cdot \frac{1}{1+e^{c-bx}},$$ (9.12d)

Model V: $$p(x) = \frac{1}{1+e^{a+bx}} \cdot \frac{1}{1+e^{c+dx}}.$$ (9.12e)

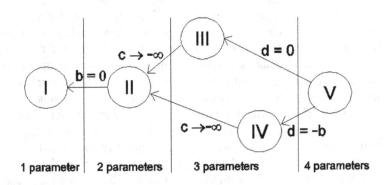

Figure 9.2. The relation between the five models fitted.

Here, x is the 10-logarithm of seed fresh mass and $p(x)$ is the probability that a seed with log-mass x is removed. Model I describes a constant probability, independent of the seed mass. Model II describes a probability that increases gradually from 0 to 1 (or decreases from 1 to 0). Model III describes a gradual increase (or decrease) of the probability from 0 to some intermediate value. Finally, Models IV and V both describe an optimum relationship, Model IV being symmetric and Model V asymmetric. Note that the models are functions of the log of the seed mass, so that the symmetry of Model IV is in the log of the seed mass, not in the seed mass itself.   Figures of the five models are given in Appendix A, and the relationship between them is shown in Figure 9.2.

We used the AICc criterion to select the model best describing the data. Each data point is denoted as (x_i, y_i), where x_i is the log of the seed mass and y_i equals 1 if the seed is removed, and 0 if it is not. For each model, the likelihood L of the data is

$$L = \prod p_\theta(x_i)^{y_i}(1 - p_\theta(x_i))^{1-y_i}, \qquad (9.13)$$

and the log-likelihood $\ln(L)$ is

$$\ln(L) = \sum y_i \ln(p_\theta(x_i)) + (1 - y_i)\ln(1 - p_\theta(x_i)). \qquad (9.14)$$

The value of p depends on the variable x and on the parameter value θ. The ML estimator of θ is the value of θ that maximizes the likelihood or log-likelihood.

We fitted the five models at four levels: (1) to the pooled data; (2) to poor and rich years separately; (3) to years separately; and (4) to individual trials. Furthermore, we also fitted the models as random effect models. That is, we assumed that for each trial the parameters were independent drawings from a normal distribution and estimated the mean and standard deviation of these parameters. If the model had more parameters, we assumed that the parameters were independent. Random effect models were fitted at three levels: (1) to the pooled data; (2) to poor and rich years separately; and (3) to years separately. Note that some trials showed no variation because all seeds were removed.

We also fitted some mixed effect models to the same three levels as the random effect models. Here, we assumed that the slope parameters (b and d) had fixed values. For each trial, the parameters determining the position of the model (a and c) were randomly drawn from some normal probability distribution. Finally, we fitted special versions of Models II and IV. Here, we assumed that the slope parameter b and the maximum M (Model IV only) were constant. This was done: (1) for all trials; (2) for the trials in poor and rich years separately; or (3) for the trials in one year. The position of the inflection point (Model II, $-a/b$) or top (Model IV, $(c-a)/2b$) was fitted for each trial separately. The random effect, mixed effect and special effect models were only considered with the AICc as selection criterion.

We wanted to distinguish certain basic relations between seed mass and the probability of seed removal. The five hierarchical models allow us to assess: (1) whether any such relationship exists (Model II versus Model I); (2) whether there is an upper limit < 1 to the probability of seed removal (Model III versus Model II); and (3) whether the probability of seed removal is maximal at intermediate seed mass or rather monotonously increasing or decreasing with seed mass (Models IV or V versus Model II). These relations are only of interest within the normal range of seed masses, i.e. 3-50g in our example.

Analysis

The size of the effect to be detected

First, we determined which effect we wanted to be able to detect. This rather arbitrary process lead to the following choices:
- Model II versus Model I: If the differences in log-odds at the smallest and largest seed mass is greater than 2, we wish to be able to assess an increasing (or decreasing) trend in probability. Then the slope parameter b should be less than -1.64. Also, we are not interested in assessing a monotone increase if it is an increase from almost never to very rarely (the maximum probability should be over 0.3) or an increase from in most cases to almost always (the minimum probability should be at most 0.7). For the minimum detectable slope this leads to $a \in (-0.07, 3.63)$. Figure 9.3a shows three versions of Model II, which we wish to be able to distinguish from Model I.
- Model III versus Model II: We wish to be able to assess whether the upper limit of the probability is at most 0.8 ($c > -1.39$), if that upper limit is approached sufficiently closely and if the conditions under which Model II can be distinguished from Model I are met. "Approaching the upper limit sufficiently closely" is operationalized as a log-odds distance from the upper limit of less than 0.5, i.e. if the upper limit is 0.8, the maximum probability reached in the range of possible seed masses is at least about 0.7. In Figure 9.3b some possible versions of Model III are given, which we wish to be able to distinguish from Model II.

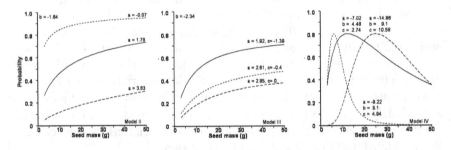

Figure 9.3. Examples of Model II (Figure 9.3a), Model III (Figure 9.3b) and Model IV (Figure 9.3c) which we wish to be able to distinguish from simpler models or models with the same number of parameters.

- Models IV or V versus Model II: We wish to be able to recognize a maximum in the probability if the differences is greater than two between the log-odds of that maximum, as attained at intermediate seed mass, and the log-odds of the probabilities at the two limits of the seed mass range. Furthermore, the top should be well within the range of the seed mass, say between 6 and 25 g. The minimum probability at both borders of the seed mass range should be less than 0.7 and at the top at least be 0.3. In Figure 9.3c three versions of Model IV are drawn. We wish to be able to distinguish these from Model II.

Levels of lumping data

Our first analysis was to compare fitting results of all trials lumped together, and the trials lumped for poor and rich years separately. Fitting Model II, for example, to trials lumped for poor and rich years separately, can be considered as fitting the model to the complete data set with an extra factor for poor or rich years. Model II then becomes

$$p(x) = \frac{1}{1 + e^{a_1 + a_2 z + (b_1 + b_2 z)x}} \tag{9.15}$$

where z is the factor for the year type ($z = 1$ in rich years, and $z = 0$ in poor years) and x is the log of the seed mass. Figure 9.4 shows the data for the probability of seeds being removed and the corresponding best fitting models. The AICc values for all models are given in Table 9.1a (first two lines).

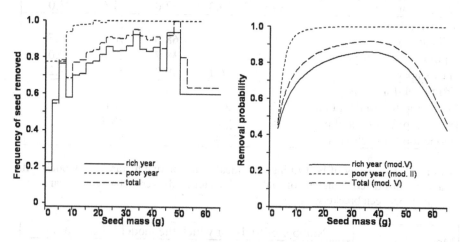

Figure 9.4. (a) The frequency of seed removal per size class, with all trials lumped and with trials lumped for rich and poor years separately. The size classes have a width of at least 5g. Size classes with less than 10 observations were lumped. (b) The corresponding models that gave the best fit.

Table 9.1. AICc values for the hierarchical set of Models I-V fitted to seed removal data with different levels of data lumping and model types. AICc values are given for the fixed effect models (a), the random effect models (b), the mixed effect models (c), and the special models (d). Levels of lumping were: all trials lumped together, trials lumped for poor and rich years separately, trials lumped for all five years separately and trials all considered separately. Note that AICc values were standardized by subtracting the smallest AICc value (the special version of model II with rich and poor years fitted separately). The smallest AICc values for each level are printed in bold.

AICc	Model I	Model II	Model III	Model IV	Model V
(a) Fixed effect					
All data together	712.5	619.5	620.2	618.5	**613.2**
Split in rich/poor years	502.1	**410.4**	413.2	412.1	411.4
Years apart	422.4	**318.5**	325.1	324.7	330.5
All trials apart	95.0	**44.0**	156.9	162.8	299.9
(b) Random effect					
All data together	175.1	72.4	76.6	**70.3**	76.7
Split in rich/poor years	146.3	38.8	39.0	**37.7**	49.3
Years apart	140.0	**39.3**	52.0	51.6	73.0
(c) Mixed effect					
All data together		89.0	**68.5**	76.1	70.5
Split in rich/poor years		49.8	**38.7**	42.0	42.0
Years apart		**46.1**	49.2	51.4	59.2
(d) Special models					
Slope/top for all data together		9.3		181.2	
Slope/top for rich and poor years		**0.0**		23.7	
Slope/top for each year		3.4		8.5	

Table 9.2. Frequency of best-fitting individual trials for five hierarchical models. All trials are considered. Numbers of trials in which all seeds were removed (no variance) are given between brackets.

Model	Number of trials for which the model is best fitting	
	Removed seeds	
	Poor year	Rich year
I	17 (17)	26 (20)
II	4	15
III	0	3
IV	0	1
V	0	0
Total	21 (17)	45 (20)

Clearly, the parameters in the poor and rich years do not have the same values. Moreover, the best fitting model differs between all years lumped together and years grouped into poor and rich years. The best model for all trials lumped shows an optimum seed size for removal. Consideration of rich and poor years separately however, reveals that an optimum seed mass for removal exists only in rich years. Poor years show rather an exponential rise to a maximum removal probability.

We then investigated how further reduction of the level of trial lumping affected the results. We extended the models with dummy variables, as in equation (9.15), to find out whether the parameters differed between years or even between individual trials. Especially the latter increased the number of parameters considerably.

The results are shown in Table 9.1a (lines 3-4). Clearly, fitting the models to trials separately yields considerably lower AICc values than fitting to lumped trials, despite the large number of extra variables involved. Seed removal is best described at the trial level by Model II, indicating higher probability of removal with increasing seed mass.

Subsequently, we determined which of the five models best fitted each trial individually. The distribution of best fitting models among trials is given in Table 9.2. Simply counting how many times each of the models turns out to give the best fit would have resulted in a constant probability per trial to be removed (Model I). However, this does not guarantee that it is the best model (Hemerik *et al.*, 2002; Hemerik and van der Hoeven, 2003). None of the five models will always be identified as the best fitting model even for data simulated with that very model (see Appendix A).

Random and mixed effect models

Another approach to account for differences between trials is to fit the five models as random effect or mixed effect models. In random effect modelling, we assume that all parameters for each trial are independent drawings from a normal distribution. In mixed effect modelling, we assumed that the parameters a and c, which determine the position of the model, were randomly drawn for each trial from some normal probability distribution, while the slope parameters (b and d) had fixed values. The resulting AICc values are given in Tables 9.1b and 9.1c, respectively.

The best fitting random effect model was Model IV with trials lumped for rich and poor years. The AICc value was even lower for this model than for the fixed effect Model II in which trials were treated separately. Figure 9.5 shows the envelopes containing 80% and 95% of the probabilities according to this model. In contrast the mixed effect models performed poorly. They never fitted better than the random effect models (Table 9.1), and rarely better than the best fitting fixed effects model. Mixed effect models had lower AICc values than fixed effect models only with trials lumped for rich and poor years.

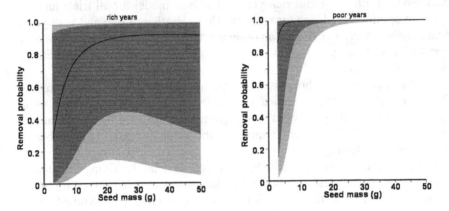

Figure 9.5. The probability of seed removal as a function of seed fresh mass according to the random effect version of Model IV. The black line indicates the removal probability with all parameters at their mean value. The dark grey envelope contains 80% of all possible realizations of parameter combinations, the grey area 95%.

Some special models

We have now seen that the best fixed effect description is obtained by fitting models to each trial separately. For the removal data, a slightly better but not very informative fit is reached by the random effect version of Model IV fitted to the data of rich and poor years separately. Our main question however, is whether there is a general relationship between seed size and the probability of seed removal (and subsequent dispersal). The two logical alternative relationships are an increase and an optimum. To investigate this, we fitted special versions of Models II and IV. Here, we assumed that the slope parameter b and the maximum M (Model IV only) were the same for all trials, while the position of the inflection point (Model II, $-a/b$) or the optimum (Model IV, $(c-a)/2b$) were fitted for each trial separately.

The AICc values for these special models are given in Table 9.1d. The lowest AICc values by far were for Model II with slope parameter b (–3.1), or even better, with slope parameter for rich ($b = -2.7$) and poor years ($b = -8.6$) separately. The models for each trial are shown in Figure 9.6 (a and b). For six out of the 45 rich trials and 17 of the 21 poor trials, the inflection point is way below the observable range of seed mass (3 to 60 g), leading to a removal probability of nearly 1, independent of the seed mass.

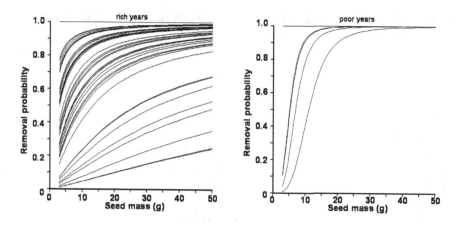

Figure 9.6. The probability of seed removal according to the special version of Model II with a fixed slope parameter for all trials in rich and in poor years respectively (Figures 9.6a and 9.6b). The upper horizontal line represents trials in which all seeds were removed, six trials in rich years (Figure 9.6a) and 17 trials in poor years (Figure 9.6b).

Information loss through fixed effect modelling with lumped data?

To investigate how much information was lost by lumping data, we calculated what percentage of the variance explained by the best fitting fixed effect model was also explained at higher levels of lumping (Burnham and Anderson, 2002). We calculated the ratio of (1) the difference between twice the log-likelihoods of an intermediate model and the simplest model (Model I with all data lumped) and (2) the difference between the best fitting model (all trials separated, Model II) and the simplest model (equation (9.16)). Let $\ln(L_b)$ be the log-likelihood of the best fitting model, $\ln(L_s)$ the log-likelihood of the most simple model and $\ln(L_i)$ the log-likelihood of the intermediate model. Then the multiple coefficient of determination, R^2 is

$$R_i^2 = \frac{2\ln(L_i) - 2\ln(L_s)}{2\ln(L_b) - 2\ln(L_s)}. \tag{9.16}$$

R_i^2 can be interpreted as the fraction of the structural information in the best fitting model, which is also contained in the intermediate model (i).

Calculating the R_i^2 for the best fitting model gives 17%, 53% and 68% explained for complete lumping, lumping in poor and rich years, and lumping per year, respectively. These percentages indicate that lumping trials in rich and poor years conserves about 50% or more of the information. Figure 9.7

shows the best fitting models for the probability of removal with at least 50%
of the information retained.

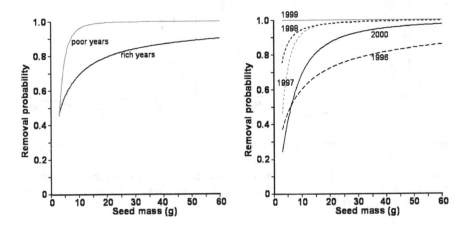

Figure 9.7. The best fitting fixed effect models with data lumped into rich and poor
years (a) with data lumped per year (b). Black lines represent rich years, grey lines
poor.

The Likelihood-Ratio approach for the fixed effect models

An alternative for using the AICc criterion is a stepwise test of a simpler
model against a one-step more complex fixed effect model. Figure 9.8 shows
all possible pathways of hypothesis testing in the case of our five models.
There are two main pathways. The first (sequence 1) is to test whether the data
can be split into groups. Subsequently, if further splitting is not significant and
thus not allowed, models are tested in order of increasing complexity. The
second (sequence 2) is to test the models in sequence of increasing complexity
for the lumped data, and then, for the most complex model allowed, test
whether the data can be split into groups.

Note that two alternatives are tested against Model II. Testing both at the
5% significance level will lead to a larger than 5% probability that Model II is
rejected under the null hypothesis. Here, we have chosen to ignore this fact
because a standard Bonferroni type correction would be far too conservative.

Both main sequences lead to the same conclusion, viz. that the best model
is Model II for all trials separately (Figure 9.8). Note however, that Model II is
rejected in favour of Model V ($p = 0.0058$) when tested at level (1) (all data
lumped).

Here, we do not explore the LR approach further because our aim is to find the model that best describes our data, rather than to choose the simplest possible model.

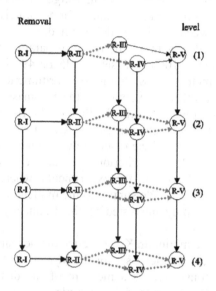

Figure 9.8. Pathways of pairwise testing for selection of the best-fitting model for seed removal by scatterhoarding rodents. Each model is tested against a one step more extended version using the χ^2 approximation of a likelihood-ratio test. Note that the extension can either be in the direction of a more complex relationship (horizontal, for example Model II instead of Model I), or in the direction of splitting the data in extra classes (vertical). Levels of lumping are: (1) all trials lumped; (2) trials lumped within rich and poor years; (3) trials lumped per year; and (4) all trials separately. Dotted grey arrow: simpler model cannot be rejected ($\alpha = 0.05$), thin arrow: the significance level between 0.05 and 0.005, intermediate arrow: the significance level between 0.005 and 0.0005, fat arrow: the significance level less then 0.0005. Left: LR test for seed removal, right: LR test for seeds being found cached.

9.4 SIMULATION

Methods

We have seen that models selected on the basis of lumped trials may differ considerably from models selected at the trial level. The best fitting fixed effect models at the trial level only indicated a simple increase of seed removal with seed mass, whereas the fixed models indicate an optimum when all data are lumped together, and the random effect models indicate an optimum both when all data are lumped together as well as when the data are split in rich and poor years. Other research has indicated that rodents discriminate against both small and large seeds, resulting in an optimum seed size for dispersal (Jansen *et al.*, 2002; Jansen, 2003). We therefore investigated whether trials with as few as 25 or 49 observations are at all suitable for accurately discriminating at the single trial level between the five models studied.

We used the five models with a wide range of parameter values. For each combination, we simulated 1000 data sets of 25 or 49 observations (seeds) with masses in a geometric series (log-masses arithmetic). The log-masses were centred and the width of the series of the log-masses was taken from -0.6109 to +0.6109. Then we determined which of the five models fitted best to each of these simulated data sets, using the AICc criterion, and counted how often the fitted model was indeed the model by which the data were generated. We used the AIC(c) as selection criterion because we wish to compare the results with the model selection in the experiments, where none of the five models will be completely true. We simulated the models with fixed parameter values because we were only interested in the frequency of correct model selection in one single trial.

We also calculated the minimum K-L distance of each simpler model to the simulated model for each parameter set for the Models II, III, IV and V. This minimum distance is considered as a measure of the distance between the simulated model and the other model. Examples of simulation and fitting results are given in Appendix A.

Simulation results

The simulations showed that 25 or even 49 seeds per plot provide too few observations per trial to accurately distinguish the five models for realistic values of the parameters. The following points emerge:

- Models II and III can be distinguished from Model I more easily if the slope parameter b is larger (in absolute value) and if the point of inflection is more in the centre of the data.
- Model IV can be distinguished more easily if the slope parameter b is large.
- Model I is chosen more often if the top of Model IV is closer to the median of the data points ([$a–c$] small), whereas Model II is chosen more often if the top moves farther away from the median of the data point (abs($a - c$) becomes larger).
- Models IV and V are chosen only rarely if a simpler model (or Model III) is true.
- If Model IV is true, the best fitting model is often Model III instead of Model IV. Only if the slope parameter b is very large, will Model IV be chosen as best model more often than Model III.

- Model I was erroneously chosen less often in the simulations with 49 observations than in those with 25 observations.

We compared the distribution of best fitting models to the experimental data (see Table 9.3) with the distribution of best fitting models to simulations with Model I both for $n = 25$ and $n = 49$. Parameter a was chosen nearest to the estimate of a for all data together. We used $a = -2$ for seed removal. The observed distribution differed significantly from the simulated one ($p = 0.01$ to 0.015, Kruskal-Wallis test).

Discrimination of the models

How different should alternative models be to be accurately discriminated? To answer this question we used the Kullback-Leibler (K-L) distance as a measure for the discrepancy between two models, and between a model and the data (see Section 9.2). The K-L distance of model A to the "real" model (the simulated one) depends on the parameter values of model A. If the real model is nested in model A, the parameters can always be chosen so that the distance is 0. In other cases, the distance will have some positive value, depending on the parameters of model A. The parameters minimising the K-L distance can be determined, and these parameters belong to the version of model A best fitting to the real model. This minimum K-L distance of model A to the real model will be indicated as the K-L distance of model A to the real model. If this distance is small, the difference between model A and the real model is small, and in model selection model A will often be preferred over the real model, i.e. the simulated one. If the distance is calculated in a limited number of data points, the distance will depend on the values of the independent variables (the seed masses) at these data points. The more observations (seeds) are used, the larger the K-L distance between models will become (see Section 9.2).

Figure 9.9 gives an impression of the K-L distance to the real model and the best approximating versions of Models I to IV for the real model being Model V.

In Section 9.3, we showed some versions of the Models II, III and IV that we wanted to distinguish from simpler models (Figure 9.3). For 25 observations, with mass geometrically spaced, the minimum K-L distances of Model I to the three examples of Model II with increasing value for a are 0.530, 1.044 and 0.530, respectively (Figure 9.3a). The minimum KL distances of Model III to Model II with increasing values of a and c are 0.011, 0.009 and 0.008, respectively (Figure 9.3b). The minimum K-L distances of Model IV to Model II with increasing parameter c (decreasing a) are 1.19, 1.88 and 1.88, respectively (Figure 9.3c). Note, however, that another choice of the seed mass distribution may dramatically affect the minimum KL distance. For example, applying the actual used distributions of seed masses in trials with 25 observations to the case of Model IV with the smallest parameter value for c (a = -7.02, b = 4.48 and c = 2.74, KL distance in case of geometric spacing 1.19) the mean of the minimum K-L distances for the 43 seed mass distributions is 0.48 (min.: 0.012, max.: 1.52).

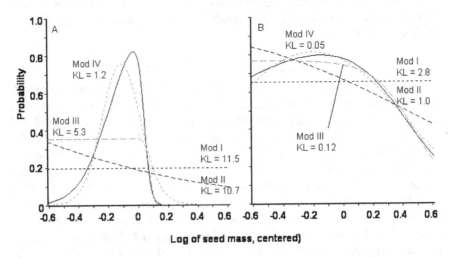

Figure 9.9. Model V (black line) and the best approximating Models I, II, III and IV. The parameter values (a,b,c,d) of Model V are in Figure 9.9A: (-2, -10, -2,50) and in Figure 9.9B:(-2, -2, -2,5). The minimum K-L distance of these models to Model V is given for 49 equidistantly spaced log-seed masses. In Figure 9.9B, Model V can be approximated reasonably well by the Models IV, III and II. In Figure 9.9A, only Model IV looks somewhat like Model V, but its distance to Model V is larger than the distance of Model II to Model V in Figure 9.9B.

The probability of the simulated model being identified as the best model increases with its difference from simpler models. For instance, if Model IV is simulated, it is chosen as the best model more often if the minimum K-L distance to the simulated model is larger for the models with less parameters (Models I and II) or with the same number of parameters (Model III). Figure 9.10 shows how the percentage of correct model choices depends on the least of all minimum K-L distances of the simpler models to the simulated model.

Figure 9.11 shows that Model I is chosen as best fitting model more often if the minimum K-L distance of Model I to the simulated model is small. For Model II, the same conclusion holds, provided that the K-L distance of Model I to the simulated model is not small too. If both Model I and Model II have a small K-L distance to the simulated model, Model I is often preferred above Model II, illustrating in fact that parsimonious models are favoured.

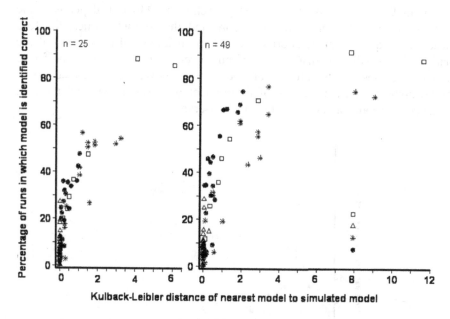

Figure 9.10. The probability of being classified as the correct model as a function of the K-L distance of the nearest simpler model to the simulated model.

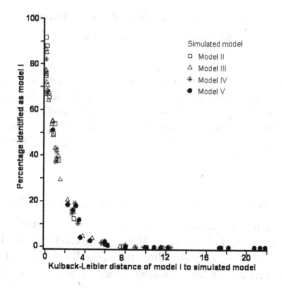

Figure 9.11. The K-L distance of Model I to the simulated model plotted against the percentage of the simulation runs in which Model I is chosen as best model using the AICc criterion. The number of observations in each simulation is 49.

Figure 9.12 shows how the frequency of choosing Model II depends on the K-L distance to Model II. In this figure and the following ones, the K-L distance is square root transformed to obtain an improved illustration of the data with a small K-L distance. The square root transformation is preferred above other possible transformations because the power of a test tends to be proportional to the square root of the number of observations, and the K-L distance is proportional to the number of observations.

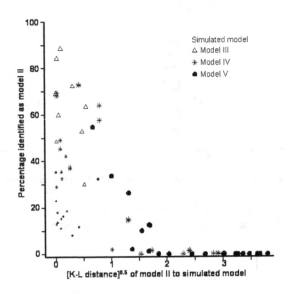

Figure 9.12. Percentage of the simulation runs in which Model II is chosen as best model (AICc criterion) as a function of the square root of the K-L distance of Model II to the simulated model. If close to the simulated model, Model I is often preferred over Model II. Markers scaled by distance of Model I to the simulated model: (1): > 2; (2): between 1 and 2; (3): between 0.5 and 1; and (4): smaller than 0.5. The number of simulated observations is 49.

Relation between K-L distance and model identification

The percentage of the simulation runs erroneously identified as Model I increases with decreasing K-L distance of the real model to Model I. Using 25 observations instead of 49 almost halves the K-L distance of Model I to the real model. The probability of choosing Model I instead of the simulated model depends only on the K-L distance between them, not on the number of simulated observations (Figure 9.13).

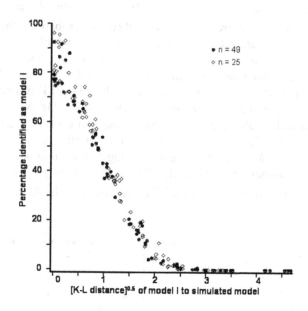

Figure 9.13. Percentage of simulation runs in which Model I is erroneously chosen as best model (AICc criterion) as a function of the square root of the K-L distance. Simulations with either 25 or 49 observations per trial

If the K-L distance for 49 observations is very small (say 0.05), halving the K-L distance will not increase the percentage of best fits of Model I considerably. On the other hand, if the K-L distance is very large, say above 10, halving the K-L distance will only slightly increase the choice for Model I. Thus, for models and parameter values with an intermediate K-L distance, a larger number of observations will reduce the K-L distance proportionally as well as the number of erroneous choices for Model I. Note that this minimum K-L distance between models is the theoretical distance, whereas the AIC calculates the observed difference for a given data set. If the minimum discrepancy to any of the simpler models is 2, the probability of choosing the extended model is about 50%.

The power of the AIC

Let us compare two nested models of which the more extended one is true but does not differ too much from the simpler one. Then twice the difference between the log-likelihoods, T, is asymptotically non-central χ^2 distributed with degrees of freedom $v = k_1 - k_2$ and non-centrality parameter λ (Cox and Hinkley, 1974). We will sketch some of the implications of the non-central χ^2

distribution of T for the power of the AIC. If T is $\chi^2_{\nu,\lambda}$ distributed, the power of the AIC to choose the more extended model (or the probability that $T > 2\nu$) can be calculated. Figure 9.14a shows the power for models with only one parameter difference ($\nu = 1$), assuming $\lambda = 2M$, where M is the minimum K-L distance of the simpler model to the more extended one. This value for λ shows a good fit to the simulation results, and Akaike (1974) has proven that the non-centrality parameter λ can be approximated by $2M$. Note, however, that for the general results, the exact value of the non-centrality parameter is irrelevant. Given $\lambda = 2M$, for $\nu = 1$ a power of about 50% is reached if $M = 1$ and of about 80% if $M = 2.5$. As long as nothing is known about the specific properties of the models, the probability $P(T > 2\nu)$ for $T \sim \chi^2_{\nu,\lambda}$ with $\lambda = 2M$ can be used as a first impression of the potential power of the AIC.

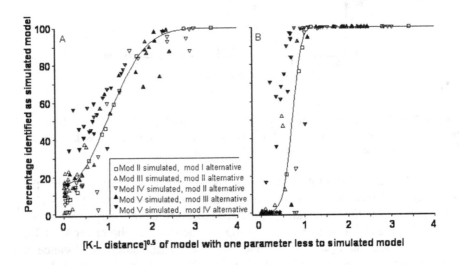

Figure 9.14. Percentage of simulation runs correctly identifying the simulated model (AICc criterion) as a function of the square root of the K-L distance to a single alternative model with one parameter less. Simulations with 49 observations per trial. The best fit is determined for separate runs (a), and for 25 runs combined (b). The drawn black lines show the theoretical prediction of the power if $\lambda = 2M$.

An increase in the number of observations will result in a more or less proportional increase of M, and thus of the non-centrality parameter λ. Sometimes, it will not be feasible to increase the number of observations in one trial with uniform conditions, for instance in the same trial. In this case, carrying out several trials, each with their own conditions, can increase the number of observations. The differing conditions in each trial may necessitate estimating a different set of model parameters for each trial. Let each trial contain n observations and let the number of trials be r. In this case, $T \sim \chi^2_{rv,\lambda'}$ with $\lambda' = r\lambda$, and the more extended model is chosen if $T > 2rv$. Figure 9.15 shows the relation between the number of trials and the power $\beta = P(T > 2rv | T \sim \chi^2_{rv,r\lambda})$ for $\lambda = 0, 0.5, 1, 2, 3$ and 4. For $\lambda = 0$ (the simpler model is true) and $\lambda = 0.5$ the power decreases with an increasing number of trials. Increasing the number of trials leads to an increase in power only for $\lambda = 2, 3$ and 4.

To illustrate how the K-L distance is related to the power of the AIC, we composed a set of 25 simulated trials for each simulated model by randomly drawing (with replacement) 25 runs (trials). For the 25 runs combined, the simulated model was tested against a model with one parameter less. This was repeated 1000 times. Figure 9.14b shows the relation between the percentage of correctly identified models and the K-L distance to the simpler model for a single simulation run. The theoretically expected power is also shown.

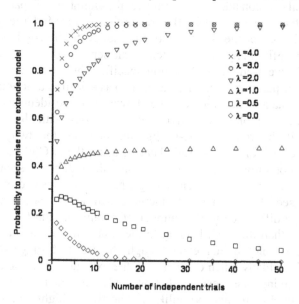

Figure 9.15. Power to correctly select a more extended model rather than a simpler one (AICc criterion) as a function of the number of independent trials. The difference between the maximum of the log-likelihoods for each trial is non-central χ^2 distributed with non-centrality parameter λ and degrees of freedom v.

9.5 CONCLUSIONS AND DISCUSSION

Experimenters must sometimes find a balance between what is statistically desirable and what is biologically realistic and feasible. One example of such a situation is formed by ecological experiments that consist of many replicate trials, each containing a limited number of observations. This approach may be the only way to obtain sufficient data, or be more informative for answering the ecologically relevant question. However, the variation between trials should be accounted for statistically. We have shown that rigorous application of this principle may make it impossible to distinguish any pattern present in ecological data.

In ANOVA types of problems it is a generally accepted rule that two groups can only be combined if they do not differ significantly. We applied this rule also for combining data in model selection and model parameter estimation. Using fixed effect models, the number of parameters is proportional to the number of fits of the model to separate groups of data sets. So if each trial is considered separately, the number of parameters is proportional to the number of trials. Before combining data in larger groups, it should be tested whether the fit of any of the models to the combined data is better than the best fit of any of the models to the separate data. To identify the best fitting model we used the AICc (a non-specific robust adaption of the AIC to large number of parameters or small data sets) as selection criterion.

If lumping of data is not allowed, we showed that the identifiability of a true model against simpler alternatives only increases with the number of replicate trials if the minimum K-L distance between the two models is sufficiently large ($> 1/2$) in each trial. That is, using the AIC(c) the power of model identification only increases by increasing the number of separate trials if the power in each trial is sufficiently large. Note that if instead of the AICc the LR test is used, the power would increase with an increasing number of trials. For a large difference in the number of parameters the LR criterion is less conservative than the AIC, leading to the situation that the AIC criterion may prefer the simpler model even though the LR test suggests that the simpler model is unlikely when compared to the extended one. The AIC leads to the model giving the best prediction and the LR test to the most parsimonious model being not too unlikely. The BIC is highly biased against the more extended models, whereas its claimed consistency is only relevant for very large numbers, and therefore totally uninteresting for most biological experiments with a relatively small number of observations.

To test whether data of several trials can be lumped, the fit of fixed and random effect versions of the models at each level of data lumping should be compared. At a given level of data lumping, the parameters of the fixed effect models have a fixed value, whereas the parameters of the random effect model are drawn for each trial from a probability distribution with fixed parameters at that level of lumping. The random effect models are the analogue of random effect models in ANOVA. In our example, the random effect model with data lumped in rich and poor years appears to fit best.

Next to fixed and random effect models, the fit of some mixed effect models might also be considered. For instance, we fitted models with a constant value for the slope parameter and trial-specific values for the other parameter(s). The fit of such mixed models have to be compared with the fit of models with parameters at only one level of data lumping. Note however, that such a mixed model may seem to show a general trend, for instance a relation with a maximum, but that this might imply for some trials a uniformly increasing trend and for others an uniformly decreasing trend, depending on which part of the curve is observed. The supposed general trend would in such a case be based on extrapolation and thus ecologically irrelevant.

Before starting an experiment it is sensible to investigate whether the statistical power is sufficient to answer the research questions. If one of the aims is to distinguish different models, the minimum K-L distance (MKLD) between alternatives can be used as an indication for the power of the AIC(c) decision procedure. To calculate the MKLD to a model, the parameters of that model and the values (or distribution) of the independent variables have to be specified. The parameter values can be based on previous experience, and can also reflect the minimum deviation of the complex model from the simpler one for which the complex model still merits consideration. The values of the independent variable(s) may be chosen to maximize the MKLD to the more complex model. The MKLD should be sufficiently large to be able to distinguish the more complex model from the simpler one. Increasing the number of observations by increasing the number of separate trials will only increase the power if the power in each separate trial is already sufficiently high.

The method described above can be summarized as:

1. Select models describing the hypothetical relationships that are of ecological importance.

2. Choose for each model realistic parameter ranges and decide for which parameter range a model should be distinguishable from the simpler alternatives.

3. Consider the range of independent model variables applicable in each experiment.

4. Calculate the MKLD of simpler models to more extended ones for the realistic parameter values (point 2) and independent variables (point 3).

5. If the MKLD is less than 1/2, models are not distinguishable without data lumping. In advance answer the question whether data lumping might be acceptable. If MKLD is sufficiently larger than 1/2, increasing the number of trials will increase the power of model identification. Calculate the number of replicate trials necessary to reach the desired power.

6. Perform the experiments.

7. Fit models to data of each trial and to the lumped data. Use fixed and random effect models. If necessary, also use mixed effect models.

8. Select the model with the lowest AIC(c), in this way both selecting the model type and the level of lumping. Remember that small differences between AIC(c)'s are not very informative.

Although an extensive body of literature exists on model selection, no guidelines are given on how to deal with data collected in a large set of separate trials as in our example. To our knowledge this is the first attempt to provide guidelines to facilitate the use of model selection methods in ecological applications and incorporate the intended model selection into the experimental set-up.

ACKNOWLEDGEMENTS

We thank Elizabeth van Ast for correction of the English, and Johan Grasman and two anonymous reviewers for critical comments on an earlier version of the manuscript. P.A.J. was supported by the Netherlands Foundation for the Advancement of Tropical Research (NWO-WOTRO grant 84-407).

REFERENCES

Akaike, H. (1974). A new look at the statistical model identification. IEEE Transactions on Automatic Control 19: 716-723.

Borowiak, D. S. (1989). Model Discrimination for Non-Linear Regression Models. Marcel Dekker Inc., New York.

Burnham, K. P. and D. R. Anderson (2002). Model Selection and Inference. A Practical Information-Theoretic Approach. Springer, New York.

Cox, D. R. and D. V. Hinkley (1974). Theoretical Statistics. Chapman and Hall, London.

Hallwachs, W. (1994). The Clumsy Dance between Agoutis and Plants: Scatterhoarding by Costa Rican Dry Forest Agoutis (*Dasyprocta punctata*: Dasyproctidae: Rodentia). PhD thesis, Cornell University, New York.

Hemerik, L. and N. van der Hoeven (2003). Egg distributions of solitary parasitoids revisited. Entomologia Experimentalis et Applicata 107: 81-86.

Hemerik, L., N. van der Hoeven and J. J. M. Van Alphen (2002). Egg distributions and the information a solitary parasitoid has and uses for its oviposition decisions. Acta Biotheoretica 50: 167-188 .

Hilborn, R. and M. Mangel (1997). The Ecological Detective. Confronting Models with Data. Princeton University Press, Princeton.

Huisman, J., H. Olff and L. F. M. Fresco (1993). A hierarchical set of models for species response analysis. Journal of Vegetation Science 4: 37-46.

Hurvich, C. M. and C.-L. Tsai (1989). Regression and time series model selection in small samples. Biometrika 76: 297-307.

Jansen, P. A., M. Bartholomeus, F. Bongers, J. A. Elzinga, J. Den Ouden and S. E. Van Wieren (2002). The role of seed size in dispersal by a scatter-hoarding rodent. pp. 209-225. In: Levey, D., W. R. Silva, and M. Galetti (Eds) Seed Dispersal and Frugivory: Ecology, Evolution and Conservation. CAB International, Wallingford.

Jansen, P. A. (2003). Scatterhoarding and Tree Regeneration. Ecology of Nut Dispersal in a Neotropical Rainforest. PhD thesis, Wageningen University, The Netherlands.

Kullback, S. and R. A. Leibler (1951). On information and sufficiency. Annals of Mathematical Statistics 22: 79-86.

Linhart, H. and W. Zucchini (1986). Model Selection. John Wiley and Sons, New York.

Schwarz, G. (1978). Estimating the dimension of a model. Annals of Statistics 6: 461-464.

Smith, C. C. and O. J. Reichman (1984). The evolution of food caching by birds and mammals. Annual Review of Ecology and Systematics 15: 329-351.

Umbach, D. M. and A. J. Wilcox (1996). A technique for measuring epidemiologically useful features of birthweight distributions. Statistics in Medicine 15: 1333-1348.

Van der Hoeven, N., (in press). A general method to calculate the power of likelihood-ratio based tests to choose between two nested models. Journal of Statistical Planning and Inference.

Vander Wall, S. B. (1990). Food Hoarding in Animals. Chicago University Press, Chicago.

Vander Wall, S. B. (2003). Effects of seed size of wind-dispersed pines (Pinus) on secondary seed dispersal and the caching behavior of rodents. Oikos 100: 25-34.

Nelly van der Hoeven,
Department of Theoretical Biology, Leiden University

Lia Hemerik
Biometris, Department of Mathematical and Statistical Methods,
Wageningen University

Patrick A. Jansen
Forest Ecology and Forest Management Group, Wageningen University

APPENDIX A

Results of simulations with Models I to V (see equation (9.13)) and different sets of parameter values (1000 runs per set). Numbers indicate the number of runs at which each of the five models was selected as the best-fitting according to the AICc criterion. The best fitting model if all 1000 runs are combined is also given. The models for each parameter combination are illustrated in figures.

Number of observations	Model	Parameters				Model number					Best model, all simulations together	Figures with examples of model
		a	b	c	d	I	II	III	IV	V		
						Number of runs in which model was chosen as best.						
25	I	-1				809	117	45	15	14	1	
25	I	-2				794	158	35	10	3	1	
25	I	-5				963	37	0	0	0	1	
49	I	-1				773	128	68	17	14	1	
49	I	-2				792	136	50	8	14	1	
49	I	-5				925	74	0	1	0	1	
25	II	0	-1			620	242	100	16	22	1	
25	II	0	-2			376	474	95	30	25	2	
25	II	0	-5			4	854	109	19	14	2	
25	II	-2	-1			742	200	29	19	10	1	
25	II	-2	-2			572	366	36	19	7	2	
25	II	-2	-5			45	882	59	10	4	2	
49	II	0	-1			494	358	84	37	27	2	
49	II	0	-2			123	707	92	32	46	2	
49	II	0	-5			0	878	58	28	36	2	
49	II	-2	-1			654	257	46	27	16	1	
49	II	-2	-2			378	541	45	23	13	2	
49	II	-2	-5			3	914	50	27	6	2	

Number of seeds	Model	Parameters				Model number					Best model, all simulations together	Figures with examples of model
		a	b	c	d	I	II	III	IV	V		
						Number of runs in which model was chosen as best						
25	III	0	-2	0		588	221	124	36	31	1	Model III, parameter b = -2
25	III	0	-2	-5		360	488	111	28	13	2	
25	III	-2	-2	0		721	153	82	16	28	1	
25	III	-2	-2	-5		611	321	45	13	10	2	
49	III	0	-2	0		432	355	138	31	44	2	
49	III	0	-2	-5		151	692	92	28	37	2	
49	III	-2	-2	0		669	154	107	27	43	1	
49	III	-2	-2	-5		398	486	70	30	16	2	
25	IV	-2	-5	0		310	118	267	162	143	3	Model IV, parameter a = -2, b = -5
25	IV	-2	-5	-2		384	39	174	266	137	3	
25	IV	-2	-5	-5		139	601	110	97	53	2	
49	IV	-2	-5	0		109	146	298	301	146	3	
49	IV	-2	-5	-2		191	17	178	462	152	4	
49	IV	-2	-5	-5		17	577	125	194	87	2	
25	V (IV)	-2	-2	-2	2	722	130	81	38	29	1	Model V, parameter a = -2, b = -2, c = -2
25	V	-2	-2	-2	5	347	283	229	90	51	2	
25	V	-2	-2	-2	10	16	299	422	178	85	3	
25	V (IV)	-2	-10	-2	10	59	0	39	542	360	4	
25	V	-2	-10	-2	20	17	1	44	579	359	4	
25	V	-2	-10	-2	50	22	2	54	584	338	4	
49	V (IV)	-2	-2	-2	2	684	119	115	52	30	1	Model V, parameter a = -2, b = -10, c = -2
49	V	-2	-2	-2	5	160	337	305	150	48	3	
49	V	-2	-2	-2	10	0	124	554	261	61	3	
49	V (IV)	-2	-10	-2	10	0	0	10	723	267	4	
49	V	-2	-10	-2	20	0	0	5	552	443	5	
49	V	-2	-10	-2	50	0	0	11	318	671	5	

10

Resilience and Persistence in the Context of Stochastic Population Models

Johan Grasman, Onno A. van Herwaarden and Thomas J. Hagenaars

ABSTRACT

The resilience of an ecological system is defined by the velocity of the system as it returns to its equilibrium state after some perturbation. Since the system does not arrive exactly at the equilibrium within a finite time, the definition is based on the time needed to decrease the distance to the equilibrium with some fraction. In this study it is found that for stochastic populations this arbitrarily chosen function disappears because the equilibrium point can be replaced by a small (confidence) domain containing the equilibrium. The size of this domain is a measure for the (local) persistence of the system. This method is fully worked out for the stochastic logistic equation as well as for a prey-predator system.

Keywords: resilience, persistence, ecosystem, Fokker-Planck equation, logistic equation, prey-predator system, epidemiology.

10.1 INTRODUCTION

In the mathematical modelling of ecological systems, stability of an equilibrium state can be quantified in several ways. The stability definition, used in the mathematical theory of dynamical systems, is a qualitative one: an equilibrium is asymptotically stable if for all initial states near the equilibrium the system returns to this equilibrium state as time tends to infinity. For the purpose of quantifying how fast the system returns to the equilibrium, the notion of resilience has been introduced in ecology. Although resilience has since obtained a wider meaning in part of the literature, we will here be primarily concerned with its original version. This notion of (engineering) resilience (Holling, 1996) measures the speed of return: a highly resilient ecological system rapidly restores its equilibrium if a deviation occurs. Its mathematical definition is based on the assumption that deviations from equilibrium are small so that the dynamic behaviour can be approximated by a linear system of differential equations. In this case the speed of return to the

T.A.C. Reydon and L. Hemerik.,(eds.), Current Themes in Theoretical Biology,
267-280.
© 2005 *Springer. Printed in the Netherlands.*

equilibrium state is governed by the real part of the eigenvalue nearest to the imaginary axis in the complex plane. Resilience is then defined by (the absolute magnitude of) this real part being the reciprocal of the time in which deviations from equilibrium shrink by a factor $1/e$ (DeAngelis, 1992). The latter time is referred to as the *return time* (although obviously it is not literally the time of return to exact equilibrium, which would always be infinitely long). We note that the deterministic return time is intrinsically only defined up to a multiplicative constant, as is reflected by the arbitrariness in the choice of the $1/e$ reduction factor. The return time that corresponds with this factor is also called the relaxation time in physics. In other applications a factor $1/2$ is taken (half-time).

In the ecological literature resilience has obtained a wider meaning than we sketched above (Holling, 1996). In order to make a distinction between the definition completely based on the eigenvalues, the expression "ecological resilience" is used in the biological literature (see Peterson *et al.*, 1998). This resilience definition covers two more elements compared with the one based on the stability of an equilibrium. First it stresses the continuation of a state in which all populations remain present and not the equilibrium state in particular. It is remarked that this creates some overlap with the existing idea of persistence. A second element, that is added, deals with the role of biodiversity. For a resilient ecosystem certain ecological functions have to be fulfilled. If more than one species takes care of this, a higher degree of resilience (persistence) will be guaranteed.

In this study we will extend the notion of "engineering resilience", being the expression used for the eigenvalue analysis. It is the result of a stochastic modelling of interacting biological populations. An important difference with deterministic systems is that here the close-to-equilibrium behaviour is typically only *quasi*-stationary. In the presence of an absorbing unstable equilibrium state with one species being extinct, stochastic exit will occur with certainty. Although the system may persist for very long times close to the (deterministically) stable internal equilibrium it will eventually reach a neighbourhood of this unstable equilibrium, slow down and will during one of those visits be absorbed. As a consequence, in the process of defining resilience for stochastic models we naturally touch upon the notion of persistence. We propose a definition of local persistence that differs from the dynamical systems literature, see Holling (1973), Ives (1995) and Grimm and Wissel (1997). Our definition connects closely to approaches adopted to quantify persistence in population dynamics (e.g. Roozen, 1987) and in epidemiology (e.g. Nåsell, 1999). The above situation of a dynamical system with a stable internal equilibrium and unstable equilibria at the boundary, such as in the prey-predator system, is a special case of the general configuration with one or more stable internal attractors (equilibria, limit cycles and strange attractors) with each having its domain of attraction. Persistence of a stable state can be seen as the ability of the system to stay in a domain of attraction under external perturbations (Carpenter *et al.*, 2001). Hitting an external boundary, resulting in the extinction of a species, reduces the state space dimension by one. In defining local persistence we consider a neighbourhood of a stable internal equilibrium. The method can in principle be extended to other types of attractors.

The method we present to quantify resilience and also local persistence applies to systems consisting of arbitrarily many interacting biological populations, but for simplicity of presentation we concentrate on two low-dimensional systems: the logistic model and a predator-prey model. Let x_1 and x_2 be scaled population variables representing respectively the prey and predator population, then the deterministic predator-prey model is given by the following system of differential equations containing three parameters α_1, α_2 and δ:

$$x_1'(t) = \alpha_1\{1 - x_1(t) - x_2(t)\}x_1(t), \tag{10.1a}$$

$$x_2'(t) = \alpha_2\{-1 + \frac{x_1(t)}{1-\delta}\}x_2(t), \qquad 0 < \delta < 1. \tag{10.1b}$$

For the *predator-prey system* (10.1) we analyse the behaviour near the internal equilibrium $\underline{x} = (1 - \delta, \delta)$ by substituting $x(t) = \underline{x} + v(t)$, giving

$$v'(t) = Bv(t) \quad \text{with} \quad B = \begin{pmatrix} -\alpha_1(1-\delta) & -\alpha_1(1-\delta) \\ \alpha_2\delta/(1-\delta) & 0 \end{pmatrix}. \tag{10.2ab}$$

The eigenvalues of B are

$$\lambda_{1,2} = -\frac{1}{2}\alpha_1(1-\delta) \pm \frac{1}{2}\sqrt{\alpha_1^2(1-\delta)^2 - 4\alpha_1\alpha_2\delta}. \tag{10.3}$$

Their real part is used to quantify resilience as described above.

In a similar manner for the *logistic equation*, Equation (10.1a) with $x_2 = 0$, we may derive that for the system linearized at the equilibrium $x_1 = 1$ the single eigenvalue equals $-\alpha_1$.

In the quasi-stationary state of the stochastic counterparts to these models, the system remains in a well-defined neighbourhood of the stable equilibrium most of the time. The return time to this neighbourhood turns out to be similar to the one of the deterministic system. The only difference is that now the multiplicative constant, that was undetermined in the deterministic case, is expressed in the size of the stochastic component of the system. More surprisingly, the linearization that is made in the deterministic approach does not have to be made in the stochastic model. The way the return time depends on the size of the stochastic component reduces considerably its use for describing ecological stability and it leads us to critically examine the value of a resilience definition based on the eigenvalues in Section 10.2.

In Section 10.3 we will introduce the concept of local persistence. It is argued that the diameter of the domain where the stochastic system remains most of the time, is a measure of how a system close to equilibrium copes with stochastic perturbations and persists in remaining close to its equilibrium state. We also make an excursion to a related problem, that of the duration of the endemic period of an infectious disease, for which a model similar to a prey-predator system applies.

10.2 STOCHASTIC DYNAMICS AND RETURN TIME

In this section we introduce the concept of return time for a stochastic model of a system of interacting populations. We make the assumption that the random fluctuations are small. The probability that the system is in the state space outside some neighbourhood of the equilibrium is (asymptotically) very small with respect to the size of the random perturbation. This neighbourhood of the equilibrium is the domain in state space where the system can most likely be expected and where it acts regularly. When speaking of the return time in the stochastic model we mean the time needed to return to this domain. The system may leave this domain from two types of actions: either by a sudden shock (Pimm, 1993) or by a sequence of small stochastic fluctuations.

Table 10.1. The transition probabilities of the prey-predator system.

Transition	Probability
$N_1 \rightarrow N_1 + 1$	$b_1 N_1 \Delta t$
$N_1 \rightarrow N_1 - 1$	$(d_1 + \dfrac{\alpha_1}{K_1} N_1 + \dfrac{\alpha_1}{K_2} N_2)$
$N_2 \rightarrow N_2 + 1$	$(b_2 + \dfrac{\alpha_2}{K_1(1-\delta)} N_1) N_2$
$N_2 \rightarrow N_2 - 1$	$d_2 N_2 \Delta t$

Let $N_1(t)$ and $N_2(t)$ be respectively the size of the prey and the predator population at time t. The transition probabilities over the time interval $(t, t + \Delta t)$ are given in Table 10.1, where

$$\alpha_1 = b_1 - d_1 > 0, \quad \alpha_2 = -b_2 + d_2 > 0 \quad \text{and} \quad 0 < \delta < 1 \qquad (10.4)$$

with b_i and d_i, $i = 1,2$, the birth and death rates of the two populations. Furthermore the parameters K_1 and K_2 have been used to obtain the scaled deterministic system (10.1) and show up when scaling back again. The probability of the event that either population increases or decreases with 1 in a time interval of length Δt is given. The probability of two or more events taking place in that time is of the order $O((\Delta t)^2)$ and is not taken into consideration.

Using the scaling

$$N_1 = K_1 x_1 \quad \text{and} \quad N_2 = K_2 x_2 \tag{10.5}$$

we find for the change Δx_i over the time interval $(t,\ t + \Delta t)$:

$$E\{\Delta x_1\} = \alpha_1(1 - x_1 - x_2)x_1 \Delta t$$

$$E\{\Delta x_2\} = \alpha_2(-1 + \frac{x_1}{1 - \delta})x_2 \Delta t$$

and

$$E\{(\Delta x_1)^2\} = \frac{1}{K_1}(\beta_1 + \alpha_1 x_1 + \alpha_1 x_2)x_1 \Delta t$$

$$E\{(\Delta x_2)^2\} = \frac{1}{K_2}(\beta_2 + \frac{\alpha_2 x_1}{1 - \delta})x_2 \Delta t$$

where

$$\beta_1 = b_1 + d_1 \quad \text{and} \quad \beta_2 = b_2 + d_2.$$

Taking into account only terms of order $O(\Delta t)$, we obtain for the variance

$$\text{var}\{\Delta x_1\} = E\{(\Delta x_1)^2\} \quad \text{and} \quad \text{var}\{\Delta x_2\} = E\{(\Delta x_2)^2\}.$$

For the corresponding diffusion process we derive the Fokker-Planck equation for the time-dependent probability density function $p(t, x)$ (Grasman and Van Herwaarden, 1999):

$$\frac{\partial p}{\partial t} = \sum_{j=1}^{2}\left\{ -\frac{\partial}{\partial x_j}(b_j(x)p) + \frac{1}{2K}\frac{\partial^2}{\partial x_j^2}(a_j(x)p)\right\} \tag{10.6a}$$

with

$$K = \frac{1}{2}(K_1 + K_2) \tag{10.6b}$$

and the drift terms

$$b_1(x) = \alpha_1(1 - x_1 - x_2)x_1, \quad b_2(x) = \alpha_2(-1 + \frac{x_1}{1 - \delta})x_2. \tag{10.7ab}$$

From the variances we derive the diffusion terms

$$a_1(x) = (\varepsilon_1 + \delta_1 x_1 + \delta_1 x_2)x_1, \quad a_2(x) = (\varepsilon_2 + \frac{\delta_2 x_1}{1 - \delta})x_2, \tag{10.8ab}$$

where

$$\varepsilon_j = \beta_j K/K_j \quad \text{and} \quad \delta_j = \alpha_j K/K_j, \quad j = 1,2. \tag{10.9ab}$$

The above diffusion approximation allows us to define the return time and to obtain (approximate) results for it. For that purpose asymptotic methods are used.

In our asymptotic approximation of the return time we use the fact that K_1 and K_2 are large and of a comparable order of magnitude (and so is K). For any given initial distribution the system tends to a stationary distribution $p^{(s)}(x)$ concentrated in a $1/\sqrt{K}$-neighbourhood of the stable internal equilibrium. A point in state space lies within this neighbourhood depending on its distance d to the equilibrium and the value of K. For $K \to \infty$ the distance d of points in the $1/\sqrt{K}$-neighbourhood of the equilibrium should satisfy the condition that $d\sqrt{K}$ remains bounded or using the Landau order symbol $d = O(1/\sqrt{K})$. Actually the distribution can only be *quasi*-stationary as the probability density is slowly leaking away at a boundary. The function $p^{(s)}(x)$ satisfies the stationary Fokker-Planck equation

$$\sum_{j=1}^{2} \left\{ -\frac{\partial}{\partial x_j}(b_j(x)p^{(s)}) + \frac{1}{2K}\frac{\partial^2}{\partial x_j^2}(a_j(x)p^{(s)}) \right\} = 0, \qquad (10.10)$$

see (10.6). To compute the expected return time $T(x_0)$ if starting at a point x_0 in state space we first have to compute the expected arrival time $T(x_0, x)$ at an arbitrary point x. Next the expected return time follows from

$$T(x_0) = \int\int T(x_0,x)\,p^{(s)}(x)\,\mathrm{d}x_1\mathrm{d}x_2. \qquad (10.11)$$

This expression for the return time applies to any stochastic dynamical system. Van Kampen (1995) gives the very illustrative example of magnetotactic bacteria that take the direction of the magnetic south pole. In an experiment this direction is changed and the process of re-orientation is then observed. Because the bacteria model is a one-dimensional stochastic dynamical system, one is in the position to do all the calculations. For higher dimensional systems this would become quite cumbersome. We therefore simplify the definition of the return time slightly. We define it by the time needed to arrive in the $1/\sqrt{K}$-neighbourhood. Since outside the $1/\sqrt{K}$-neighbourhood the drift prevails over the diffusion the expected return time is approximated very well by that of the corresponding deterministic system.

The stochastic logistic system

We perform the calculations for the stochastic logistic system by analysing the one-dimensional stochastic birth-death process for $N_1(t)$ given in Table 10.1 with $N_2(t) = 0$. It has the probability density function $p(t, x_1)$ satisfying the Fokker-Planck equation

$$\frac{\partial p}{\partial t} = -\frac{\partial}{\partial x_1}\{\alpha_1(1-x_1)x_1 p\} + \frac{1}{2K_1}\frac{\partial^2}{\partial x_1^2}\{(\beta_1 x_1 + \alpha_1 x_1^2)p\}.$$

We approximate $p^{(s)}(x_1)$ near $x_1 = 1$ asymptotically by

$$p^{(s)}(x_1) = \frac{1}{\sqrt{2\pi\sigma_1^2}} \exp\left[-\frac{(x_1-1)^2}{2\sigma_1^2}\right], \quad \sigma_1^2 = \frac{\alpha_1 + \beta_1}{2\alpha_1 K_1}, \quad K_1 \gg 1. \quad (10.12)$$

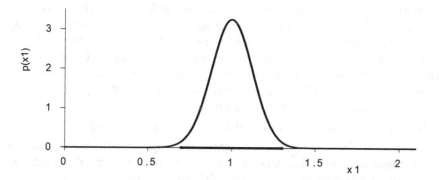

Figure 10.1. The quasi-stationary probability density approximated by (10.12). The $1/\sqrt{K_1}$-neighbourhood of the equilibrium point is indicated by the solid line on the x_1-axis.

We next need to indicate more precisely what it means that the state of the system is in a $1/\sqrt{K_1}$-neighbourhood of the equilibrium $x_1 = 1$. For this purpose we introduce an arbitrary large constant L independent of K_1 with $1 \ll L \ll \sqrt{K_1}$. Then the $1/\sqrt{K_1}$-neighbourhood is for this system defined as

$$1 - L/\sqrt{K_1} < x_1 < 1 + L/\sqrt{K_1},$$

see Figure 10.1. Selecting an arbitrary point x_1 on the positive x_1-axis away from the boundary $x_1 = 0$ and outside the $1/\sqrt{K_1}$-neighbourhood of $x_1 = 1$, we approximate the expected arrival time at this domain. Since outside the domain the drift prevails over the random motion, we may take the deterministic return time as the approximation

$$T(x_1) = \int_{x_1}^{1-L/\sqrt{K_1}} \frac{1}{\alpha_1 s(1-s)} ds \quad \text{for } x_1 < 1 - L/\sqrt{K_1} \qquad (10.13a)$$

and

$$T(x_1) = \int_{x_1}^{1+L/\sqrt{K_1}} \frac{1}{\alpha_1 s(1-s)} ds \quad \text{for } x_1 > 1 + L/\sqrt{K_1}. \qquad (10.13b)$$

Both formulas have the same asymptotic behaviour for large K_1:

$$T(x_1) = \frac{1}{\alpha_1} \ln \sqrt{K_1} (1 + o(1)) \qquad (10.14)$$

where $o(1)$ denotes a term that contains x_1, L and K_1 and that tends to zero for $K_1 \to \infty$. Consequently, the return time is asymptotically independent of the starting point. This is due to the fact that the system slows down near the equilibrium, so that most of the time is spent in the last part of the return path. Thus, the stochastic return time can be defined very naturally without resorting to a linearization assumption as is done in the deterministic resilience definition. It holds for any starting point within the domain of attraction of the equilibrium not too close to the boundary. Moreover, the arbitrary constant in the deterministic return time is absent now. Note that also the constant L is not present in the leading term of the return time.

It is remarked that for $K_1 \to \infty$ the $1/\sqrt{K_1}$-neighbourhood shrinks to the equilibrium of the corresponding deterministic logistic differential equation and that then the return time tends to infinity. We also note that differences in return time (resilience) between systems differing in carrying capacity K_1 need to be interpreted with care. As the domain of regular action (the $1/\sqrt{K_1}$-neighbourhood) increases with decreasing K_1 (smaller populations exhibiting relatively stronger stochastic fluctuations), the distance to this domain and thus the return time shortens. At the same time however, the increasing relative magnitude of the perturbation term will increase the frequency at which perturbations lead to extinction, i.e. will reduce persistence. This illustrates our view that, in order to obtain a full insight into the ecological stability of a system, one should not use the resilience definition introduced above as the sole measure, but simultaneously study a measure of persistence. To this end we introduce the notion of local persistence in Section 10.3.

The stochastic prey – predator system

For the two-dimensional prey-predator system ($N_2 \neq 0$) satisfying the Fokker-Planck equation (10.6) the probability density is concentrated in a $1/\sqrt{K}$-neighbourhood of the equilibrium $\underline{x} = (1-\delta, \delta)$. The probability density function $p^{(s)}(x)$ can be approximated by a bivariate normal distribution satisfying the stationary Fokker-Planck equation corresponding with (10.6) linearized at the equilibrium $\underline{x}$:

$$p^{(s)}(x) = \frac{K}{2\pi\sqrt{(1-\rho^2)\sigma_1^2\sigma_2^2}} \exp[-\frac{1}{2}(x-\underline{x})^T S^{-1}(x-\underline{x})]$$

with covariance matrix

$$S = \frac{1}{K} \begin{pmatrix} \sigma_1^2 & \rho\sigma_1\sigma_2 \\ \rho\sigma_1\sigma_2 & \sigma_2^2 \end{pmatrix}$$

satisfying the matrix equation

$$K^{-1}A + BS + SB^T = 0,$$ (10.15)

where B is given by (10.2b) and A is a diagonal matrix with $a_i(\underline{x})$, $i = 1,2$ on the diagonal. Solving this matrix equation we obtain, using (10.9),

$$\sigma_1^2 = \frac{a_1(\varepsilon_2 + \delta_2)(1-\delta) + a_2(\varepsilon_1 + \delta_1)}{2\alpha_1\alpha_2},$$ (10.16a)

$$\sigma_2^2 = \frac{a_1 a_2 \delta(\varepsilon_2 + \delta_2)(1-\delta) + a_2^2 \delta(\varepsilon_1 + \delta_1)}{2\alpha_1^2 \alpha_2 (1-\delta)^2} + \frac{(\varepsilon_2 + \delta_2)(1-\delta)}{2\alpha_2},$$ (10.16b)

$$\rho\sigma_1\sigma_2 = -\frac{(\varepsilon_2 + \delta_2)(1-\delta)}{2\alpha_2}.$$ (10.16c)

Now the $1/\sqrt{K}$-neighbourhood is a domain with an ellipsoidal boundary

$$(\underline{x} - \underline{x})^T S^{-1}(\underline{x} - \underline{x}) = L$$ (10.17)

with L again an arbitrary number independent of K with $1 \ll L \ll K$ so that $L/K \ll 1$. These domains with L as parameter can be seen as the confidence domains.

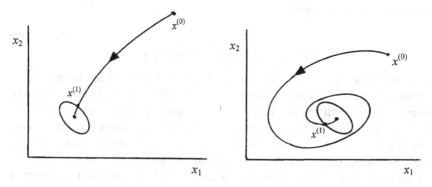

(a) λ_1 is a real eigenvalue.

(b) $\lambda_{1,2}$ are complex conjugated eigenvalues.

Figure 10.2. Return of a system of populations to the domain of regular operation, i.e. the $1/\sqrt{K}$-neighbourhood of the equilibrium. In (10.18) the leading term in the asymptotic approximation of the return time is given.

The way this ellipsoid is approached depends on the eigenvalue(s) of B nearest to the imaginary axis in the complex plane. If this is a real one, $\lambda_1 = -\alpha$, then the trajectories approach along the corresponding eigenvector $v^{(1)}$. If the nearest eigenvalues are complex $\lambda_{1,2} = -\alpha \pm i\beta$, then the trajectories spiral towards the equilibrium. Let $x^{(0)}$ be the starting point of a trajectory, away from the ellipsoidal domain containing the equilibrium, and $x^{(1)}$ the point where the trajectory enters the domain, see Figure 10.2. Then the time needed to travel from $x^{(0)}$ to $x^{(1)}$ is mostly spent at the last part of the interval, where the distance to the equilibrium $r(t) = |x(t) - \underline{x}|$ approximately satisfies $r' = -\alpha r$. Thus

$$T \approx \int_{|x^{(0)} - \underline{x}|}^{|x^{(1)} - \underline{x}|} \frac{dt}{dr} dr \approx \frac{-1}{\alpha} \int_{|x^{(0)} - \underline{x}|}^{|x^{(1)} - \underline{x}|} \frac{1}{r} dr$$

$$= \frac{-1}{\operatorname{Re}\lambda_{1,2}} \ln\sqrt{K}(1 + o(1)) \qquad (10.18)$$

with $\operatorname{Re}\lambda_{1,2}$ given by (10.3). It is noted that the starting point and the direction of approach do not play any role in the first order approximation. Moreover, it is seen that the form of the expression remains unchanged compared with the one-dimensional system except for the possibility of complex conjugated eigenvalues. It is easily deduced that this expression also holds for systems with a dimension higher than two.

10.3 LOCAL PERSISTENCE

Persistence is a notion different from, but still closely related to resilience, as we already mentioned in our introduction. A highly resilient system quickly recovers from a perturbation that puts it in state space far away from the equilibrium in which it regularly operates. A persistent system is characterised by the ability to avoid fluctuations away from the equilibrium that are large enough to cause exit from the domain of attraction of that equilibrium. Persistence of an ecological system is quantified in literature by the expected time until it collapses resulting in the extinction of one or more species. In contrast to resilience, extinction can be studied only in the context of a *stochastic* population model. The randomness can be of demographic origin, such as random birth and death, or arise from the environment. The definition for local persistence we will give differs from the ones given in the dynamical-systems literature, see Grimm and Wissel (1997).

In a resilient system there is a rapid return to the equilibrium state because in the complex plane the eigenvalues of the linearized deterministic system are sufficiently far away from the imaginary axis, while in a system that lacks persistence the random motion, that drives it away from the equilibrium, is felt. It is clear that persistence depends on the balance between these random forces and the size of the real part of the eigenvalues nearest to the imaginary axis. The size of the domain of regular operation is determined by this balance of opposite tendencies. In the calculation of the return time we were in the fortunate position that the arbitrary large constant L in the expression for the domain boundary dropped out in the first order approximation. If we quantify the size of the ellipsoidal domain of regular operation by the maximal diameter of the domain, this would lead to an expression for this size containing the constant L that now remains present. The ellipsoid with $L = 1$ indicating the size of the variance σ^2 in all directions is a good measure for the size of the domain of regular operation. Thus we have to find the maximal diameter $2\sigma_{max}$ of the ellipsoid

$$x^T S^{-1} x = 1. \tag{10.19}$$

The transformation to y-coordinates coinciding with the axes of the ellipsoid is of the form $x = Uy$ with U a unitary matrix ($U^T = U^{-1}$) with

$$U^{-1} S U = D, \tag{10.20}$$

where D is a diagonal matrix; the largest element on the diagonal equals σ_{max}^2.

For the logistic model $\sigma_{max}^2 = \sigma_1^2$ is given in (10.12). For the prey-predator system we have, using (10.16),

$$\sigma_{max}^2 = \frac{1}{2K}(\sigma_1^2 + \sigma_2^2) + \frac{1}{2K}\sqrt{(\sigma_1^2 + \sigma_2^2)^2 - 4\sigma_1^2\sigma_2^2(1-\rho^2)}. \tag{10.21}$$

Thus, it is proposed that the maximal variance is used as measure for the (lack of) persistence of an equilibrium state. It is defined by the local stochastic dynamics of the system.

Persistence in epidemiology

A closely related persistence measure, useful in the context of epidemiology, is the quasi-stationary coefficient of variation in the number of the infected individuals (Hagenaars et al., 2003). In applications where persistence is quantified by the expected extinction time, the smallest ellipsoid is selected that is tangent to one of the boundaries of the positive quadrant of the x_1, x_2-plane. The value of $p^{(s)}$ at the point of tangency determines the expected extinction time. This method breaks down when the linearization at the internal equilibrium fails to produce a good approximation near the axes. It has been successfully applied to calculate the duration of the endemic period of infectious diseases with the infectious period much shorter than the host life span (Nåsell, 1999). An alternative approach is based on the value of $p^{(s)}$ near an unstable equilibrium at the boundary. Such an approximation gives good results for problems with extremely large extinction times, see Grasman et al. (2001) and Van Herwaarden and Van der Wal (2002).

It is noted that in practice persistence is not always solely determined by exit behaviour from a quasi-stationary internal equilibrium state. In the epidemiological regime studied by Nåsell (1999) for example, where the host life expectancy is much longer than the infectious period of the agent, the introduction of an infection in a population typically sets off a major outbreak that may 'self-extinguish' before reaching quasi-stationarity (Van Herwaarden, 1997). As a consequence, the expected duration of outbreaks is in such a regime only partly determined by the expected exit time conditional on quasi-stationarity.

10.4 CONCLUDING REMARKS

The study of resilience in a population model with stochastic elements has led to an expression for the return time that on the one hand is consistent with the one for deterministic systems. It even extends the range of validity of the formula to starting points far away from the equilibrium but within the domain of attraction. On the other hand a return time defined as the time needed to arrive at the domain of regular operation has the peculiar property that an increase of noise decreases the return time because of the expansion of the domain of "regular" operation. It is concluded that for studying the stability of a system of interacting biological populations the return time should not be used as the sole measure. As a complementary measure local persistence is defined by the size of the domain of regular operation at an equilibrium state: the smaller the domain the better the system persists in this state.

The wish to come to a more coherent description is also found in other publications. Ludwig *et al.* (1997) include other limit states than only a single equilibrium in their analysis. Cropp and Gabric (2002) employ for the purpose of consistency a metric based upon thermodynamic laws. With our approach we stay close to the definition of (engineering) resilience as it is mostly used in literature. In addition it turns out that in case studies the essential element of our approach, the ellipsoidal domain of regular operation, is already playing an important role: we refer to problems in epidemiology, see Nåsell (1999) and Hagenaars *et al.* (2003), population biology (Roozen, 1987) and in fishery (Grasman and Huiskes, 2001).

Ecological resilience as presented by Peterson *et al.* (1998) stresses aspects of the stability of ecosystems different from our expressions. It is believed that if ecological resilience is made more quantitative in its analysis of large systems of interacting populations, that more of the methodology we developed can be incorporated in this description of local persistence. Making an estimate of all parameters in a higher dimensional model will probably be a larger problem than solving the matrix equation (10.15) forming the key to the problem of determining the diameter of the confidence ellipse. The problem of extinction that is studied in the quantitative literature mostly deals with the extinction of only one species that is not crucial for the continued existence of the remaining population (species-deletion stability), see e.g. Grasman *et al.* (2001). If this population were crucial, the collapse of the ecosystem, as a result of the extinction of this keystone species, is still open to a quantitative analysis of the type we made by considering separately the extinction of each species that follows next.

REFERENCES

Carpenter, S., B. Walker, J. M. Anderies and N. Abel (2001). From metaphor to measurement: Resilience of what to what? Ecosystems 4: 765-781.

Cropp, R., and A. Gabric (2002). Do ecosystems maximize resilience? Ecology 83: 2019-2026.

DeAngelis, D. L. (1992). Dynamics of Nutrient Cycling and Food Webs. Chapman & Hall, London.

Grasman, J., and O. A. van Herwaarden (1999). Asymptotic Methods for the Fokker-Planck Equation and the Exit Problem in Applications. Springer-Verlag, Heidelberg.

Grasman, J., and M. J. Huiskes (2001). Resilience and critical stock size in a stochastic recruitment model. Journal of Biological Systems 9: 1-12.

Grasman, J., F. van den Bosch and O. A. van Herwaarden (2001). Mathematical conservation ecology: a one predator – two prey system as case study. Bulletin of Mathematical Biology 6: 259-269.

Grimm, V., and C. Wissel (1997). Babel, or the ecological stability discussions: An inventory and analysis of terminology and a guide for avoiding confusion. Oecologia 109: 323-334.

Hagenaars, T. J., C. A. Donnelly, and N. M. Ferguson (2003). Spatial heterogeneity and the persistence of infectious diseases. preprint.

Holling, C. S. (1973). Resilience and stability of ecological systems. Annual Review of Ecology and Systematics 4: 1-23.

Holling, C. S. (1996). Engineering resilience vs. ecological resilience. In: Schulze, P. C. (Ed.). Engineering within Ecological Constraints. National Academy Press, Washington DC. pp. 31-43.

Ives, A. R. (1995). Measuring resilience in stochastic systems. Ecological Monographs 65: 217-233.

Ludwig, D., B. Walker and C. S. Holling (1997). Sustainability, stability and resilience. Ecology and Society 1:
    http://www.ecologyandsociety.org/vol1/iss1/art7/.

Nåsell, I. (1999). On the time to extinction in recurrent epidemics. Journal of the Royal Statistical Society B 61: 309-330.

Pimm, S. L. (1993). Nature's short sharp shocks. Current Biology 3: 288-290.

Peterson, G., C. R. Allen and C. S. Holling (1998). Ecological resilience, biodiversity and scale. Ecosystems 1: 6-18.

Roozen, H. (1987). Equilibrium and extinction in stochastic population dynamics. Bulletin of Mathematical Biology 49: 671-696.

Van Herwaarden, O. A. (1997). Stochastic epidemics: the probability of extinction of an infectious disease at the end of a major outbreak. Journal of Mathematical Biology 35: 793-813.

Van Herwaarden, O. A., and N. J. van der Wal (2002). Extinction time and age of an allele in a large finite population. Theoretical Population Biology 61: 311-318.

Van Kampen, N. G. (1995). The turning of magnetotactic bacteria. Journal of Statistical Physics 80: 23-34.

Johan Grasman
Biometris, Department of Mathematical and Statistical Methods
Wageningen University

Onno A. van Herwaarden
Biometris, Department of Mathematical and Statistical Methods
Wageningen University

Thomas J. Hagenaars
Division of Infectious Diseases, Animal Sciences Group
Wageningen University

11

Evolution of Specialization and Ecological Character Displacement: Metabolic Plasticity Matters

Martijn Egas

ABSTRACT

An important question in evolutionary biology, especially with respect to herbivorous arthropods, is the evolution of specialization. In a previous paper the combined evolutionary dynamics of specialization and ecological character displacement was studied, focusing on the role of herbivore foraging behaviour. In this paper the robustness of these results is examined with respect to the assumption about the (metabolic) feeding efficiency function, changing it from a fixed to a plastic response. For low specialization costs, the model yields qualitatively similar results. Through the process of evolutionary branching, the herbivore population radiates into many specialized phenotypes for basically any level of sub-optimal foraging (where plant utilization is to some degree determined by the relative growth rate on each plant type). However, for an increased cost for specialization, the model loses its primary evolutionary equilibrium point. In this part of the parameter space there is run-away selection towards the ultimate generalist strategy. Under the conditions for evolutionary branching, the model predicts host race formation and sympatric speciation in herbivorous arthropods when mating is host-plant associated.

Keywords: evolution, specialization, resource gradient, ecological character displacement, foraging behaviour, adaptive radiation, metabolic plasticity.

11.1 INTRODUCTION

In insect-plant biology, host plant specialization has received a lot of attention over the decades (e.g. Dethier, 1954; Ehrlich and Raven, 1964; Levins and MacArthur, 1969; Bernays and Graham, 1988; Thompson, 1994; Berenbaum, 1996; Schoonhoven *et al.*, 1998) because so many herbivorous insects are strongly specialized (Futuyma and Gould, 1979; Chapman, 1982; reviews in Strong *et al.*, 1984; Bernays and Chapman, 1994). Since the seminal paper by Fox and Morrow (1981), it is widely recognized that even generalist species can represent a collection of locally specialized populations. Moreover, studies on the genetic structure in insect populations frequently

T.A.C. Reydon and L. Hemerik.,(eds.), Current Themes in Theoretical Biology,
281-304.

show discrete groups (demes) at a sometimes surprisingly small spatial scale: insects feeding on a tree species may even adapt to different individual trees (Edmunds and Alstad, 1978; Mopper, 1996).

Observations of insect species on host plants in the field are the basis for the above-mentioned pattern that large numbers of herbivorous insects are specialized. This type of data does not exclude the possibility that many insect species can feed on a wide range of host plants. For such a species its natural distribution suggests that it is very specialized, but when resource utilization (the generally accepted criterion for host plant specialization) is considered, it is not.

The observed pattern of specific host use in the field can be explained by foraging theory. Given a choice between several host plants that vary in quality (measured as fitness when utilising this host), a herbivore chooses the highest quality species (see e.g. Levins and MacArthur, 1969; Rosenzweig, 1981, 1987). Optimal foraging theory predicts that the ideal herbivorous arthropod will occur on only a specific subset of plant species if their quality exceeds some marginal value.

However, foraging theory is not an explanation of the evolution of specialization in herbivorous arthropods. It is designed to explain animal behaviour, not evolutionary change, and several aspects of the evolution of specialization still require explanation. First, there is usually a good correlation between specificity according to occurrence in nature and specificity according to utilization: herbivorous arthropods are also specialized when the criterion is based on resource utilization. Second, there are ample examples of sister species in a clade of herbivorous arthropods that seem to have radiated onto a clade of plant species (e.g. Farrell and Mitter, 1990, 1994; Futuyma and McCafferty, 1990; Futuyma et al., 1995; Farrell, 1998; Termonia et al., 2001). If the group of sister species as a total is able to adapt to a group of hosts, why does each sister species utilize only its own specific selection of these hosts?

From an ecological perspective there is not much theoretical work providing an insight into these aspects of specialization in herbivorous arthropods. Basically, for evolution towards specialized resource (host) utilization in a species, individuals competing for two resources should be subject to a strong fitness trade-off between the two resources (Levins, 1962). This prediction, based on the assumption of a random distribution of consumers over two discrete resources, states that specialization evolves when generalists have greatly reduced fitness compared to specialists, such that the fitness benefit of specializing on one resource is higher than the fitness cost on the alternative resource.

However, herbivorous arthropods are not generally known to be randomly distributed over their resources. On the contrary, they generally display foraging behaviour. Studies on the evolution of specialization that have incorporated foraging behaviour used the theory of optimal foraging. These studies predict that for the evolution of specialization the strength of the fitness trade-off does not matter: any trade-off suffices (Rosenzweig, 1981, 1987). Through optimal foraging behaviour a specialist can feed exclusively on the host it is specialized on, thereby avoiding the cost of lower fitness from feeding on the alternative resource.

Optimal foraging *sensu stricto* is in itself a biologically unrealistic case, because it requires omniscience and unlimited mobility. Real foragers need to sample their environment: herbivorous arthropods need time to select their host plant, exhibit stage-dependent mobility, and mothers may not always select the best host for their (immobile) offspring (e.g. Wainhouse and Howell, 1983; Rauscher, 1983; Rauscher and Papaj, 1983; Whitham, 1983; Robertson, 1987; Moran and Whitham, 1990; Valladares and Lawton, 1991; Underwood, 1994). Evidence is accumulating that herbivorous arthropods can approximate optimal foraging by adaptive learning (Dukas and Bernays, 2000; Egas and Sabelis, 2001; Egas *et al.*, 2003; Nomikou *et al.*, 2003). Therefore, analyses of the evolution of specialization in herbivorous arthropods should focus on adaptive but sub-optimal foraging behaviour.

Recently, Egas *et al.* (2004b) studied the evolutionary dynamics of herbivore specialization, taking into account sub-optimal foraging behaviour of the herbivores, a quality gradient of plant types and explicit plant population dynamics. Herbivore adaptation can occur with respect to two metric characters: the level of specialization in feeding efficiency and the point on the plant quality gradient at which the herbivore's feeding efficiency is highest. This makes it possible to link the evolution of specialization to ecological character displacement. The model yields broad conditions for the adaptive radiation of a herbivore population into many specialized phenotypes through evolutionary branching, for basically any level of sub-optimal foraging (where plant utilization is to some degree determined by the relative gain of each plant type). Because the process of evolutionary branching was incorporated, both the number of phenotypes in the model, their level of specialization and the amount of character displacement among them are the result of the evolutionary dynamics, which in turn depends on the characteristics of the species. Lower levels of sub-optimal foraging lead to lower degrees of specialization. Foraging costs also influence the level of specialization and hence limit the number of species that can be "packed" in the resource gradient, whereas the amount of character displacement evolving between the different phenotypes depends on the level of sub-optimal foraging, but not on foraging costs.

In this paper, one of the assumptions in the above model is relaxed, dealing with the metabolic efficiency with which herbivores can process plant food into herbivore mass. In the original model this was a fixed function: herbivores are not phenotypically plastic with respect to digestion rate of plant food. Here the herbivores are assumed to display phenotypic plasticity. Herbivores can reallocate metabolic effort for plant types that are not present in the environment. Both assumptions are biologically plausible for herbivorous arthropods; but do they have similar effects on the evolution of specialization?

11.2 PLANT-HERBIVORE POPULATION DYNAMICS

In this section, the population dynamics of plants and herbivores and the foraging behaviour of the herbivores are defined (see also Egas et al., 2004b).

Dynamics along the plant gradient

Assume that the plant types in the model can be arranged along a quality axis q. The growth rate of the density of plants $p(q)$ with quality q is described by a standard Lotka-Volterra differential equation,

$$\frac{d}{dt} p(q) = r(q) \cdot p(q) \cdot \left[1 - \frac{p(q)}{K(q)}\right] - c_f \cdot h \cdot u(q) \cdot p(q). \tag{1}$$

Here, $r(q)$ and $K(q)$ are, respectively, the intrinsic growth rate and carrying capacity of plants with quality q; the constant c_f is the feeding rate per plant per herbivore, h is the number of herbivores and $u(q)$ is the resource utilization spectrum, describing the proportional utilization of plants with quality q.

Utilization spectrum, derived from foraging behaviour

As the growth rate of the herbivore will be assumed to depend on the total food intake rate it achieves (defined below, in the section *Herbivore growth*), this is the currency on which to base foraging behaviour (Stephens and Krebs, 1986). The gain $g(q)$ that an individual herbivore can extract from plants of quality q is then defined as its intake rate when feeding on those plants only. This is a product of plant quality, feeding rate, and feeding efficiency ($e(q)$),

$$g(q) = q \cdot c_f \cdot p(q) \cdot c_c \cdot e(q). \tag{2}$$

Here, c_f is the feeding rate constant of (Equation 1) and c_c is a scaling constant for conversion of plant biomass into herbivore biomass. Feeding efficiency describes how well a herbivore type can digest a specific resource (see below). The parameter q enters this function to scale the gain of plant biomass with respect to its quality.

The herbivores are assumed to forage dynamically, so that the distribution of herbivores over the plants can change continuously in time (Krivan, 1997). In this way, optimally foraging herbivores will always feed on plants of the quality or qualities that yield(s) the highest gain. To describe sub-optimal levels of foraging, the utilization spectrum $u(q)$ of plants of quality q is taken to be proportional to $g(q)^\alpha$,

$$u\ (q) = \frac{g\ (q)^{\alpha}}{\int g\ (q)^{\alpha} dq} \tag{3}$$

with the denominator ensuring proper normalization of the probability distribution $u\ (q)$. The parameter α can take any value in the range $[0, +\infty)$, and be viewed as describing the herbivore's foraging accuracy. Setting α to zero yields non-selective feeding, whereas when α goes to infinity, the utilization function $u\ (q)$ describes dynamical optimal foraging. Values of α in between these two extremes result in selective but sub-optimal foraging.

Feeding efficiency

In exploiting different plant types, the herbivores face a trade-off for feeding efficiency on these plants. For instance, the metabolic machinery of herbivores consists of a complex assembly of digestion enzymes, which together determine the efficiency with which a certain food type can be digested (here called feeding efficiency). Herbivorous arthropods cannot digest all plant types with the highest efficiency because the amino acids and/or the amount of energy needed to make the enzymes are limited. In insect-plant biology it is often assumed that this trade-off is of a physiological nature (e.g. Bernays and Graham, 1988; Futuyma and Moreno, 1988; Jaenike, 1990). However, for the model considered here, the trade-off can stem from any constraint that affects feeding efficiency, e.g. involving morphological characters of the insect's mouthparts and plant leaf surface. Therefore, the interpretation of the results of our study need not be restricted to physiological trade-offs.

The trade-off is modelled by assuming a Gaussian distribution of feeding efficiency along the resource gradient. Hence, the efficiency distribution over plant qualities $e(q)$ can be described by the two parameters of a Gaussian distribution: its mean μ and standard deviation σ. Herbivores can either increase their efficiency of using a narrow range of plant qualities, thereby leaving a large part of the resources unused, or they can increase the range of plant qualities utilized, but thereby decreasing their feeding efficiency on each of these plant types. Note that this definition of $e(q)$ means that any herbivore type has a defined efficiency for plant qualities up to infinity and hence will have some degree of pre-adaptation to plant qualities not occurring on the plant gradient interval considered.

In this paper, the herbivore is assumed to reallocate metabolic effort for plant qualities not present in the considered plant community. This is the only difference with the model in Egas et al. (2004b). The efficiency distribution, therefore, is normalized for the range of plant qualities available (i.e. $\int e(q)dq$ is constant for all values of the trait $[\mu, \sigma]$). In terms of the models with a trade-off between two resources or habitat types, this kind of fitness trade-off would be termed a linear trade-off, i.e. the border case between a weak and a strong trade-off. Under a weak trade-off, $\int e(q)dq$ would decrease with increasing specialization; under a strong trade-off, $\int e(q)dq$ would increase with increasing specialization.

Herbivore growth

The total intake rate I of a herbivore is the product of gain and proportional utilization, integrated over the resource gradient,

$$I = \int g(q) \cdot u(q)\, dq. \tag{4}$$

The per capita birth and death rates of the herbivores are functions of this intake rate. Individuals can produce offspring if their intake rate exceeds a threshold I_{min}. Above this threshold, the birth rate $b(I)$ is a saturating function of intake. The death rate $d(I)$ is asymptotically declining from the starvation mortality d_{max} towards a background mortality d_{min}. In the model, the functions used are: $b(I) = \max(b_{max}(1 - \exp[c_b(I_{min} - I)]),0)$ and $d(I) = d_{min} + (d_{max} - d_{min}) \cdot \exp[-c_d I]$. Here, b_{max} is the maximum birth rate, and c_b is a scaling constant determining the slope of increase. For the mortality rate c_d is a scaling constant determining the slope of decrease. The herbivore per capita growth rate is defined as $b(I) - d(I)$ and the population-level growth rate is obtained by summing the per capita growth rates of all individuals. Note that this growth rate is zero for an intake rate $\hat{I}$ that satisfies $b(\hat{I}) = d(\hat{I})$.

Specialization cost

A foraging cost for specialists is modelled as a decrease in the utilization of plants when σ decreases. This describes situations in which specialists need more time to search for the specific host plants they are specialized on and therefore spend less time exploiting those host plants. The cost is simply modelled by multiplying the utilization spectrum $u(q)$ with a Holling type II function for σ, $C(\sigma) = \sigma/(\sigma + c_s)$, with the half-saturation constant c_s measuring the cost of specialization.

11.3 ADAPTIVE DYNAMICS

The evolution of the shape of the efficiency function is studied and therefore coevolution of the two parameters μ and σ is modelled. The two parameters describe two different aspects of exploitation. The mean μ represents the mean of the plant quality range on which the herbivore phenotype is focused. The standard deviation represents the level of specialization: a low value of σ implies a relatively high efficiency in a narrow range of the resource gradient, whereas a high value results in a relatively low efficiency, but in a wide range of the plant spectrum.

The theory of adaptive dynamics (Metz *et al.*, 1996; Dieckmann and Law, 1996; Dieckmann, 1997; Geritz *et al.*, 1997, 1998) is used to analyze evolutionary dynamics. In addition to individual-based models, the investigation is based on a deterministic approximation, the canonical equation of adaptive dynamics (Dieckmann and Law, 1996). In this approximation, population dynamics are treated deterministically and the evolutionary change that arises from small mutational steps deterministically follows the local gradient of the fitness landscape around a resident phenotype. Assuming that mutations occur with low probability (mutation-limited evolution), adaptation can be studied by evaluating the fate of mutants in populations that consist of one or more resident phenotypes at their ecological equilibrium.

Attractors of evolutionary dynamics can be of several types (Dieckmann, 1997; Geritz *et al.*, 1998); two of which are encountered in this study. A continuously stable strategy (CSS) is a strategy that is an evolutionary attractor (i.e. a singular point that is convergence stable under the canonical equation) and that also cannot be invaded by any neighbouring phenotype (i.e. the singular strategy is locally evolutionarily stable). In contrast, an evolutionary branching point (EBP) is an evolutionary attractor that is not evolutionarily stable (Metz *et al.*, 1992, 1996; Geritz *et al.*, 1998). Mutants close to such a singular phenotype have a chance to mutually invade each other's populations and form a stable dimorphism. The resulting dimorphic adaptive dynamics allows for two resident phenotypes in the population and describes the initial phenotypic divergence and subsequent evolutionary change in the two subpopulations. Such a pair is again expected to converge to a singular point, the evolutionary stability of which can once more be determined by evaluating whether mutants of small effect around the singular point are able to invade.

Whereas the deterministic approximation is a versatile tool for investigating the adaptive dynamics under low degrees of polymorphism (monomorphic and dimorphic evolution) it becomes tedious thereafter. The individual-based model therefore has two advantages. First, it allows checking the robustness of conclusions obtained from the deterministic approximation when relaxing the simplifying assumption that this approximation is based on. Second, it is naturally suited for investigating adaptive dynamics involving multiple evolutionary branching events, which give rise to higher degrees of polymorphism.

In the individual-based model, all individual herbivores can be assigned different phenotypes (μ, σ) and demography as well as mutations are treated stochastically. Based on herbivore birth and death rates, the waiting times for the next birth or death event to take place are drawn from an exponential probability distribution (Van Kampen, 1981; Dieckmann *et al.*, 1995). During a birth event, the phenotype of offspring individuals can either be faithfully inherited from their parent or it can be affected by mutation. The latter occurs at probability p_m and the mutant phenotype is then drawn from a normal distribution with standard deviation s_m around the parental phenotype.

11.4 RESULTS

In this section, the evolutionary dynamics of specialization under sub-optimal foraging (i.e. $0 < \alpha < \infty$) are examined. The two extreme cases of non-selective foraging ($\alpha = 0$) and optimal foraging ($\alpha = \infty$) yield results equal to those in Egas *et al.* (2004b). In short, for the case of non-selective foraging, evolution results in one phenotype specialized on plant types that yield the maximum value of $q \cdot K(q)$ (which reflects the maximum potential intake $g[q]$). In this case, there is always one plant type with the highest gain and it always pays to specialize on that type, even though non-selective herbivores cannot help but use the entire gradient of plant phenotypes. The unrealistic case of optimal foraging behaviour leads to unrealistic evolutionary outcomes: an infinite diversity of extreme specialists. In this particular situation there exists no evolutionary singular point for any herbivore phenotype. This is due to the interplay of optimal foraging and evolution: any phenotype can maximize its fitness through optimal foraging, but any resident (monomorphic) population can always be invaded by any mutant of small effect, since a mutation in the efficiency function always leads to a higher efficiency for at least one plant quality.

Deterministic approximation

Cases of sub-optimal foraging (i.e. $0 < \alpha < \infty$) are analyzed by numerical simulation. For this purpose, the resource axis is discretized. First, we describe a representative example of the evolutionary dynamics, where we have set $\alpha = 1$ and $c_s = 0$ (no costs for specialization; as in Egas *et al.*, 2004b). The original model predicts evolution towards a generalist strategy which is an evolutionary branching point (EBP) for reasonably low costs of specialization. This primary branching point always allows for the herbivore population to split up into a specialist subpopulation on heavily exploited plants of high quality and a generalist subpopulation on plant qualities that are less extensively used. Subsequent evolution reveals two characteristic phases in the adaptation of herbivore phenotypes: a fast 'character displacement' phase, and a slow 'coevolutionary niche shift' phase (Egas *et al.*, 2004b). As described below, the qualitative predictions are upheld in the alternative model analyzed here but the details of the process are different.

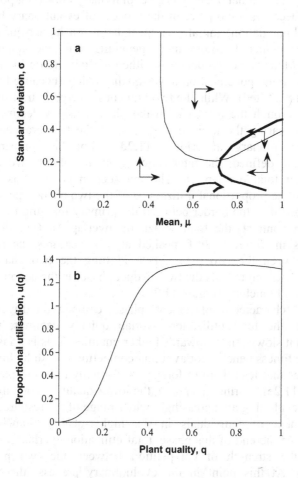

Figure 11.1. Evolution of the efficiency strategy in a monomorphic population under sub-optimal foraging. (a) Evolutionary phase portrait, with isoclines and directions of evolutionary change. There is one isocline for the standard deviation σ (thin line) and two for the mean μ (thick lines). The singular point, where two of the different isoclines cross, is a branching point. (b) The utilization spectrum along the resource gradient of a herbivore population with the singular strategy and at ecological equilibrium. (Parameters: $K(q) = 50 + 10 \cdot N_{0.1}(q - 0.5)$, $r(q) = 1$, $b_{max} = 1.0$, $c_f = 0.0025$, $c_c = 400$, $c_b = 1.0$, $c_d = 3.0$, $c_s = 0$, $d_{max} = 1.0$, $d_{min} = 0.02$, $I_{min} = 0.2$, $\alpha = 1.0$.)

In a single-phenotype population (a monomorphic population) evolutionary change leads to a globally attractive singular strategy (Figure 11.1a). At the ecological equilibrium for this singular strategy, the herbivores homogeneously utilize the higher-quality part of the resource gradient (Figure 11.1b). The singular strategy is an evolutionary branching point, hence the population becomes dimorphic. In the process of evolutionary branching, one branch (subpopulation) initially is specializing on higher-quality plants while the other branch becomes more generalist on lower-quality plants (Figure 11.2a). Mutants can do better by either specializing on higher-quality plants or by becoming generalist on lower-quality plants (because low-quality plants are underexploited). While the two branches diverge in trait space, their mutual impact through the feedback on the plant densities declines. This is apparent by comparing the utilization spectra of the two branches as they change over evolutionary time (Figure 11.2d-f). For this comparison, two characteristics are defined: the intersection point $q*$ of the two utilization spectra and their total overlap Δu. The intersection point $q*$ is that plant quality at which the proportional utilizations of the two phenotypes are equal, and can be regarded as the border between the primary foraging ranges of the two spectra. In contrast, the total utilization overlap Δu (i.e. the roughly triangular areas in Figure 11.2d-f, peaked at $q*$) measures the amount of competition between the two phenotypes. Plotting the two characteristics against each other again reveals the two distinct phases in the adaptive process after evolutionary branching (Figure 11.2b).

In the first, 'character displacement' phase, competition drives the two branches apart: the total utilization overlap quickly declines, while the intersection point slowly shifts towards higher qualities. This is a fast process, indicating large fitness benefits for avoiding competition. Both herbivore types evolve strategies that lead them to forage on distinctly different plant quality ranges (Figure 11.2e). During this phase, the lower-quality branch increases in abundance, by exploiting an increasingly wider range of the resource gradient, while the number of individuals in the higher-quality branch declines (Figure 11.2c). At the end of this phase, total utilization overlap is minimized (and with it the strength of competition between the two phenotypes, Figure 11.2b,e). At this point in the evolutionary process, the benefit of avoiding competition is counterbalanced in the higher-quality branch by the disadvantages of specializing even further on higher plant qualities and giving up foraging on even more of the resource gradient.

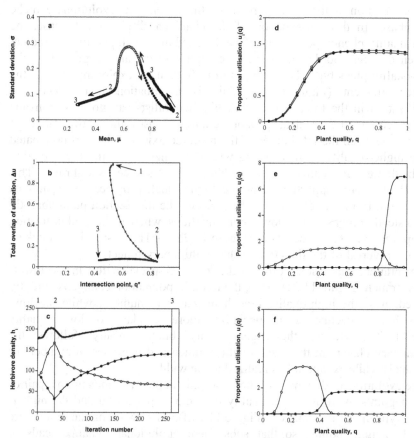

Figure 11.2. Evolutionary dynamics of the dimorphic population (open dots: low-quality branch; closed dots: high-quality branch). (a) Evolutionary trajectories of the two resident phenotypes. Starting from the branching point, arrows indicate the direction of subsequent evolutionary change. Utilization spectra from positions 1, 2 and 3 are presented in panels (d)-(f), respectively. (b) Change over evolutionary time of the total utilization overlap between the two phenotypes (Δu) and the intersection point of the two utilization spectra (q^*). The first, fast phase shows character displacement, as evidenced by the strongly decreasing utilization overlap. The second, slow phase is characterized by coevolutionary niche shift: the intersection point shifts to lower plant qualities, while the total utilization overlap stays roughly constant. (c) Resultant changes in herbivore densities (total herbivore density is shown by the thick line). (d) Initial utilization spectra of the two phenotypes in the population [corresponding to position 1 in panels (a)-(c)]. (e) Utilization spectrum of the two phenotypes at the end of the character displacement phase [corresponding to positions 2 in panels (a)-(c)]. (f) Utilization spectra of the two phenotypes at the dimorphic singular strategies, i.e. at the coevolutionary equilibrium [corresponding to positions 3 in panels (a)-(c)]. Parameters are as in Figure 11.1.

This situation is the starting point of the second, 'coevolutionary niche shift' phase. In the end-situation of the character displacement phase, the higher-quality phenotype finds itself sandwiched in a small part of the gradient between the right border of the resource axis and the area exploited by the lower-quality phenotype. The latter, in turn, finds itself exploiting most of the resource gradient (Figure 11.2e). This is a situation that selects for specialization in the lower-quality branch. Generalists can no longer invade that branch, as the resident phenotype already uses the entire range of resources between the left border of the resource axis and the part dominated by the higher-quality branch. Along with specialization in the lower-quality branch, though, its realized niche width on the resource axis shrinks. This gives evolutionary opportunity for the higher-quality branch to expand its niche width. Both are reflected by the fact that the intersection point abruptly starts a steady progression to lower plant qualities, whereas the total utilization overlap remains fairly constant (Figure 11.2b). The result is a marked, complete reversal of the direction in the evolutionary trajectory of the higher-quality branch, and a steep drop in the level of specialization in the lower-quality branch (Figure 11.2a). Also, the trend in population change is abruptly reversed: now the high-quality branch increases in number, while the low-quality branch declines as it specializes more on plants of lower quality (Figure 11.2c). Hence, the higher-quality branch neatly follows the evolutionary change in the lower-quality branch and increases in numbers of individuals while expanding its realized niche width.

The evolutionary dynamics of the dimorphic population ends up in a pair of singular points, with the lower-quality type more specialized and the higher-quality type more generalized (Figure 11.2a,f). The two singular points are again branching points, so that subsequent evolutionary change leads to trimorphic and quadrimorphic populations.

The evolutionary dynamics in this example describes a general pattern that applies to a wide range of sub-optimal foraging under low specialization costs. First, evolution in a monomorphic population always leads to a singular point at which evolutionary branching can take place. Second, the initial direction of dimorphic evolution after branching is always characterized by one branch becoming more specialized on higher plant qualities (towards the extensively used range of the resource gradient) and the other branch becoming more generalized on lower plant qualities (towards the under-exploited range of the resource gradient). Subsequently, the two phases of evolutionary change in the dimorphic population result in a generalist branch on high plant qualities and a specialist branch on low plant qualities.

Tracking the primary branching point through trait space for different values of α showed that for increasing α-values the singular strategy for a monomorphic population is more and more generalized (Figure 11.3). When considering a monomorphic population of herbivores with a closer-to-optimal foraging behaviour, we allow the herbivores to focus their feeding efforts more on the plants that yield the highest gains. Therefore such herbivores need be less specialized to exploit the full spectrum of plant qualities.

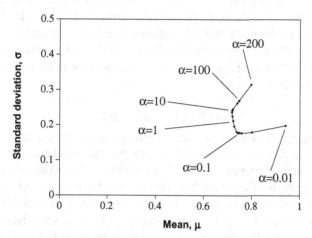

Figure 11.3. Singular strategies (μ, σ) for various values of the level of sub-optimal foraging, α. Other parameters as in Figure 11.1.

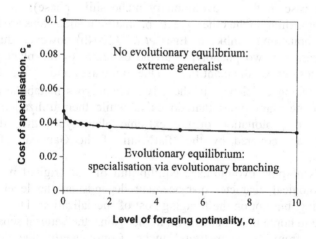

Figure 11.4. Critical values of the cost of specialization, c_s for different levels of sub-optimal foraging, α. Below the line, the singular point is an evolutionary branching point: the population splits up in two or more phenotypes with increasing degrees of specialization. Above the line, the evolutionary equilibrium has vanished; instead, there is run-away selection towards extreme generalization. Other parameters as in Figure 11.1.

An unexpected effect occurs when the cost of specialization is increased from zero (Figure 11.4). Increasing the cost of specialization makes the σ-isocline (see Figure 11.1a) move upwards (as there is an increasing penalty on low values of σ) whereas the μ-isocline is hardly affected. Below a critical cost evolutionary branching results in specialized subpopulations of herbivores, as described above. Above the critical cost the evolutionary equilibrium point of the monomorphic population is lost, resulting in run-away selection towards the ultimate generalist by ever-increasing values of both μ and σ. The critical cost of specialization is lower if α is higher, i.e. if the herbivores are better foragers.

Individual-based model

The evolutionary dynamics of the herbivore population after secondary branching as described above is analyzed with the individual-based model (for the parameter space in Figure 11.4 where evolutionary branching occurs). The results show a pattern of repeated branching (i.e. adaptive radiation; Schluter, 2000a,b) in the μ-dimension, leading to a large number of specialized types in the population (Figure 11.5 gives an example). The individual-based model also allows us to check the deterministic results under small but finite mutation steps. As in Egas *et al.* (2004b), the 'coevolutionary niche shift' phase is generally absent: branching occurs in one or both of the two branches before this phase sets in. When the fitness gradient is very small in one direction (which is the case in the 'coevolutionary niche shift' phase), and favours evolutionary branching in the other direction, small mutation steps are likely to produce such branching. Unlike in Egas *et al.* (2004b) however, during the radiation process the two phenotypes at the extremes of the plant gradient evolve μ-values outside the plant range. Due to the assumption of phenotypic plasticity in feeding efficiency, in these two phenotypes metabolic effort is reallocated to the plant types that do exist, while their utilization spectra allows improved exploitation of the extreme plant types they dominate (because they are covered by the 'flat' tail of the Gaussian efficiency distribution).

In the polymorphic evolutionary equilibrium of the original model, the degree of ecological character displacement depends on the level of sub-optimal foraging, but not on the foraging cost of specialization. The better the forager is able to home in on plants with higher gains, the better it separates its realized niche from its competitors' niche. Consequently, the degree of ecological character displacement in the end will be lower for more discriminate foragers. Costs of foraging influence the level of specialization that is eventually evolving and hence limit the number of phenotypes that 'fit' in the gradient. Such costs however, do not influence the degree of ecological character displacement, because they do not affect the ability of the phenotypes to minimize competition with other phenotypes and forage in distinctly separate ranges of the resource gradient. In the alternative model analyzed here, the results are qualitatively similar.

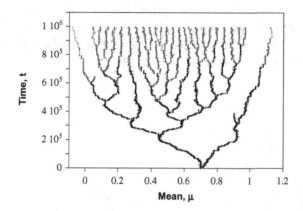

Figure 11.5. Adaptive radiation of phenotypes over the resource gradient, involving repeated events of evolutionary branching (black branches consist of more than 200 individuals, dark grey branches 100-200 individuals, and light grey branches less than 100 individuals). Specialization increases with the number of branches. Parameters: $s_m = 0.005$, $p_m = 0.0003$, $K(q) = 50 + 10 \cdot N_{0.1}(q - 0.5)$, $r(q) = 3$, $b_{max} = 1.0$, $c_f = 0.0025$, $c_c = 400$, $c_b = 1.0$, $c_d = 3.0$, $c_s = 0$, $d_{max} = 1.0$, $d_{min} = 0.02$, $I_{min} = 0.2$, $\alpha = 1.0$.

A higher degree of sub-optimal foraging α allows for more phenotypes to coexist in the population (Figure 11.6a). This is because more selective foragers can focus their foraging effort more, even when they are specialized. At the same time, the phenotypes are less specialized when they are more selective foragers (Figure 11.6b). Increasing the cost of specialization decreases the number of branches that can coexist along the resource gradient (Figure 11.6a). The number of types that 'fit' in the resource gradient decreases, because the herbivores evolve a lower degree of specialization when costs of specialization are higher (Figure 11.6b).

The degree of ecological character displacement can be expressed by the so-called d/w ratio (May and McArthur, 1972). Here, d stands for the distance between two adjacent phenotypes as measured by the means μ_i of their efficiency distributions, $d = |\mu_1 - \mu_2|$. In the d/w ratio, the distance d is then considered relative to the width w of the efficiency distribution of the focal phenotype, which is here taken as the standard deviation σ of the efficiency distribution. In this way, each comparison of a pair of adjacent phenotypes gives two values of the d/w ratio (one for each phenotype) and the total number of d/w ratios is $2 \cdot$(number of branches $- 1$). The d/w ratio is an expression of the displacement in the fundamental niches of the phenotypes, not of their realized niches (which are kept roughly separate due to their foraging behaviour). In the polymorphic evolutionary equilibrium this measure of character displacement is fairly constant along the gradient. Only the two extreme phenotypes yield higher d/w ratios, because they maintain larger distance with the neighbouring branch. This has a negligibly small effect on the average d/w ratios found, hence the results on ecological character displacement are presented using the average ratio.

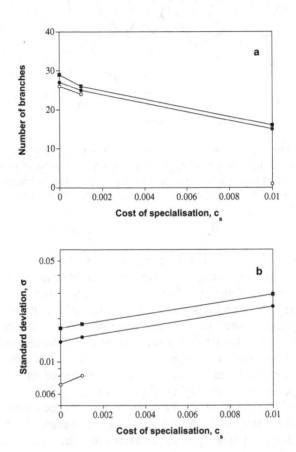

Figure 11.6. Characteristics of the polymorphic evolutionary equilibrium for different levels of sub-optimal foraging, α, and cost of specialization, c_s. (a) Number of branches (open dots: $\alpha = 0.1$; solid dots: $\alpha = 0.5$; solid squares: $\alpha = 1.0$). Note that for $\alpha = 0.1$ evolutionary branching only occurs for $c_s = 0$ and $c_s = 0.001$. (b) Average level of specialization σ among the branches. (c) See next page.

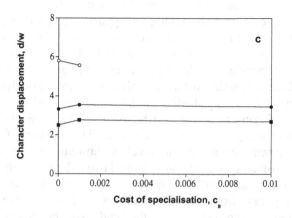

Figure 11.6 continued. Characteristics of the polymorphic evolutionary equilibrium for different levels of sub-optimal foraging, α, and cost of specialization, c_s. (c) Average character displacement d/w among the branches. Other parameters as in Figure 11.5.

Values for d/w increase from roughly 2.5 to 6 as α decreases from 1 to 0.1, but are not significantly affected by the foraging cost of specialization c_s (Figure 11.6c). The cost of specialization does not affect foraging ability (determined by the value of α), so that the degree of separation between phenotypes along the resource axis remains the same when c_s is increased. Hence, the degree of ecological character displacement depends critically on the degree of sub-optimal foraging behaviour.

11.5 DISCUSSION

An important question in evolutionary biology, especially with respect to herbivorous arthropods, is the evolution of specialization. Recently, Egas *et al.* (2004b) studied the evolutionary dynamics of herbivore specialization, taking into account sub-optimal foraging behaviour of the herbivores, a quality gradient of plant types and explicit plant population dynamics. They have shown that the evolutionary dynamics of herbivore efficiency results in evolutionary branching of the herbivore population into specialized subpopulations with distinctly different exploitation strategies. The driving mechanism behind this evolutionary branching process is frequency-dependent selection deriving from sub-optimal foraging.

In this paper, the assumption of fixed digestion efficiency of plants was relaxed: the metabolic effort for plant types not present in the model was reallocated to existing plant types. The results show that metabolic plasticity matters: it yields qualitatively different predictions for the evolution of specialization and ecological character displacement. Evolutionary branching still occurs for the whole range of foraging behaviour ($0 < \alpha < \infty$). However, for relatively low costs of specialization (Figure 11.4), the primary branching point is lost and run-away selection for extreme generalization ensues. Repeated branching (when the cost of specialization is low enough to allow for primary branching) leads to adaptive radiation of the herbivores over the resource gradient (Figure 11.5; see also Schluter, 2000a,b), with the level of specialization increasing within the branches. The degree of ecological character displacement depends on the level of sub-optimal foraging, but not on the foraging cost of specialization (Figure 11.6). The better the forager is able to home in on plants with higher gains, the better it is at keeping its realized niche separate from its competitors. Consequently, the degree of ecological character displacement in the end will be lower for more discriminate foragers. Costs of foraging do not influence the d/w ratio, because they do not affect the ability of the phenotypes to minimize competition with other phenotypes and forage in distinctly separate ranges of the resource gradient. However, such costs do affect the level of specialization that eventually evolves and hence limit the number of phenotypes that 'fit' in the gradient.

The next section discusses the effects of assuming phenotypic plasticity in digestion efficiency that are contrary to the original model. The last section focuses on the implications of the model results for host race formation and sympatric speciation in herbivorous arthropods.

Feeding efficiency: fixed or flexible

The most striking effect of making feeding efficiency flexible is the loss of the evolutionary attractor for relatively low costs of specialization. In the original model, there is a transition from an evolutionary branching point (EBP) to a continuously stable strategy (CSS) for high values of this cost. This pattern describes a gradual decrease in the number of phenotypes in the polymorphic evolutionary equilibrium (with ever decreasing levels of specialization) until there is only one phenotype predicted (the CSS), which is increasingly generalist with increasing cost. In the current model, there is a discontinuous transition from an evolutionary end state with multiple specialist phenotypes to one extremely generalist phenotype.

With respect to the process of ecological character displacement, the current model predicts qualitatively similar outcomes to Egas et al. (2004b). Most importantly, the 'ecological character displacement' phase and the 'coevolutionary niche shift' phase in the deterministic approximation are retained as the basic pattern of dimorphic divergence. However, the evolutionary process in the second phase is reversed: the high-quality branch evolves to become the generalist (with the higher density), and the low-quality branch ends up as the specialist (with the lower density). This direction of evolution makes sense when the utilization spectra are considered (specifically, Figure 11.2e), which are in part determined by the assumption of fixed or flexible feeding efficiency.

A last effect of metabolic plasticity discussed here is the evolution of 'border phenotypes'. At both extremes of the plant quality gradient, a phenotype is present in the polymorphic evolutionary equilibrium, with an efficiency function focused outside the boundary of the gradient (see Figure 11.5). Their existence is fully due to the fact that metabolic effort for plants not available in the system is reallocated to digestion efficiency of available plants. Hence, if such phenotypes would be shown to exist in experimental systems of adaptive radiation (e.g. Farrell and Mitter, 1990, 1994; Futuyma and McCafferty, 1990; Futuyma et al., 1995; Farrell, 1998; Termonia et al., 2001), this would indicate metabolic plasticity. It makes sense that experimental biologists would not normally check whether extreme phenotypes in a radiation are actually better at using resources not naturally occurring in their habitat.

In the current model, no cost of phenotypic plasticity is assumed. When such a cost is considered, the amount of metabolic effort reallocated, or the range of plant types for which reallocation occurs can also be subject to evolutionary change. In that case, it would make intuitive sense that evolution of specialization creates selection for decreased phenotypic plasticity.

Do herbivorous arthropods display phenotypic plasticity in digestion? To the author's knowledge, there is not much data to address this question. There are reports of extremely specialized insect herbivores that refuse to eat any other food but the host plant they are specialized on (see e.g. references in Szentesi and Jermy, 1990), but also reports of herbivorous arthropods being able to improve performance on a novel diet (e.g. for the two-spotted spider mite, Chatzivasileiadis et al., 2001; Agrawal et al., 2002; Magowski et al., 2003). However suggestive, though, both cases do not reveal whether metabolic efficiency is fixed or flexible over a range of plant types. Since this property can have significant effects on the evolution of specialization and ecological character displacement, it would be useful to have more insight into the metabolic plasticity of herbivorous arthropods.

Host race formation and sympatric speciation in herbivorous arthropods

Increasing attention has been paid to the notion that ecological feedback may lead to host race formation and sympatric speciation (e.g. Orr and Smith, 1998; Dieckmann and Doebeli, 1999; Geritz and Kisdi, 2000; Doebeli and Dieckmann, 2000, 2003; Schluter, 2001; Via, 2001). Two conditions are recognized as necessary for these processes to unfold: frequency-dependent selection, leading to stable coexistence of phenotypes, and a degree of reproductive isolation, usually assumed to emerge from assortative mating (Bush, 1994). Evolutionary branching yields the first of these two conditions and, in the model considered here and in Egas *et al.* (2004b), a population of herbivores easily splits up into two (and usually more) types.

Because the model is based on asexual reproduction we cannot elaborate on the second condition, assortative mating. Felsenstein (1981) pointed out that sexual reproduction works against the divergence of sympatric populations. Even though frequency-dependent selection may initially produce linkage disequilibrium, recombination (through sexual reproduction) tends to destroy this association. Felsenstein thus predicted that unrealistically strong selection pressures would be needed for sympatric speciation to proceed. However, other studies have shown that in models with sexual reproduction, assortative mating may readily evolve in the course of evolutionary branching (Dieckmann and Doebeli, 1999; Geritz and Kisdi, 2000; Doebeli and Dieckmann, 2003). Moreover, as pointed out by Rice (1987), this negative interaction between the processes of selection and reproduction may be circumvented when assortative mating can occur as a by-product of selection. This relaxes the conditions for sympatric speciation (Dieckmann and Doebeli, 1999; see also Drossel and McKane, 2000). For herbivorous arthropods, assortative mating may well be a by-product of host plant specialization, producing host-associated mating (Bush 1975, 1994). The results in this paper and in Egas *et al.* (2004b) show that host plant specialization is the expected evolutionary outcome when herbivores forage at least to some extent selectively (see also Egas *et al.*, 2004a,c). Thus, these results make host race formation and sympatric speciation plausible in herbivorous arthropods, through specialization and ecological character displacement.

ACKNOWLEDGEMENTS

Parts of the results described in this paper were obtained during the YSSP 1998 visit of ME to the International Institute for Applied Systems Analysis (IIASA) in Laxenburg, Austria, for which funding by NWO is gratefully acknowledged. I thank Ulf Dieckmann for supervision of this work during my visits to IIASA, and André Noest, Mikko Heino, Maurice Sabelis, Steph Menken and two anonymous reviewers for comments on the work presented in this manuscript.

REFERENCES

Agrawal, A. A., F. Vala and M. W. Sabelis (2002). Induction of preference and performance after acclimation to novel hosts in a phytophagous spider mite: adaptive plasticity? American Naturalist 159: 553-565.

Berenbaum, M. R. (1996). Introduction to the symposium: on the evolution of specialization. American Naturalist 148: S78-S83.

Bernays, E. and M. Graham (1988). On the evolution of host specificity in phytophagous arthropods. Ecology 69: 886-892.

Bernays, E. A. and R. F. Chapman (1994). Host-Plant Selection by Phytophagous Insects. Chapman & Hall, New York.

Bush, G. L. (1975). Modes of animal speciation. Annual Review of Ecology and Systematics 6: 339-364.

Bush, G. L. (1994). Sympatric speciation in animals: new wine in old bottles. Trends in Ecology and Evolution 9: 285-288.

Chapman, R. F. (1982). Chemoreception: the significance of receptor numbers. Advances in Insect Physiology 16: 247-356.

Chatzivasileiadis, E. A., M. Egas and M. W. Sabelis (2001). Resistance to 2-tridecanone in *Tetranychus urticae*: effects of induced resistance, cross-resistance and heritability. Experimental and Applied Acarology 25:717-730.

Dethier, V. G. (1954). Evolution of feeding preferences in phytophagous insects. Evolution 8: 33-54.

Dieckmann, U. (1997). Can adaptive dynamics invade? Trends in Ecology and Evolution 12: 128-131.

Dieckmann, U. and R. Law (1996). The dynamical theory of coevolution: A derivation from stochastic ecological processes. Journal of Mathematical Biology 34: 579-612.

Dieckmann, U. and M. Doebeli (1999). On the origin of species by sympatric speciation. Nature 400: 354-357.

Dieckmann, U., P. Marrow and R. Law (1995). Evolutionary cycling in predator–prey interactions: Population dynamics and the red queen. Journal of Theoretical Biology 178: 91–102.

Doebeli, M. and U. Dieckmann (2000). Evolutionary branching and sympatric speciation caused by different types of ecological interactions. American Naturalist 156: S77-S101.

Doebeli, M. and U. Dieckmann (2003). Speciation along environmental gradients. Nature 421: 259-264.

Drossel, B. and A. McKane (2000). Competitive speciation in quantitative genetic models. Journal of Theoretical Biology 204: 467-478.

Dukas, R. and E. A. Bernays (2000). Learning improves growth rate in grasshoppers. Proceedings of the National Academy of Sciences USA 97: 2637-2640.

Edmunds, G. F. and D. N. Alstad (1978). Coevolution in insect herbivores and conifers. Science 199: 941-945.

Egas, M. and M. W. Sabelis (2001). Adaptive learning of host preference in a herbivorous arthropod. Ecology Letters 4: 190-195.

Egas, M., D.-J. Norde and M. W. Sabelis (2003). Adaptive learning in arthropods: spider mites learn to distinguish food quality. Experimental and Applied Acarology 30: 233-247.

Egas, M., U. Dieckmann and M. W. Sabelis (2004a). Evolution restricts the coexistence of specialists and generalists – the role of trade-off structure. American Naturalist: 163: 518-531.

Egas, M., M. W. Sabelis and U. Dieckmann (2004b). Evolution of specialization and ecological character displacement of herbivores along a gradient of plant quality. Evolution, submitted.

Egas, M., M. W. Sabelis, F. Vala and I. Lesna (2004c). Adaptive speciation in agricultural pests. In: Dieckmann, U., J. A. J. Metz, M. Doebeli and D. Tautz (Eds), Adaptive Speciation. Cambridge University Press, Cambridge, in press.

Ehrlich, P. R. and P. H. Raven (1964). Butterflies and plants: a study in coevolution. Evolution 18: 586-608.

Farrell, B. D. (1998). "Inordinate fondness" explained: Why are there so many beetles? Science 281: 555-559.

Farrell, B. D. and C. Mitter (1990). Phylogenesis of insect plant interactions – have *Phyllobrotica* leaf beetles (Chrysomelidae) and the Lamiales diversified in parallel. Evolution 44: 1389-1403.

Farrell, B. D. and C. Mitter (1994). Adaptive radiation in insects and plants: time and opportunity. American Zoologist 34: 57-69.

Felsenstein, J. (1981). Skepticism towards *Santa Rosalia*, or why are there so few kinds of animals? Evolution 35: 124-138.

Fox, L. R. and P. A. Morrow (1981). Specialization: species property or local phenomenon? Science 211: 887-893.

Futuyma, D. J. and F. Gould (1979). Associations of plants and insects in a deciduous forest. Ecological Monographs 49: 33-50.

Futuyma, D. J. and S. S. McCafferty (1990). Phylogeny and the evolution of host plant associations in the leaf beetle genus *Ophraella*. Evolution 44: 1885-1913.

Futuyma, D. J. and G. Moreno (1988). The evolution of ecological specialization. Annual Review of Ecology and Systematics 19: 207-233.

Futuyma, D. J., M. C. Keese and D. J. Funk (1995). Genetic constraints on macroevolution: the evolution of host affiliation in the leaf beetle genus *Ophraella*. Evolution 49: 797-809.

Geritz, S. A. H. and É. Kisdi (2000). Adaptive dynamics of diploid, sexual populations and the evolution of reproductive isolation. Proceedings of the Royal Society London B Biological Sciences 267: 1671-1678.

Geritz, S. A. H., J. A. J. Metz, É. Kisdi and G. Meszéna (1997). Dynamics of adaptation and evolutionary branching. Physical Review Letters 78: 2024–2027.

Geritz, S. A. H., É. Kisdi, G. Meszéna and J. A. J. Metz (1998). Evolutionary singular strategies and the adaptive growth and branching of the evolutionary tree. Evolutionary Ecology 12: 35-57.

Jaenike, J. (1990). Host specialization in phytophagous insects. Annual Review of Ecology and Systematics 21: 243-273.

Krivan, V. (1997). Dynamic ideal free distribution: effects of optimal patch choice on predator-prey dynamics. American Naturalist 149: 164-178.

Levins, R. (1962). Theory of fitness in a heterogeneous environment. I. The fitness set and adaptive function. American Naturalist 96: 361-373.

Levins, R. and R. H. MacArthur (1969). An hypothesis to explain the incidence of monophagy. Ecology 50: 910-911.

Magowski, W., M. Egas, J. Bruin and M. W. Sabelis (2003). Intraspecific variation in induction of feeding preference and performance in a herbivorous mite. Experimental and Applied Acarology 29: 13-25.

May, R. M. and R. H. MacArthur (1972). Niche overlap as a function of environmental variability. Proceedings of the National Academy of Sciences USA 69: 1109-1113.

Metz, J. A. J., R. M. Nisbet and S. A. H. Geritz (1992). How should we define "fitness" for general ecological scenarios? Trends in Ecology and Evolution 7: 198–202.

Metz, J. A. J., S. A. H. Geritz, G. Meszéna, F. J. A. Jacobs and J. S. van Heerwaarden (1996). Adaptive dynamics, a geometrical study of the consequences of nearly faithful reproduction. In: Van Strien, S. J. and S. M. Verduyn Lunel (Eds), Stochastic and Spatial Structures of Dynamical Systems. pp. 183-231. North Holland, Amsterdam.

Mopper, S. (1996). Adaptive genetic structure in phytophagous insect populations. Trends in Ecology and Evolution 11: 235-238.

Moran, N. A. and T. G. Whitham (1990). Differential colonization of resistant and susceptible host plants: *Pemphigus* and *Populus*. Ecology 71: 1059-1067.

Nomikou, M., A. Janssen and M. W. Sabelis (2003). Herbivore host plant selection: whitefly learns to avoid host plants that are unsafe for her offspring. Oecologia 136: 484-488.

Orr, M. R. and T. B. Smith (1998). Ecology and speciation. Trends in Ecology and Evolution 13: 502-506.

Rauscher, M. D. (1983). Alteration of oviposition behavior by *Battus philenor* butterflies in response to variation in host-plant density. Ecology 64: 1028-1034.

Rauscher, M. D. and D. R. Papaj (1983). Demographic consequences of discrimination among conspecific host plants by *Battus philenor* butterflies. Ecology 64: 1402-1410.

Rice, W. R. (1987). Speciation via habitat specialization: the evolution of reproductive isolation as a correlated character. Evolutionary Ecology 1: 301-314.

Robertson, H. G. (1987). Oviposition site selection in *Cactoblastis cactorum* (Lepidoptera): constraints and compromises. Oecologia 73: 601-608.

Rosenzweig, M. L. (1981). A theory of habitat selection. Ecology 62: 327-335.

Rosenzweig, M. L. (1987). Habitat selection as a source of biological diversity. Evolutionary Ecology 1: 315-330.

Schluter, D. (2000a). The Ecology of Adaptive Radiation. Oxford University Press, Oxford.

Schluter, D. (2000b). Ecological character displacement in adaptive radiation. American Naturalist 156: S4-S16.

Schluter, D. (2001). Ecology and the origin of species. Trends in Ecology and Evolution 16: 372-380.

Schoonhoven, L. M., T. Jermy and J. J. A. van Loon (1998). Insect-Plant Biology: From Physiology to Evolution. Chapman and Hall, London.

Stephens, D. W. and J. R. Krebs (1986). Foraging Theory. Princeton University Press, Princeton.

Strong, D. R., J. H. Lawton and R. Southwood (1984). Insects on Plants – Community Patterns and Mechanisms. Blackwell Scientific Publications, Oxford.

Szentesi, A. and T. Jermy (1990). The role of experience in host plant choice by phytophagous insects. In: Bernays, E. A. (Ed.). Insect-Plant Interactions. Vol. II. CRC Press, Boca Raton. pp 39-74.

Termonia, A., T. H. Hsiao, J. M. Pasteels and M. C. Milinkovitch (2001). Feeding specialization and host-derived chemical defense in Crysomeline leaf beetles did not lead to an evolutionary dead end. Proceedings of the National Academy of Science USA 98: 3909-3914.

Thompson, J. N. (1994). The Coevolutionary Process. University of Chicago Press, Chicago.

Underwood, D. L. A. (1994). Intraspecific variability in host plant quality and ovipositional preferences in *Eucheira socialis* (Lepidoptera: Pieridae). Ecological Entomology 19: 245-256.

Valladares, G. and J. H. Lawton (1991). Host-plant selection in the holly leaf-miner: does mother know best? Journal of Animal Ecology 60: 227-240.

Van Kampen, N. G. (1981). Stochastic Processes in Physics and Chemistry. North-Holland, Amsterdam.

Via, S. (2001). Sympatric speciation in animals: the ugly duckling grows up. Trends in Ecology and Evolution 16: 381-390.

Wainhouse, D., and R. S. Howell (1983). Intraspecific variation in beech scale populations and in susceptibility of their host *Fagus sylvatica*. Ecological Entomology 8: 351-359.

Whitham, T. G. (1983). Host manipulation of parasites: within plant variation as a defense against rapidly evolving pests. In: Denno, R. F., and M. S. McClure (Eds). Variable plants and herbivores in natural and managed systems. Academic Press, New York. pp. 15-41.

Martijn Egas
Institute for Biodiversity and Ecosystem Dynamics, University of Amsterdam

List of Contributors

Dullemeijer, P.: Piet Dullemeijer is Emeritus Professor in Functional Morphology from Leiden University. He has been Managing Editor of *Acta Biotheoretica* in the periods 1966 – 1969 and 1987 – 1998.
Address: Mariahoevelaan 3, 2343 JA Oegstgeest, The Netherlands

Egas, M.: Martijn Egas studied biology and physical geography at the University of Amsterdam. He obtained his Ph.D. degree *cum laude* with a thesis on the evolution of specialisation in phytophagous arthropods and its implications for host race formation and sympatric speciation. Currently, his main research interests are the combined study of experimental and theoretical evolutionary ecology and evolutionary game theory, with a keen side-interest in population dynamics and food web dynamics. Also, he is working on the interaction between *Wolbachia* bacteria and the animal hosts they infect.
Address: Department of Population Biology
                      Institute for Biodiversity and Ecosystems Dynamics
                      University of Amsterdam
                      P.O. Box 94084, 1090 GB Amsterdam, The Netherlands
                      Email: egas@science.uva.nl

Etienne, R. S.: During the 1990s Rampal Etienne studied Foundations of Physics, Philosophy of Science (University of Utrecht) and Environmental Science (University of Nijmegen). In 1997 he developed and taught a course in Environmental Modeling for a year (University of Nijmegen). In February 1998 he started his Ph.D. research on the question of how mathematical models and methods can be used in nature management of metapopulations (Wageningen University). After obtaining his degree in March 2002, he started postdoctoral research at Wageningen University to explain large-scale biodiversity patterns (species-abundance distribution, species-area curve) with mathematical models. In September 2002 he moved to the University of Groningen where he continued working on these issues.
Address: Community and Conservation Ecology Group
                      University of Groningen
                      P.O. Box 14, 9750 AA Haren, The Netherlands
                      Email: r.etienne@biol.rug.nl

T.A.C. Reydon and L. Hemerik.,(eds.), Current Themes in Theoretical Biology,
305-310.
© 2005 *Springer. Printed in the Netherlands.*

Grasman, J.: The main interest of Johan Grasman is in the mathematical education of students in the applied sciences. In research he mostly works on differential equation models of physical and biological processes. These differential equations range from ordinary non-linear equations to partial differential equations. Through the study of a class of linear diffusion equations (the Fokker-Planck equation), also stochastic differential equations are being included. This type of mathematical modeling is applied in various fields of science such as meteorology, hydrology, ecology and epidemiology.

Address: Biometris
Department of Mathematical and Statistical Methods
Wageningen University
P.O. Box 100, 6700 AC Wageningen, The Netherlands
Email: johan.grasman@wur.nl

Hagenaars, T. J.: Thomas Hagenaars was trained as a Theoretical Physicist (doctoral degree from Utrecht University in 1995) before moving into the field of theoretical biology. His main current interest is in the epidemiology of infectious diseases, an interest rooted in postdoctoral work at the University of Oxford and at Imperial College London.

Address: Quantitative Veterinary Epidemiology
Division of Infectious Diseases, Animal Sciences Group
Wageningen University
P.O. Box 65, 8200 AB Lelystad, The Netherlands
Email: thomas.hagenaars@wur.nl

Hemerik, L.: Lia Hemerik is a Theoretical Biologist at the Department of Mathematical and Statistical methods at Wageningen University in the Netherlands. She received her Ph.D. in Ecology in 1991 from Leiden University after finishing both her M.Sc. in Mathematics and in Biology. Her current work focuses on the application of Cox's regression model and modeling population dynamical processes in behavioural ecology, conservation biology, aquatic ecology or soil biology. She has been Managing Editor of *Acta Biotheoretica* from 1999 to 2001.

Address: Biometris
Department of Mathematical and Statistical Methods
Wageningen University
P.O. Box 100, 6700 AC Wageningen, The Netherlands
Email: lia.hemerik@wur.nl

Hengeveld, R.: Rob Hengeveld has an Honorary Professorship at the Free University in Amsterdam. He is interested in stochastic spatial processes as part of ecological adaptation. As this assumes the organisms to be rigid, he also concentrates on the possible age of certain ecologically relevant traits, as well as on the mechanisms keeping these traits the same. A consequence of continuous spatial adaptation, though, is that spatial units like populations and communities can no longer apply; one has to concentrate on processes among individual organisms. Part of his work is therefore also methodological and historical analysis of the current demographic paradigm.

Address: Department of Ecotoxicology, Vrije Universiteit
De Boelelaan 1087, 1081 HV Amsterdam, The Netherlands
Email: rob.hengeveld@ecology.falw.vu.nl,
rhengeveld@wish.net

Jansen, P. A.: Dr Patrick Jansen obtained his Ph.D. degree in ecology from Wageningen University (The Netherlands) in 2003. After that, he was at Alterra (Wageningen University and Research Centre) as a research scientist in international nature conservation. He is now working at the Smithsonian Tropical Research Institute in Panama – on seed dispersal by agoutis – as a postdoctoral fellow with the University of Groningen.

Address: Community and Conservation Ecology Group
University of Groningen
P.O. Box 14, 9750 AA Haren, The Netherlands
Email: p.a.jansen@biol.rug.nl

Kornet, D. J.: Diedel Kornet is Professor of Philosophy of Biology at Leiden University since 1995.

Address: Philosophy of the Life Sciences Group
Institute of Biology, Leiden University
P.O. Box 9516, 2300 RA Leiden, The Netherlands
Email: kornet@rulsfb.leidenuniv.nl

Kooijman, S. A. L. M.: Professor Dr S. A. L. M. (Bas) Kooijman graduated in biology (1974), and obtained his Ph.D. degree (1974) on the statistical analysis of point patterns at Leiden University. During this period he taught in and advised on applications of mathematics in biology. During 1977-1985 he was affiliated with the TNO laboratories in Delft as Study Director, and supervised research in ecotoxicology. Since 1985 he has been appointed as Professor and Head of the Department of Theoretical Biology of the Vrije Univesiteit and was Advisor at the TNO laboratories till 1995. He (co)authored about 200 articles, book chapters, including three books. Since 1979 most work relates to the Dynamic Energy Budget theory that he created; this theory is about quantitative aspects of metabolic organisation. Extensive information about his professional activities can be found at the department's home-page (http://www.bio.vu.nl/thb/).

Address: Department of Theoretical Biology, Vrije Universiteit
De Boelelaan 1087, 1081 HV Amsterdam, The Netherlands
Email: bas@bio.vu.nl

McAllister, J. W.: James W. McAllister is university lecturer at the Faculty of Philosophy, Leiden University, and Visiting Professor of Philosophy of Technology and Culture at Delft University of Technology. He studied natural sciences and philosophy at Cambridge and Toronto, gaining his Ph.D. at the Department of History and Philosophy of Science, Cambridge, in 1989. He has held visiting appointments at the Institute for Advanced Study, Princeton, and the Center for Philosophy of Science, University of Pittsburgh. He is the author of *Beauty and Revolution in Science* (Ithaca, N.Y.: Cornell University Press, 1996) and editor of the Routledge journal *International Studies in the Philosophy of Science*.

Address: Faculty of Philosophy, Leiden University
P.O. Box 9515, 2300 RA Leiden, The Netherlands
Email: j.w.mcallister@let.leidenuniv.nl

Reydon, T. A. C.: Thomas Reydon holds Master's degrees in physics and philosophy of science (both from Leiden University). He is now working toward a Ph.D. in philosophy of biology (also at Leiden University) with a dissertation on the species problem. His research interests encompass issues in general philosophy of science as well as in the philosophical foundations of particular scientific disciplines. He has been Managing Editor of *Acta Biotheoretica* since the beginning of 2002.

Address: Philosophy of the Life Sciences Group
Institute of Biology, Leiden University
P.O. Box 9516, 2300 RA Leiden, The Netherlands
Email: reydon@rulsfb.leidenuniv.nl

Van der Hoeven, N.: Nelly van der Hoeven has studied mathematical biology at Leiden University and obtained her Ph.D. in 1985 from Groningen University. Next, she has been working for 10 years at the TNO Institute of Environmental Sciences, and presently she is the Head of ECOSTAT, a consultancy agency for statistics in ecology and ecotoxicology. She is President of the Netherlands working group on Statistics and Ecotoxicology and Board Member of the Biometric Section of the Netherlands Society for Statistics and Operational Research and the Netherlands section of the International Biometric Society.

Address: Department of Theoretical Biology
Institute of Biology, Leiden University
P.O. Box 9516, 2300 RA Leiden, The Netherlands
Present address: ECOSTAT
Vondellaan 23, 2332 AA Leiden, The Netherlands
Email: nvdh@ecostat.nl

Van der Steen, W. J.: Wim J. van der Steen is Emeritus Professor of Philosophy of Biology, Department of Biology and Faculty of Philosophy, Vrije Universiteit Amsterdam. He has published books and articles in biology, biomedicine, philosophy of science and ethics.

Address: Maluslaan 3, 1185 KZ Amstelveen, The Netherlands
Email: wjvds@wanadoo.nl

Van der Weele, C. N.: Cor van der Weele studied biology at the University of Utrecht and philosophy at the universities of Utrecht and Groningen. Her Ph.D. thesis (1995, Vrije Universiteit, Amsterdam) with the title *Images of Development; Environmental Causes in Ontogeny* dealt with developmental biology in the border area of biology, philosophy of science and ethics. After that she worked in various places with ethics as a constant theme. In her present job at Wageningen University she returns to the border area of biology and philosophy, with a special interest in metaphor and selective attention, and genetics and genomics again as important topics.

Address: Department of Applied Philosophy
                 Wageningen University
                 P.O. Box 8130, 6700 EW Wageningen, The Netherlands
                 Email: cor.vanderweele@wur.nl

Van Herwaarden, O. A.: Onno van Herwaarden has studied mathematics at Utrecht University. In 1996 he finished his Ph.D. with a thesis "Analysis of unexpected exits using the Fokker-Planck equation" at Wageningen University. His main fields of interest are the education of mathematics, in particular the use of computer algebra, and the study of exit problems, mainly in biological applications.

Address: Biometris
                 Department of Mathematical and Statistical Methods
                 Wageningen University
                 P.O. Box 100, 6700 AC Wageningen, The Netherlands
                 Email: onno.vanherwaarden@wur.nl

Wouters, A. G.: Arno Wouters studied biology (Wageningen University) and philosophy (Groningen University). He obtained his Ph.D. in the Philosophy of Science from Utrecht University. His research interests include explanation in functional biology, the relation between the different levels of organization in biology, and the philosophical impacts of the recent genomics research trend. The article was written as part of postdoc project at the Department of Philosophy of Nijmegen University. Arno Wouters is currently affiliated as a Guest Researcher to the Philosophy of the Life Sciences Group of the Institute of Biology, Leiden University.

Address: Philosophy of the Life Sciences Group
                 Institute of Biology, Leiden University
                 P.O. Box 9516, 2300 RA Leiden, The Netherlands
                 Email: arno.wouters@rulsfb.leidenuniv.nl